Nikolai M. Plakida
High-Temperature Superconductivity

Nikolai M. Plakida

High-Temperature Superconductivity

Experiment and Theory

With 87 Figures

 Springer

Professor Dr. Nikolai M. Plakida

Joint Institute for Nuclear Research, Laboratory of Theoretical Physics
141980 Dubna, Moscow Region, Russia

Translator

Dr. Alexander N. Ermilov

Academy of Sciences, Steklov Mathematical Institute, 42 Vavilov St.
1117966 Moscow GSP-1, Russia

Title of the original Russian edition: *Vysokotemperaturnye sverkhprovodniki*
© Fizmatlit Publishing House (in preparation)

ISBN-13:978-3-642-78408-8   e-ISBN-13:978-3-642-78406-4
DOI: 10.1007/978-3-642-78406-4

Preface

At the beginning of 1986, Karl Alex Müller and Johan Georg Bednorz, who were research associates at the IBM Research Division, Zurich Research Laboratory, discovered the occurrence of superconductivity in lanthanum and barium copper oxides at temperatures below 35 K. This discovery, for which the authors received the Nobel Prize in Physics in 1987, has caused an unprecedented wave of scientific activity in study of superconductivity. Early in 1987, it was observed that the compounds obtained by the Zurich group after replacing La by Y became superconducting at 90 K. New compounds based of bismuth and thallium copper oxides with transition temperatures up to 120 K were discovered at the beginning of 1988. Researchers are now actively looking for new compounds of oxide superconductors type. There is hope that these activities will lead to new discoveries and the creation of high-temperature superconductors with material properties appropriate for technical applications.

As a result of the enormous research effort of large number of physicists as well as chemists and material scientists, high-quality samples of high-temperature superconductors have been obtained and their main physical properties have been studied. It has emerged that these compounds possess a number of unusual properties due to a complicated interplay of electronic, spin, and lattice degrees of freedom. To study these properties, researchers have had to make use of various experimental methods. In view of the complicated character of the interplay, any theory of oxide superconductors encounters a number of difficulties. Despite the powerful modern methods of statistical physics, the study of various microscopic models has not so far resulted in an unambiguous interpretation of all the physical phenomena and the mechanism for formation of the superconducting state.

The extensive investigations of oxide superconductors have been reflected in in a huge flux of publications in the form of journal articles and reports at numerous conferences. A number of reviews have recently been published in which the results of studies obtained in separate fields or on the basis of different methods are discussed. At the same time there are very few publications in which the essential properties of high-temperature oxide superconductors are reviewed and their theoretical interpretation is discussed. The purpose of the present book is to achieve this aim in a form accessible to a wide circle of readers, both beginners in the high-temperature superconductivity field and experts interested in a general overview.

As follows from the contents of the book, the main physical properties of the new oxide superconductors are discussed in a concise form in Chap. 2–6, and the essential theoretical models for the high-temperature superconductivity are considered in Chap. 7. The discussions are given at the text-book level, but it is assumed that the reader is familiar with the main concepts of the conventional theory of superconductivity.

A major source of the interest in high-temperature superconductors is definitely due to the possibility of wide technical applications. In this respect a large number of investigations of electrodynamic and magnetic properties of oxide superconductors have been made; their results have important bearing from the point of view of possible applications. We shall not discuss these properties, since this field of applied studies constitutes a separate large branch in the physics of high-temperature superconductors and goes beyond the scope of this book.

The author expresses his deep gratitude to a large number of colleagues for illuminating discussions and for sending their preprints. The author apologizes to all colleagues whose works have been unintentionally omitted. Several recent papers, published after the manuscript was completed are given as footnotes.

I am especially indebted to Prof. P. Fulde, who read the book and made a number of important remarks. His hospitality during my stay at the Max-Planck Institute in Stuttgart where a part of this book was prepared is highly appreciated.

The name of Academician N. N. Bogolubov, my teacher, should also be mentioned; he supported our research activities in the field of superconductivity for many years. Useful discussions with colleagues in the Joint Institute for Nuclear Research at Dubna are acknowledged. Special thanks are due to Mrs. L. Gavrilenko and Mr. V. Udovenko who helped to prepare this book for publication.

Dubna
September 1994   *N. Plakida*

Contents

1. Introduction

Ever since 1911 when H. Kamerlingh Onnes first discovered superconductivity, physicists have been interested to find out why the temperature of the transition to the superconducting state T_c is so low compared to the temperatures of other phase transitions. The temperatures of the transitions to ferromagnetic or antiferromagnetic states in metals are hundreds of degrees Kelvin, while for most common superconductors T_c lies around 10 K. This fact seems surprising, since in the both cases the phase transitions take place in the electron subsystem of the crystals and are ultimately due to electron-electron interaction. To answer this question, we shall discuss in this chapter the history of Bednorz and Müller discovery of the new oxide superconductors [1.1, 2]. We shall also try to explain why they can be singled out as a separate class of superconductors and what are the principle features which distinguish them from conventional superconductors.

1.1 The Discovery of High-Temperature Superconductivity

A superconductor is usually referred to as a high-temperature superconductor, if the transition temperature T_c exceeds 90 K, i.e., if the superconducting state is attained by cooling in liquid nitrogen. A large number of papers (see, for example, [1.3]) are devoted to the general problem of designing high-temperature superconductors. In order to discuss the factors which influence the temperature of the superconducting transition, we shall consider the simple model of Bardin, Cooper, and Schrieffer (BCS). This model contains only three parameters which specify the interaction of electrons. They are the density of electron states at the Fermi surface $N(0)$ (per spin direction), an effective attraction coupling constant V and an energy shell of order $\hbar\omega$ around the Fermi surface. In the weak coupling limit when

$$\lambda = N(0)V \ll 1,$$

the transition temperature according to the BCS theory is (see, for example, [1.3])

$$kT_c \simeq \hbar\omega \exp(-1/\lambda). \tag{1.1}$$

This expression determines the temperature below which the normal state of electrons becomes unstable with respect to the formation of a condensate of electron pairs with opposite spins and zero orbital momentum (singlet s-wave pairing). The nature of the forces responsible for the effective attraction in the energy shell $\hbar\omega$ is of no importance for derivation of (1.1). The BCS theory and the relation (1.1) can therefore be used to treat superconductivity in the case of other, i.e., non-phononic, mechanisms of attraction for the formation of singlet electron pairs. In this cases one speaks of a generalized BCS pairing theory.

It is interesting to note that, besides the superconducting electron pairing, in the electron liquid an instability with respect to the formation of electron-hole pairs below a certain temperature T_0 is possible. In certain cases (for example, when the electron (hole) Fermi surface has perfect nesting), this temperature is determined by the same relation (1.1). In this case the attraction V between an electron and a hole in the enegy shell $\hbar\omega$ may be of purely Coulomb nature. For typical Coulomb energies $\hbar\omega \simeq 1\,\mathrm{eV}$, according to (1.1) we obtain $T_0 = 100\,\mathrm{K}$ even for weak coupling $\lambda \leq 0.2$. The bounded states of electron-hole pairs result in the formation of charge density waves and an metal-insulator transition or in the formation of spin density waves and a ferromagnetic or antiferromagnetic transition of the metal. These phase transitions occur at higher temperatures and often hinder the system from entering the superconducting state (see, for example, [1.3]).

The electron attraction which is responsible for superconductivity in conventional metals is due to a retarded electron-phonon interaction. It is related to the phonon energy and comes into play only in a narrow energy shell of order $\hbar\omega/k \leq 400\,K$ near the Fermi surface. The coupling constant itself depends on the phonon spectrum

$$\lambda = N(0)\langle g^2 \rangle \left\langle \frac{1}{M\omega^2} \right\rangle - \mu^*. \tag{1.2}$$

The first term is determined by the square of the matrix element g for the electron-phonon interaction averaged over the Fermi surface and by the averaged inverse lattice rigidity $\langle \Phi^{-1} \rangle = \langle (M\omega^2)^{-1} \rangle$ where M is the reduced mass of lattice ions. The second term μ^* describes an effective Coulomb repulsion of electrons. In conventional superconductors it is small, $\mu^* \simeq 0.1 - 0.2$.

A direct confirmation of the electron-phonon mechanism of pairing is the isotopic effect, i.e., the dependence of T_c on the mass M of the lattice ions, $T_c \propto M^{-\alpha}$. According to (1.1) and (1.2), the exponent α in this dependence is given by the expression

$$\alpha = -\frac{d \log T_c}{d \log M} \simeq \frac{1}{2}, \tag{1.3}$$

provided the contribution of μ^* to (1.2), which is usually small, is neglected. Only for superconductors with low T_c can the value of α be considerably lower due to large μ^* and small values of λ (1.2).

The relations (1.1) and (1.2) show that, in conventional metals with typical values of $N(0)$ and $\langle g^2 \rangle$ it is difficult to obtain high transition temperatures. Actually, if we take a rigid lattice with a high phonon frequency ω in the relation (1.1), we obtain weak coupling in (1.2), $\lambda \sim \langle \omega^{-2} \rangle$. For weak coupling, the transition temperature T_c turns out to be low even for high-frequency phonons. For example, for $\hbar\omega \simeq 400$ K and $\lambda < 0.3$ according to the relation (1.1) we obtain $T_c < 15$ K.

The coupling constant (1.2) can be increased significantly by synthesizing materials with a high density of states. There is however a limit to the increase of the coupling constant λ, since a large value in metals leads to a strong renormalization of the phonon spectrum, i.e. to its softening and to structural instability (see, for example, [1.3]). The greatest success in this direction has been attained for compounds of transition metals with the A15 structure of the A_3B type, where $A = $ Nb, V and $B = $ Sn, Si, Ge, etc. (see, for example, [1.4]). Despite great research efforts, the maximum temperature $T_c \simeq 23$ K obtained for the compound Nb_3Ge in 1973 was not exceeded till the discovery of the new class of oxide superconductors by Bednorz and Müller [1.1].

A strong coupling and high T_c can also be obtained in compounds with large value of the electron-ion interaction $\langle \bar{g}^2 \rangle$. But in conventional metals with high electron density, the matrix element of the electron-ion interaction g is considerably weakened due to strong screening and cannot attain high values. Taking into account this circumstance and the absence of progress in studying compounds of transition metals, in 1983 Bednorz and Müller addressed their attention to another class of compounds, namely, to oxide superconductors. The high polarizability of oxygen ions and poor screening of the electron-ion Coulomb interaction due to a low density of carriers may result in a strong electron-phonon coupling in these compounds. By that time conducting oxides with relatively high transition temperatures $T_c \simeq 13$ K at very low densities of electron states were already known. The most interesting was a perovskite $Ba(PbBi)O_3$ discovered by Sleight et al. in 1975 [1.5]. At a sufficiently low concentration of carriers ($n = 4 \cdot 10^{-21}$ cm^{-3}, i.e., two order of magnitude smaller than in transition metals) and therefore a small value of $N(0)$, the high value of T_c could be accounted for in the frame of the electron-phonon model (1.2) only if one assumes a large value of the electron-ion interaction. However, attempts to raise T_c in this compound by increasing density of states $N(0)$ by varying the ratio Pb : Bi failed. With increasing density of states, the compound underwent a metal-insulator transition with the formation of a charge density wave (see, for example, [1.6]).

The search for new oxide superconductors undertaken by Bednorz and Müller was based on the idea of creating conducting oxides containing so-called Jahn-Teller ions. Such ions, for example, Ni^{3+} or Cu^{2+}, are characterized by a strong interaction of electrons with local distortions of a crystal lattice. The distortion considerably decreases the electronic energy of the ion due to a lifting of the degeneracy of electron levels. (The Jahn-Teller effect for Cu^{2+} ions is considered in Sect. 5.1). The strong interaction of electrons with displacements of surrounding ions can result in the formation of polarons whose pairing in the form of bipolarons can also lead to superconductivity [1.7].

The study of nickel compounds, however, did not give positive results. In 1985 Bednorz and Müller turned to compounds of copper oxides. Among them lanthanum and barium copper oxides with metallic conductivity were known. On varying the ratio $La^{3+} : Ba^{2+}$ in these compounds it was easy to control the valence of copper and the concentration of carriers. In January 1986 when performing measurements of conductivity in compounds with various concentrations of barium Bednorz and Müller discovered a dramatic fall of the resistivity in some samples at temperatures below 35 K. The results of the measurements were published in the September issue of Zeitschrift für Physik [1.1]. Final confirmation of super-conducting nature of the phase transition in these samples was obtained after a verification of the Meissner effect [1.8].

The publication of these papers drew the attention of many scientists who, in a short period of time, confirmed the occurrence of superconductivity in the ceramics La – M – Cu – O where M = Ba, Sr, Ca [1.2]. Later on it became clear that the oxide superconductors of this type have a layered perovskite structure $La_{2-x}M_xCuO_4$ (LMCO) (see Sect. 2.1). Still higher superconducting transition temperatures were reached in January 1987 by the group of C. W. Chu at the University of Houston in collaboration with the group of M.-K. Wu at the University of Alabama. Having replaced La by Y they obtained $T_c = 90$ K in a multi-phase ceramic sample [1.9]. The superconducting phase in this compound has the layered perovskite structure $YBa_2Cu_3O_{7-y}$ (YBCO) with a deficit in oxygen (see Sect. 2.4). Thus within one year, the temperature of the superconducting transition increased several-fold compared to the value $T_c = 23$ K, the maximum known in 1973. It is very important that T_c in the new oxide superconductors exceeds the boiling point of nitrogen equal to 77 K. This has become a criterion defining a high-temperature superconductor.

The further active search for new compounds with higher values of T_c led to the discovery of superconductivity in the systems Bi – Sr – Ca – Cu – O [1.10] and Tl – Ba – Ca – Cu – O [1.11] in which T_c reached 110–120 K. A maximal value $T_c = 125$ K was found in the compound $Tl_2Ba_2Ca_2Cu_3O_{10}$. This is so far the record value of T_c verified by many laboratories [1].

The synthesis of the compound (K – Ba)BiO_3 [1.12, 13] with $T_c = 30$ K considerably exceeding the $T_c = 13$ K in Ba(Pb – Bi)O_3 [1.5] was of a great importance for understanding the mechanisms of high-temperature superconductivity. The absence of copper and the three-dimensional nature of the lattice with the standard perovskite structure $CaTiO_3$ excludes for these compounds the magnetic mechanisms of superconductivity proposed for copper-oxide superconductors (Chap. 7).

The above oxide superconductors have hole-type conductivity. Therefore, the discovery of superconductivity in the compounds $Nd_{2-x}Ce_xCuO_4$ [1.14, 15] with $T_c = 20$ K and with electron-type conductivity shows the universality and presumably a general mechanism of high-temperature superconductivity in oxide compounds.

---

[1] Even higher $T_c = 132$ K was discovered in Hg – Ba – Ca – Cu – O system [1.25] which can be further enhanced by applying external pressure. The highest $T_c \simeq 240$ K was reported for Bi- and Hg-based systems with 5–6 copper planes [1.26].

1.2 A New Class of Oxide Superconductors

There now exist a great many oxide compounds with high superconducting transition temperatures. They can be subdivided into three classes: three-dimensional compounds based on $BaBiO_3$, layered copper-oxide compounds with hole conductivity (based on La_2CuO_4, $YBa_2Cu_3O_{7-y}$, and compounds of Bi and Tl), and compounds based on Nd_2CuO_4 with electron conductivity. The main types of these compounds and their transition temperatures T_c are given in Table 1.1 [1.16]. With increasing number of CuO_2 planes in copper-oxide superconductors, a certain increase in T_c is observed. The values range from $T_c = 36\,K$ in $La_{2-x}Sr_xCuO_4$ with a single CuO_2 plane to $T_c = 125\,K$ in compounds of Tl(2223) with three CuO_2 planes [2]. It should also be taken into account that the value of T_c in these compounds is strongly dependent on the concentration of oxygen. In Table 1.1, the maximum values of T_c are shown corresponding to optimal oxidation. The dependence of T_c on the composition is discussed in greater detail in Chap. 2.

One of the most important microscopic characteristics of a superconductor which determines its transition temperature T_c (1.1) is the density of electron states on the Fermi surface. Usually, this quantity is experimentally estimated from the value of electronic specific heat at low temperatures, $C_e = \gamma T$ where $\gamma \propto N(0)$ is the Sommerfeld constant (Chap. 4). It is therefore interesting to discuss the T_c-γ plot proposed by Batlogg [1.17] and shown in Fig. 1.1. In this figure, the conventional and oxide superconductors are represented by points and circles, respectively.

As one can see from this T_c-γ plot, there is a maximum of T_c at a fixed value of γ. It is marked by a dashed line for the conventional superconductors. According to (1.1), maximum values of T_c for fixed γ can be obtained either due to a large value of the energy $\hbar\omega$ or due to a strong coupling V. A further increase of T_c is possible only via an increase of γ. For conventional superconductors, the absolute maximum is presumably attained for the A15 compounds. They have both the highest density of states for transition metals, and optimal parameters ω and V in (1.1). It should be noted that there exists another class of superconductors with heavy fermions (UPt_3, $CeCu_2Si_2$, etc.) with very high values $\gamma > 100\,mJ/mol\,K^2$ but low $T_c \sim 1\,K$. The mechanism of superconductivity in these compounds has not been yet sufficiently studied, and the simple BCS formulae (1.1) does not necessarily apply to them.

On the other hand, the new oxide superconductors lie much higher than the dashed line in Fig. 1.1. For the same density of states $N(0) \propto \gamma$, the transition temperature is several times larger than in the conventional superconductors. This fact enables the oxide superconductors to be singled out as a separate class including $BaBiO_3$-based superconductors with the cubic lattice which do not contain copper. For example, $Ba(Pb - Bi)O_3$ has the same transition temperature as the oxide compound $LiTi_2O_4$ with much larger value of γ. The unusual dependence of T_c on γ suggests the idea of a universal mechanism for the formation of the

---

[2] See footnote [1].

Table 1.1. High-temperature superconducting oxides [1.16]

Insulating Parent	HTSC Compounds	T_c (maximum) K
	Ternaries	
La_2CuO_4	$La_{2-x}A_xCuO_4$ A=Ba, Sr, Ca	36
R_2CuO_4	$R_{2-x}M_xCuO_4$ R=Pr, Nd, Sm, Eu M=Ce, Th	24
Nd_2CuO_4	$Nd_{2-x-y}Ce_ySr_xCuO_4$	20
$BaBiO_3$	$(Ba_{1-x}K_x)BiO_3$	30
	Quaternaries	
$RBa_2Cu_3O_{6+x}$ $(x < 0.4)$ R=Y, La, Nd, Sm, Eu,Gd,	$RBa_2Cu_3O_{7-y}$ $(y < 0.6)$ Dy, Ho, Er, Tm, Yb, Lu	102
	$YBa_2Cu_{3.5}O_{8-y}$	87
	$YBa_2Cu_4O_8$	80
	Quinaries	
BiRSrCuO	BiCaSrCuO 2021, 2122, 2223 [1]	110
TlRBaCuO	TlCaBaCuO 2021, 2122, 2223 1021, 1122, 1223, 1324	125
HgRBaCuO	HgCaBaCuO [2] 1021, 1122, 1223	132
$Pb_2RSrCu_3O_8$	$Pb_2ASr_2Cu_3O_{8+y}$ A=R+Sr, R+Ca	70

[1] The groups of numbers represent the composition ratio of the cations
[2] See [1.25]

superconducting state both in cubic compounds based on $BaBiO_3$ and in layered copper-oxide high-temperature superconductors. Mechanisms of superconductivity are discussed in greater detail in Chap. 7.

1.3 General Properties of Oxide Superconductors

Even the first studies of high-temperature superconductors have shown that the new oxide compounds possess many properties in common with conventional superconductors. In particular, measurements of the Shapiro steps in the Josephson effect, an observation of the flux lattice in a magnetic field, and a direct measurement of the flux quantum $\phi_0 = hc/2e$ have shown that Cooper pairs with charge 2e appear in the superconducting state. Tunnel experiments unambiguously indicate

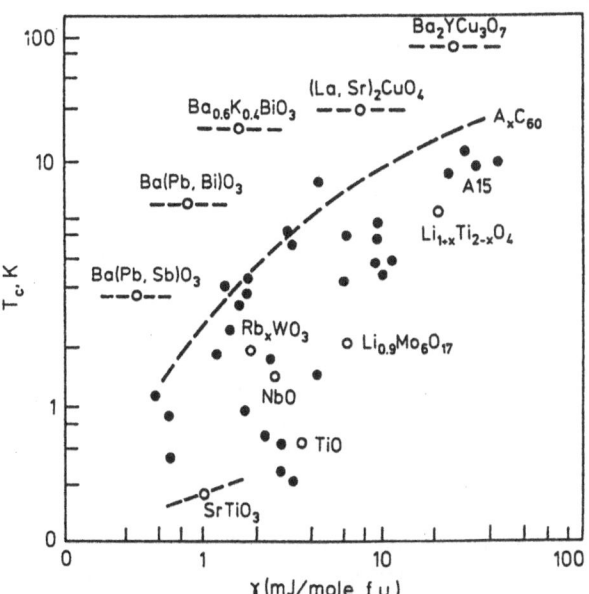

Fig. 1.1 The temperature of the superconducting transition T_c for various superconductors vs the Sommerfeld constant γ [1.17]

the formation of a gap in the spectrum of charge carriers, which also confirms the picture of Cooper pairs. The decrease in the Knight shift in the superconducting state and the temperature dependence of the penetration depth of a magnetic field point to the singlet nature of pairing as in the usual BCS theory (Chaps. 3 and 4).

At the same time the high-temperature superconductors have a number of physical properties by which they essentially differ from conventional metals. Here we discuss these properties only briefly, since they will be considered in greater detail in later chapters.

As structural studies show, the high-temperature superconductors are formed on the basis of perovskite-like structures including the cubic phase of $BaBiO_3$-based compounds and show a diversity of layered structures. In the copper-oxide superconductors, the main role is played by CuO_2 planes which are separated by layers of other ions. As a result, a high anisotropy of the electronic and, in particular, superconducting properties, which are of a quasi-two-dimensional character, is specific for the copper-oxide superconductors. The physical properties of the oxide compounds are strongly influenced by the deficit with respect to oxygen and certain variations in their composition. These problems are discussed in Chap. 2.

A peculiar distinction of the copper-oxide superconductors is a antiferromagnetic ordering of spins, which are almost localized at copper sites, in the CuO_2 planes in the insulating phase. Doping CuO_2 planes with charge carries by variation of the composition easily destroys the long-range antiferromagnetic order in the metallic state. However, strong dynamical short-range antiferromagnetic fluctuations still exist in the metallic phase. They strongly effect the properties of the copper-oxide compounds in the normal phase and may be the origin of non-phononic mechanisms of superconductivity. Chapter 3 is devoted to the description of the antiferromagnetic phase transitions and their relation to structural phase transitions. In the same chapter, experiments on neutron inelastic scattering and NMR are discussed which confirm the existence of strong dynamical antiferromagnetic fluctuations.

Studies of the thermodynamical properties of copper-oxide superconductors have shown a number of peculiarities in the temperature dependences of critical magnetic fields and specific heats. The existence of the strong anisotropy of electronic properties results in a large anisotropy of critical magnetic fields in the plane and in the direction perpendicular to this plane, as well as anisotropy of the related correlation length for the order parameter. The estimation of the correlation length shows that its value in the plane constitutes several lattice constants, while in the direction perpendicular to the plane it is approximately equal to or even smaller than the lattice constant in this direction. Such small values of the correlation length show that the number of electrons (holes) in the Cooper pair is by several order of magnitude smaller than the number of electrons n_s in conventional superconductors, where $n_s = 10^4$–10^6. A small number of electrons in Cooper pair results in considerable fluctuations which are observed in measurements of specific-heat jump at the superconducting transition. The results on the measurements of the specific heat and critical magnetic fields and the estimations of some parameters of high-temperature superconductors on the basis of the anisotropic Ginsburg-Landau theory are treated in Chap. 4.

A strong anisotropy of the electronic properties of the copper-oxide superconductors was already predicted in the first theoretical calculations of electronic band structure (Sect. 5.3). It was shown that the main contribution to states near the Fermi surface is made by strongly bonding $3d$ electron states of copper Cu^{2+} and $2p$ states of oxygen O^{2-} in a CuO_2 plane. Rather accurate experimental investigations have verified this picture only at the qualitative level and have found a considerable contribution of electron single-site correlations at copper ions which was not taken into account in the first theoretical calculations. In particular, a metal-insulator transition with the formation of the antiferromagnetic state, a considerable localization of the spin density at copper sites, the appearance of excitations related to the charge transfer, and a number of other phenomena observed in electron spectroscopy can be accounted for only by taking into consideration strong electronic correlations. Studies of electronic properties are discussed in Chap. 5 where the results on the superconducting gap $\Delta(T)$ are also considered. A considerable anisotropy of the gap and a large value of the ratio $2\Delta(0)/kT_c =$ 4–6 for the gap in the plane are specific for layered copper-oxide superconductors. This fact clearly distinguishes them from conventional superconductors, where the gap is rather isotropic and its ratio to T_c is close to the universal value 3.53 of the BCS theory.

In conventional superconductors, the electron-phonon mechanism of pairing is verified by the large value of the isotopic effect (1.3). While in three-dimensional Ba – Bi compounds the exponent α has the value close to $\alpha = 1/2$, in copper-oxide compounds the isotopic effect is suppressed, $\alpha = 0.2$–0.05, although in some cases it reveals an anomalous growth (Sect. 7.4). In this respect, studies of the phonon spectrum of oxide superconductors and the observation of manifestations of electron-phonon interaction (for example, the variation of frequency and phonon damping near T_c) are of great importance for clarifying mechanisms of high-temperature superconductivity. The results of these studies are considered in Chap. 6, where the impact of structural instability, i.e., phonon mode softening, on the value of T_c is also discussed.

Such unusual properties of high-temperature superconductors have required physicists to apply various theoretical models ranging from the standard models of a Fermi-liquid with strong electron-phonon coupling to rather exotic models of quantum spin liquids with unusual ground states (for example, flux phase with violation of spatial and time symmetry). Chapter 7 is devoted to the discussion of some of these models and of the most general mechanisms of superconductivity in high-temperature superconductors.

The results of studies of high-temperature superconductors, materials, and mechanisms of superconductivity are presented in the proceedings of various international conferences, in particular [1.18–20]. Physical properties of high-temperature superconductors are discussed in some detail in the monograph [1.21]. A number of useful reviews devoted to theoretical and experimental studies of high-temperature superconductors can be found in the books [1.22–24].

2. Crystal Structure

It is well known that, in solids, atomic structure determines the character of chemical bonding and a number of other related physical properties. Even small changes in structure often bring about considerable changes in the electronic properties of a solid, for example, the Peierls metal-insulator phase transitions. Therefore, the investigation of the crystal structure, i.e., long-range order, dependence of the structure on temperature, pressure, and composition plays an important role in studying high-temperature superconductors. This is important both for understanding the mechanisms of high-temperature superconductivity and in predicting possible ways to synthesize new superconducting compounds.

So far, about 20 topologically different types of crystal structure of layered copper-oxide superconductors have been studied [2.1]. The crystal structure of three-dimensional oxide $BaBiO_3$ based superconductors has also been studied fairly well [2.2]. These various structures can be devided into several families depending on the type of packing of a small number of structure elements, i.e., perovskite-like metal-oxide layers and layers with the structure of rock salt (see Table 1.1). Besides the type of crystal structure, the properties of oxide superconductors also depend strongly on the short-range atomic order which determines the local charge distribution in the crystal [2.3]. The aspects of the structure of oxide compounds related to crystal and chemical properties will be discussed in Sect. 5.1. In this chapter, the main crystal structures of the oxide superconductors are described. We begin with the simplest compound $(K- Ba)BiO_3$ which has the classical structure of perovskite $CaTiO_3$.

2.1 The Structure of $Ba_{1-x}K_xBiO_3$

The crystal structures of $BaBiO_3$ based compounds under replacing Ba by K or Bi by Pb can be described as small distortions of the original perovskite cubic phase $Pm\bar{3}m$ shown in Fig. 2.1 [2.2].

In the figure, thermal factors are shown as ellipsoids whose dimensions characterize the thermal fluctuations of ions in the lattice. For the oxygen ions, the strongly anisotropic thermal factors with large vibration amplitude in the plane of the cube face implies that the lattice is predisposed to structural phase transitions related to rotations of the BiO_6 octahedrons. In fact, freezing out of rotations of

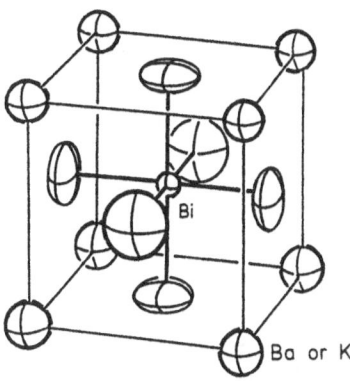

Fig. 2.1. The structure of $(K-Ba)BiO_3$ in the cubic phase [2.2]

Bi

Ba or K

the octahedrons (in antiphase to neighboring cells) around the cube axes [001] or [110] gives rise respectively to the tetragonal I4/mcm or orthorhombic Ibmm phases. The lattice constant increases by a factor of $\sqrt{2}$ in the basis x–y plane, while it doubles along the z axis. An additional freezing in of a breathing phonon mode (variation in the lengths of the Bi–O bonds) in the orthorhombic phase Ibmm decreases its symmetry to monoclinic I2/m [2.2]. In the latter case, the periodic variation in the lengths of the Bi–O bonds gives rise to a similar periodic variation in the charge on the Bi ions which can be described as a charge density wave (CDW). This monoclinic phase I2/m is just observable in the pure compound $BaBiO_3$, which turns out to be an insulator with an optical gap of $\sim 2\,eV$ due to the formation of the charge density wave.

A replacement of Ba by K suppresses the CDW. At a potassium concentration $x > 0.1$, non-equivalent positions of Bi are no longer observed. The long-range CDW disappears and the symmetry increases to Ibmm. However, metallic conductivity does not occur up to the transition into the cubic phase (at low temperatures for $x > 0.37$). Then, in the cubic phase, superconductivity with a transition temperature $T_c = 30\,K$ is observed. As the concentration increases up to $x = 0.5$ which is the solubility limit for potassium ions in a solid solution, the value of the transition temperature decreases. Under doping, the lattice constant of the pseudo-cubic lattice a_{pc} smoothly decreases. Its concentration dependence at room temperature is given by the formula [2.2]

$$a_{pc} = 4.3548 - 0.1743x \; (\text{Å}).$$

Since the ion radii of K^+ and Bi^{2+} have similar values, $1.64\,\text{Å}$ and $1.61\,\text{Å}$ respectively, the decrease in the volume of the primitive cubic cell can only be related with the decrease in the radius of the bismuth ion as the degree of its oxidation increases. The ion radii of Bi^{3+} and Bi^{5+} are $1.03\,\text{Å}$ and $0.76\,\text{Å}$ respectively.

A similar sequence of structural phase transitions is observed in $BaBi_{1-x}Pb_xO_3$ when Bi is replaced by Pb. At $x > 0.05$ the symmetry of the monoclinic phase I2/m increases to orthorhombic, although the transition to the metallic state takes place only at $x > 0.65$ when the symmetry increases to tetragonal I4/mcm. In the metallic phase, superconductivity is observed with a maximum value $T_c = 13\,K$ at

$x = 0.75$. As the lead concentration increases further, the transition temperature T_c decreases and at $x = 1$ a non-superconducting orthorhombic phase Ibmm is found with typical metallic properties.

Thus, despite the relative simplicity of the structure of cubic perovskite, the compounds $Ba_{1-x}K_xBiO_3$ and $BaBi_{1-x}Pb_xO_3$ reveal several phases with various crystal symmetry depending on the concentration x of doping ions and on the temperature. While the semiconducting properties of the original compound at $x = 0$ are readily accounted for by the formation of the charge density wave in the monoclinic phase I2/m, their persistence in orthorhombic phases where no non-equivalent positions of Bi are observed, cannot be so simply explained. A possible reason why the semiconducting gap in the electronic spectrum should remain in these phases may be the existence of local (or incommensurate) charge density waves with a small coherence length which obscure their observation in diffraction experiments. A specific feature of $BaBiO_3$ compounds is the occurrence of superconductivity near to the metal-insulator transition with decrease of the superconducting transition temperature T_c, as the number of charge carrier in the region of normal metallic properties increases. Such a non-monotonic dependence of T_c on the concentration of carriers and the suppression of superconductivity in the normal metallic phase is a feature specific to all the oxide superconductors.

2.2 $La_{1-x}M_xCuO_4$ Compounds

The structure of the copper-oxide layered compounds with the general formula

$$(Ln_{1-x}M_x)_{n+1}Cu_nO_{3n+1-m}$$

where Ln is a trivalent rare-earth ion R or Y, and M is a divalent alkaline ion, Ba, Sr, or Ca, can be characterized by the packing of CuO_6 octahedrons and the ordered system of oxygen vacancies. The number of planes constituted by interbonded oxygens of the octahedrons is charecterized by a quantity $n = 1, 2, ...$. For $n = 1$ we have the layered perovskite structure of K_2NiF_4, while for $n \to \infty$ we get a cubic perovskite ABO_3 where the octahedrons BO_6 have common oxygen ions in all the three directions. The number of oxygen vacancies is characterized by a quantity m describing the multiplicity of copper coordination. Since besides the 6-fold coordination CuO_6, copper readily allows the 4-fold, CuO_4, and 5-fold, CuO_5, coordinations, a large number of perovskite-like structures with oxygen deficiency appear. In these compounds, the copper is usually in a state with valence $2 \le v \le 2.4$, so that the number m turns out to be related to the concentration x of divalent ions M.

Figure 2.2 [2.4] shows a number of possible structures of this type. Only two of them, $YBa_2Cu_3O_7$ and $La_{1.85}Sr_{0.15}CuO_4$ become superconducting (see Table 1.1). The rest possess metallic conductivity up to helium temperatures but do not display a superconducting state despite the fact that they have the same averaged copper valence as in the superconducting compounds. In this respect, the absence

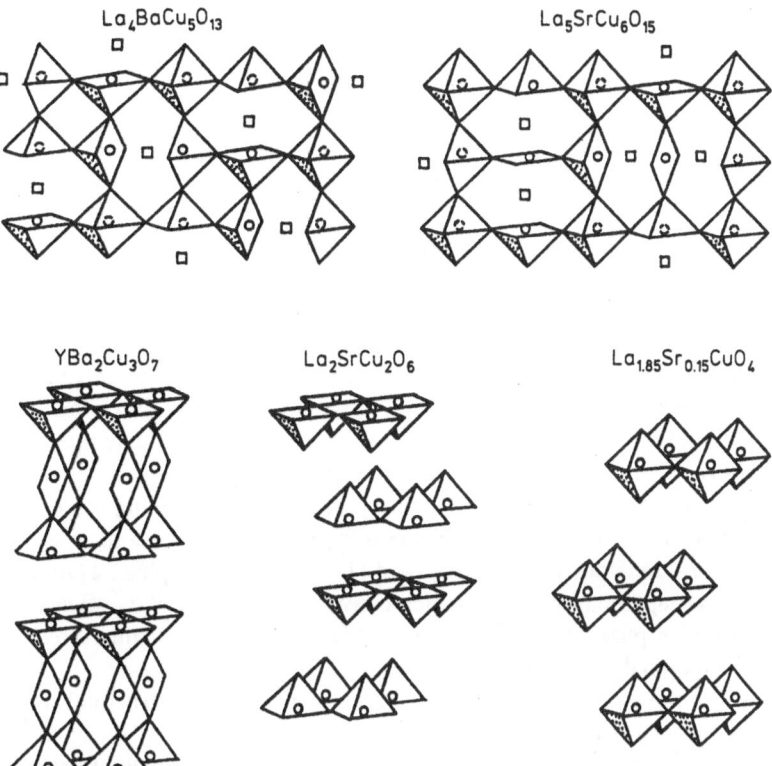

Fig. 2.2. The structure of the metallic copper-oxide compounds (Ln$_{1-x}$M$_x$)$_{n+1}$Cu$_n$O$_y$ [2.4]

of superconductivity in the simplest compound with two copper-oxygen planes La$_2$SrCu$_2$O$_6$ seems rather mysterious. The structure of this compound is close to YBa$_2$Cu$_3$O$_7$. More recent studies, however, have shown that this compound can also be made superconducting by replacing the Sr by Ca ion in the layer separating the copper planes CuO$_2$ [2.5]. It has been found that, in the layer of Sr, which is equivalent to the layer of Y in YBa$_2$Cu$_3$O$_7$, doping oxygen ions appear which can destroy superconductivity. On the other hand, the replacement of Sr by the smaller Ca ion blocks the appearance of the doping oxygen in this layer. These results demonstrate the important role played by the short-range order in the ion arrangement. The occurrence of CuO$_2$ planes with a particular degree of oxidization is apparently required for the appearance of superconductivity in copper-oxide compounds. This degree of oxidization is controlled by the distance to the neighboring metal-oxide layers and the value of their charge [2.3]. Recently high-temperature superconductivity was observed in the so called "infinite-layer structure" (Sr$_{1-x}$Ca$_x$)$_{1-y}$CuO$_2$ ($T_c = 80 - 110$ K) [2.6] and Sr$_{1-x}$Nd$_x$CuO$_2$ ($T_c = 40$ K) [2.7] which contain the CuO$_2$ planes with the Sr (Ca) or Nd layers between them.

2.2.1 The Structure of $La_{2-x}M_xCuO_{4-y}$

Let us consider in more detail the structure of the compound $La_{2-x}M_xCuO_{4-y}$ (LMCO or 2-1-4) where M = Sr, Ba, or Ca. In the high-temperature tetragonal (HTT) phase, these compounds have a structure of the type K_2NiF_4, i.e., a body-centered tetragonal lattice with one formula unit in the primitive cell (I4/mmm $-$ D_{4h}^{17}). The tetragonal unit cell of this lattice, which has two formula units, is shown in Fig. 2.3 [2.8]. Structural parameters of LMCO crystals for different compositions and their temperature dependence are discussed in the review [2.1]. A typical value of the lattice constant in the tetragonal phase is $a_t = 3.78$ Å, $c_t = 13.2$ Å. The distance Cu $-$ O1 in the plane is given by $a_t/2 = 1.89$ Å and the distance Cu $-$ O2 is 2.42 Å. The large anisotropic thermal factors for the oxygen ions are noteworthy. They indicate a large vibrational amplitude of these ions in the rotation mode (tilting type) for the CuO_6 octahedra. As the temperature decreases, a structural phase transition from the tetragonal to the low-temperature orthorhombic (LTO) phase (Cmca$-D_{2h}^{18}$) takes place, with the doubling of the unit cell. This cell is shown in Fig. 2.4 where the orthorhombic axes $b_0 > a_0 > c_0$ are chosen along the directions [001], [$\bar{1}$10], and [110] of the HTT phase I4/mmm. Another space group Bmab with the lattice constants $a_0' < b_0' < c_0'$ is often used in the orthorhombic phase. In this case, the orthorhombic axes are simply related to the initial tetragonal phase, $a_0' \simeq b_0' \simeq a_t\sqrt{2} = 5.35$ Å, $c_0' \simeq c_t$.

The study of the phonon spectrum (see [2.9]) shows that the structural phase transition HTT $\rightarrow$ LTO is due to the condensation of a soft tilting mode at the Brillouin zone boudary. In the LT0 phase, two domains appear which are related to the rotation of the CuO_6 octahedra around the tetragonal axes [110] or [$\bar{1}$10] respectively (see Figs. 2.4 and 2.8).

Under the rotation of octahedra in the orthorhombic phase (up to 5° at low temperatures [2.1]) the change in the Cu1-O1 bonds in the plane is negligible (less than 0.01%). An accompanying orthorhombic deformation increases the lattice constant along the direction perpendicular to the rotation axis, i.e., $a_0 > c_0$ under the rotation around the c_0 axis in Fig. 2.4. In the LTO phase, the compressibility of the La_2CuO_4 lattice turns out to be anisotropic. At room temperature, for the directions a_0, b_0, and c_0 (the Cmca phase) it is respectively equal to 4.1, 1.5, and $1.8(\times 10^{-4}kbar^{-1})$ [2.10].

The temperature of the HTT $\rightarrow$ LTO structural phase transition T_0 in $La_{1-x}M_xCuO_{4-y}$ rapidly decreases with increasing concentration x of doping divalent M ions. A typical dependence $T_0(x)$ for the case M = Sr is shown in Fig. 2.5 [2.7] where the regions with long-range antiferromagnetic order (the Néel state), the superconducting state, the boundary between the insulating and metallic phases, and the region of frozen spin states (the spin glass) at low temperatures are also shown.

An additional study of the dependence of T_0 on the concentration of oxygen vacancies y shows that T_0 decreases as the quantity $x - 2y$ increases, i.e., as the total number of charge carriers in a CuO_2 plane goes up. In fact, the replacement of La^{3+} by M^{2+} increases the number of holes by one, while the formation of a

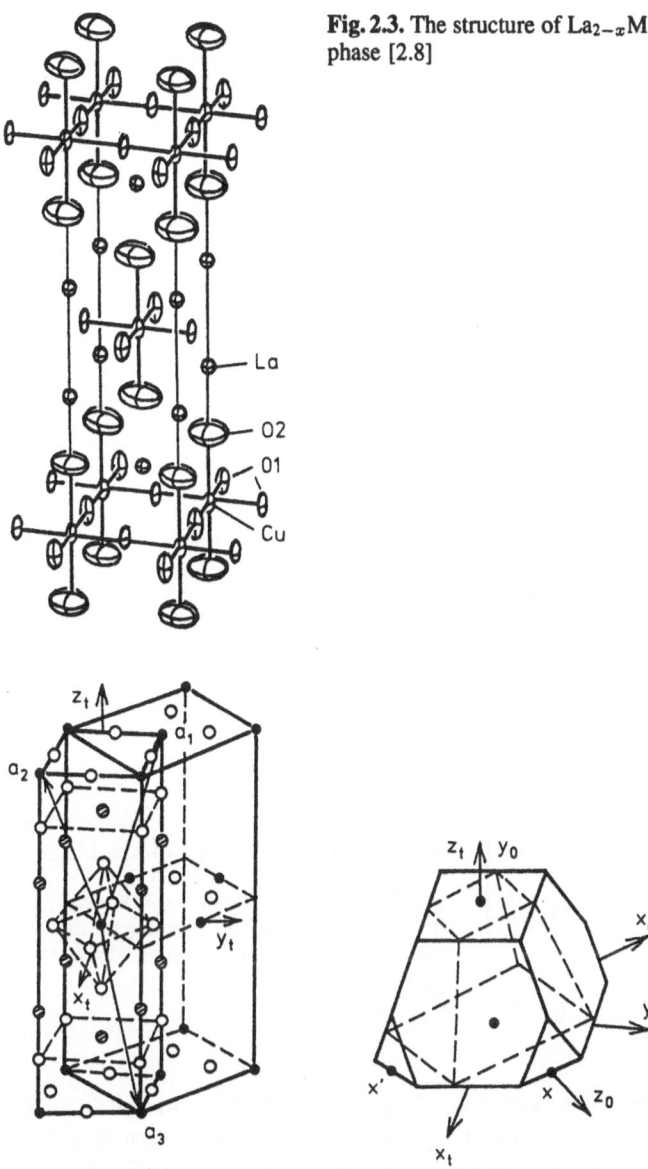

Fig. 2.3. The structure of La$_{2-x}$M$_x$CuO$_4$ in the tetragonal phase [2.8]

Fig. 2.4. The tetragonal body-centered ($a_t = b_t$, c_t) and orthorhombic ($b_0 > a_0 > c_0$) unit cells of La$_2$CuO$_4$ (**a**) and the corresponding Brillouin zones (**b**)

vacancy of O^{2-} increases the number of electrons by two [2.11]. Under applied pressure, $T_0(x)$ goes down and the orthorhombic phase disappears at pressures above the critical $p_c(x)$ (for example, $p_c = 15$ kbar for $x = 0.12$ in La$_{2-x}$Sr$_x$CuO$_4$ [2.11]).

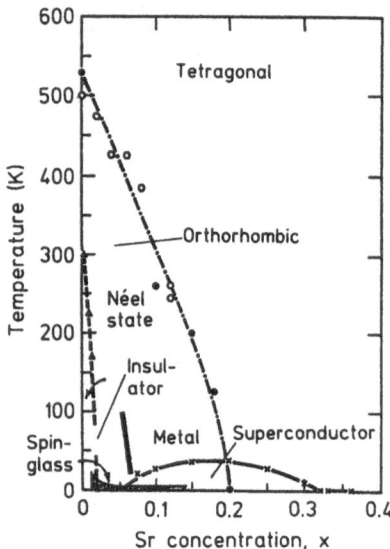

Fig. 2.5. The temperature-concentration phase diagram for $La_{2-x}Sr_xCuO_4$ [2.9]

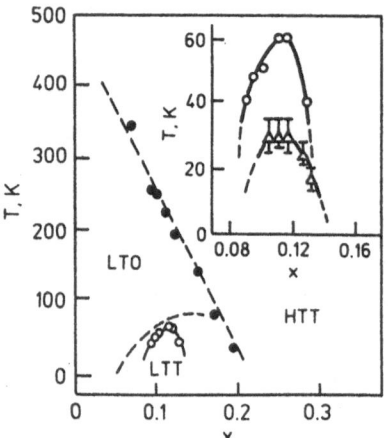

Fig. 2.6. The temperature — concentration phase diagram for $La_{2-x}Ba_xCuO_4$ [2.10]. In the insert the temperature of the structural phase transition to the LTT—phase (*dots*) and the temperature of the freezing of the copper magnetic moments (*triangular*) are shown [2.15]

More careful studies of the structure of $La_{2-x}Ba_xCuO_4$ have revealed another phase transition in the region of low temperatures. It is the phase transition from the orthorhombic to another tetragonal phase $P4_2/ncm$ (D_{4h}^{16}) [2.10–15]. The phase diagram of this compound is shown in Fig. 2.6 [2.15]. It is interesting to note that the temperature of the superconducting transition in the low-temperature tetragonal phase (LTT) has a minimum.

The sharp decrease of T_c in the LTT phase is correlated with anomalous changes in other electronic properties such as conductivity, the Hall effect, thermo electric power [2.13–15]. A freezing of the copper magnetic moments in the LTT phase was also observed by the μSR— method as it is shown in the insert on Fig. 2.6. A substitution of Th^{4+} ions for La^{3+} ions shifts the minimum of T_c in the

LTT phase to a higher concentration of Ba^{2+} ions [2.16]. This proves an electronic origin of the T_c suppression in the LTT phase.

The changes in electronic properties in the LTT phase can be related to the appearance of non-equivalent positions of the O1 oxygen ions in the plane. In the orthorhombic phase, under the rotation of CuO$_6$ octahedra around the c_0 axis (Fig. 2.4) (or a_0 for the other domain) all four O1 oxygen ions move out of the plane, as is shown in Fig. 2.8a,b. In this case, the variation of the crystal field potential in proportion to the square of the displacement of O1 ions is the same for all four ions. In the LTT phase, which can be represented as a coherent addition of the displacements in the two domains of the LTO phase, the rotation of the octahedra occurs around the tetragonal axes, i.e., the x axis in one of the CuO$_2$ planes and the y axis in the neighboring planes. Only two of the four O1 ions move out of the plane (Fig. 2.8c). This gives rise to a variation in their crystal field potential and to a local redistribution of charge. In this case, the formation of a charge density wave [2.17] or a gap in the electronic spectrum [2.18] accompanied by considerable changes in the electronic properties is possible. In the LTT phase we thus observe the influence of the structural instability of the lattice of the LMCO compounds on their superconducting properties. This is apparently a common feature for oxide superconductors.

A more complicated sequence of structural phase transitions was observed in La$_{1-y-x}$RE$_y$Sr$_x$CuO$_4$ compounds where the La^{3+} ions were replaced by the rare-earth ions RE^{3+} = Nd, Sm, Gd of a smaller radius [2.19–21]. The phase diagram for Nd– compound at $y = 0.4$ is shown in Fig. 2.7. In addition to the LTT phase which exists in a broad region of the Sr concentrations there appears a new intermediate phase – the low temperature orthorhombic phase (LTOII) (space group Pccn). In Fig. 2.7 is also shown the $T_c(x)$ dependence in the LTO phase, T_c (LTO), (at y= 0) and in the LTT phase, T_c (LTT) (at y= 0.4). As in the La–Ba–Cu–O compound $T_c(x)$ has a minimum at x= 0.12 which is more deep in the LTT phase. According to [2.21] there is no bulk superconductivity in the LTT phase. At the fixed concentration of holes (Sr^{2+}) the superconducting transition temperature $T_c(y)$ smoothly decreases in the sequence of phase transitions LTO $\rightarrow$ LTOII $\rightarrow$ LTT. In the intermediate LTOII phase Pccn the oxygen ion displacements can be represented as a sum of the displacements in the two domains in the LTO phase but with different amplitudes. Starting from one domain of the LTO phase and smoothly increasing the contribution for the oxygen displacements from the second domain one can get a continuous phase transition from LTO to LTT phase through the Pccn phase while a direct LTO $\rightarrow$ LTT phase transition is of the first order. The continious phase transition LTO $\rightarrow$ LTOII $\rightarrow$ LTT can be described as a smooth change in the direction of the rotation axis for the CuO$_6$ octahedra from [110] in LTO phase to [100] in the LTT phase (in the tetragonal HTT phase notations). As pointed out in [2.20] just these changes of the oxygen displacements can explain the supression of T_c at fixed concentration of holes (Sr^{2+}) in the considered sequence of phase transitions.

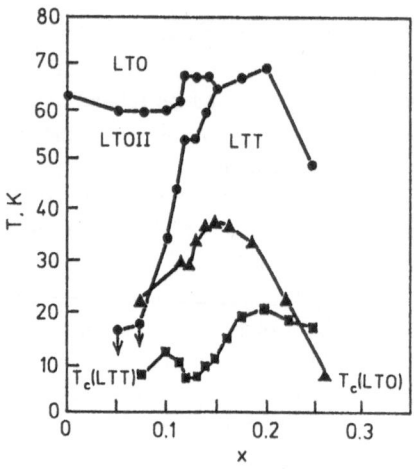

Fig. 2.7. The temperature–concentration phase diagram for $La_{2-y-x}Nd_ySr_xCuO_4$ for $y = 0.4$ and $T_c(x)$ for $y = 0$ in the LTO phase, T_c (LTO), and for $y = 0.4$ in the LTT phase, T_c (LTT) [2.19]

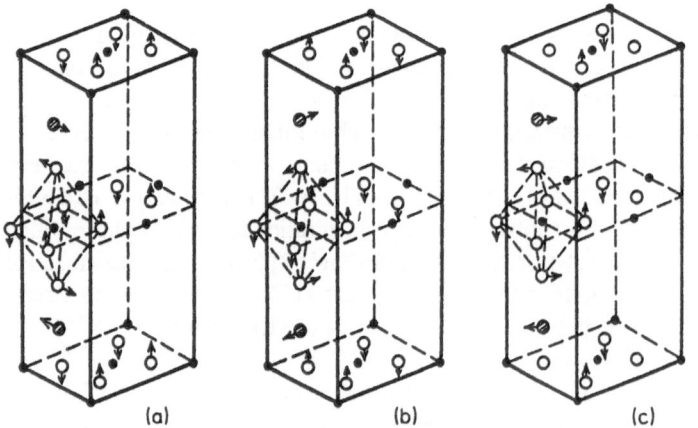

Fig. 2.8. The displacement of the O1 oxygen ions in the two domains of the orthorhombic phase (**a, b**) and in the low-temperature tetragonal phase (**c**)

2.2.2 A Phenomenological Theory of Structural Phase Transitions in La_2CuO_4- Based Compounds

Studyies of the La_2CuO_4- based copper-oxide superconductors reveal considerable anomalies in the elastic characteristics of the crystals under structural phase transitions [2.22, 23]. In these compounds, structural phase transitions may also influence their electronic and magnetic properties. In this respect, we shall consider a phenomenological theory of structural phase transitions based on a symmetry analysis and the Landau expansion for the free energy [2.24, 25]. The interplay of structural phase transitions and antiferromagnetic ordering in La_2CuO_4 is discussed in Sect. 3.1.2.

The sequence of structural phase transitions in La$_2$CuO$_4$ from the tetragonal (HTT – D$_{4h}^{17}$) to orthorhombic (LTO – D$_{2h}^{18}$, LTOII – Pccn) and low-temperature tetragonal (LTT – D$_{4h}^{16}$) phases can be described as a successive condensation of a two-component order parameter $\{C_1, C_2\}$. Namely, $C_1 \neq 0$, $C_2 = 0$ in the LTO phase $C_1 > C_2 \neq 0$ in the LTOII phase and $C_1 = C_2 \neq 0$ in the LTT phase. Figure 2.8 shows that a condensation of a soft mode related to the rotation of CuO$_6$ octahedra

$$C_1 \propto \langle R_1(\mathbf{k}_x(1)) \rangle, \quad C_2 \propto \langle R_2(\mathbf{k}_x(2)) \rangle, \tag{2.1}$$

corresponds to the two-component order parameter. In (2.1), $R_{1,2} = R_x \mp R_y$ is the rotation of the octahedra around the $[\bar{1}10]$ or $[110]$ axis for the wave vectors $\mathbf{k}_x(1,2) = (\pi/a)(\pm 1, 1, 0)$, respectively. The wave vectors $\mathbf{k}_x(1,2)$ form a two-arm star of the wave vector at the $X-$ point of the Brillouin zone of the body-centered tetragonal lattice (Fig. 2.4b).

The primitive cell of the body-centered tetragonal phase, which contains one LMCO formula unit, is given by the translation vectors

$$\mathbf{a}_1 = (-\tau, \tau, \tau_z), \quad \mathbf{a}_2 = (\tau, -\tau, \tau_z), \quad \mathbf{a}_3 = (\tau, \tau, -\tau_z),$$

where $2\tau = a_t$ and $2\tau_z = c_t$ are the parameters of the body-centered unit cell (Fig. 2.4). The reciprocal-lattice vectors are determined by the relations

$$\mathbf{b}_1 = \pi(0, \frac{1}{\tau}, \frac{1}{\tau_z}), \quad \mathbf{b}_2 = \pi(\frac{1}{\tau}, 0, \frac{1}{\tau_z}), \quad \mathbf{b}_3 = \pi(\frac{1}{\tau}, \frac{1}{\tau}, 0).$$

With this notation,

$$\mathbf{k}_x(1) = \frac{1}{2}\mathbf{b}_3 = \frac{\pi}{a}(1, 1, 0), \quad \mathbf{k}_x(2) = \frac{1}{2}(\mathbf{b}_1 - \mathbf{b}_2) = \frac{\pi}{a}(-1, 1, 0). \tag{2.2}$$

and the sum $\mathbf{k}_x(1) + \mathbf{k}_x(2)$ is equal to the vector $\mathbf{k}_z = (\pi/\tau_z)(0, 0, 1)$ to within reciprocal-lattice vector $\mathbf{b}_1$. Thus, the two-component order parameter determines two domains. Each of them is related to an irreducible representation X_3^+ for the wave vectors $\mathbf{k}_x(1)$ or $\mathbf{k}_x(2)$, respectively.

The expansion of the free energy in terms of the order parameter (OP) can be constructed in the form of a function of the corresponding invariants, $I_1 = (C_1^2 + C_2^2)$ and $I_2 = C_1^2 C_2^2$. It is convenient to write the expansion in the form

$$F_c = \frac{1}{2}r(C_1^2 + C_2^2) + \frac{1}{2}uC_1^2 C_2^2 + \frac{1}{4}v(C_1^4 + C_2^4) + \ldots \quad , \tag{2.3}$$

where $r = a(T - T_0)$, and u and v are phenomenological constants. The thermodynamic potential (2.3) for the two-component order parameter describes a wide class of structural transitions (see, for example, [2.26]). In the tetragonal phase, the strain contribution to the free energy is

$$F_\varepsilon = \frac{1}{2}C_{11}(\varepsilon_1^2 + \varepsilon_2^2) + C_{12}\varepsilon_1\varepsilon_2 + C_{13}(\varepsilon_1\varepsilon_3 + \varepsilon_2\varepsilon_3)$$

$$+ \frac{1}{2}C_{33}\varepsilon_3^2 + \frac{1}{2}C_{44}(\varepsilon_4^2 + \varepsilon_5^2) + \frac{1}{2}C_{66}\varepsilon_6^2, \tag{2.4}$$

where Voigt's notations are used for the strain tensor ε_μ and the elastic coefficients $C_{\mu\nu}$ of the crystal. The symmetrized square of the irreducible representation X_3^+ contains the invariants $(C_1^2 + C_2^2)$ and $(C_1^2 - C_2^2)$. This determines the interaction of the order parameter with the strains in the form

$$F_{C\varepsilon} = [\alpha(\varepsilon_1 + \varepsilon_2) + \beta\varepsilon_3](C_1^2 + C_2^2) + \gamma\varepsilon_6(C_1^2 - C_2^2), \tag{2.5}$$

where higher-order terms are omitted. We point out that the interaction of the type $\lambda\varepsilon_6 C_1 C_2$ is prohibited since the product $C_1 C_2$ belongs to the irreducible representation with wave vector $k_z \neq 0$.

For second-order phase transitions, equilibrium values of the order parameter are found from the conditions

$$\frac{\partial F}{\partial C_i} = 0.$$

Equilibrium values of the strains ε_μ are found from the conditions

$$\frac{\partial F}{\partial \varepsilon_\mu} = 0$$

where the full free energy is equal to the sum of the contributions (2.3)–(2.5). An analysis of these equilibrium conditions also enables one to determine the jumps in the elastic coefficients under a structural phase transition. We shall briefly summarize the results of calculations [2.24].

The phase transition HTT → LTO occurs when $(v - 4\gamma^2/C_{66}) < u$. Under this condition two domains appear, $C_1 \neq 0$ or $C_2 \neq 0$. In the orthorhombic phase the elastic coefficients then undergo the jumps

$$\Delta C_{11} = \Delta C_{12} = -2\alpha^2/v,$$
$$\Delta C_{13} = -2\alpha\beta/v, \quad \Delta C_{33} = -2\beta^2/v, \tag{2.6}$$
$$\Delta C_{66} = -2\gamma^2/v, \quad \Delta C_{44} = 0,$$

while the equilibrium strains $\varepsilon_1 = \varepsilon_2$, ε_3, and ε_6 are proportional to the square of the order parameter, for example, $\varepsilon_6 = -(\gamma/C_{66})C^2$.

The phase transition HTT → LTT occurs when $(v - 4\gamma^2/C_{66}) > u$. Under this condition $C_1 = C_2 = C \neq 0$. The strains $\varepsilon_1 = \varepsilon_2$ and ε_3 are proportional to the square of the order parameter C^2, while the elastic coefficients experience the jumps

$$\Delta C_{11} = \Delta C_{12} = -4\alpha^2/(v+u),$$
$$\Delta C_{13} = -4\alpha\beta/(v+u), \quad \Delta C_{33} = -4\beta^2/(v+u), \tag{2.7}$$
$$\Delta C_{66} = 0, \quad \Delta C_{44} = 0.$$

By comparing (2.6) and (2.7) with experimental data on the measuremed velocity of sound in the LMCO crystals and their jumps under structural phase transitions (see, for example, [2.22, 23]), one can determine the coupling constants α, β, and γ in (2.5) (see Sect. 6.3).

The phase transition LTO → LTT is not of a subgroup nature and occurs as a first-order transition. The transition temperature is determined by the equality of the free energies F in the LTO (at $C_1 \neq 0$, $C_2 = 0$) and LTT phases (at $C_1 = C_2 \neq 0$). In this case, higher powers of the invariants in the expansion of the free energy (2.3) start to play a role. For example, in [2.27], a model is proposed which contains powers of the sixth order in the form

$$\Delta F = w(I_1^2 - 4I_2)I_1 = w(C_1^2 - C_2^2)^2(C_1^2 + C_2^2). \tag{2.8}$$

In the LTO phase, this contribution differs from zero, while in the LTT phase at $C_1 = C_2$, it vanishes. At sufficiently low temperatures when equilibrium values of the order parameters become large, the LTT phase can thus be energetically more favorable than the LTO phase at $w > 0$, since the latter contains a larger positive contribution (2.8). In [2.27], assuming a certain dependence of the coefficients u and v on concentration x of doping ions, the temperature – concentration phase diagram for $La_{2-x}Ba_xCuO_4$ represented in Fig. 2.6 was been successfully described. In other models, this phase diagram was described by assuming a temperature dependence for the coefficients u and v. It is rather difficult to justify such an assumption (see, for example, [2.18, 13]).

The phenomenological expansion of the free energy (2.3), (2.5) can be obtained by calculating the free energy on the basis of a microscopic theory. Such a model microscopic theory will be discussed in Sect. 6.3. This theory consideres anharmonic vibrations of oxygen ions in a soft rotation mode and their interaction with acoustic phonons. The calculation of the ground state energy for La_2CuO_4 crystals was performed by the density-functional method in [2.18]. In the latter calculations, one derives the dependence of the ground state energy on the displacement of the oxygen ions due to the rotation of the CuO_6 octahedra in the LTO and LTT phases (Fig. 2.9).

It turned out that the LTT phase has the minimum energy. The sequence of phase transitions HTT → LTO → LTT can be described as a sequence of transitions of the order-disorder type connected with the ordering of rotations of CuO_6 octahedra. According to [2.18], the instability of the tetragonal phase with respect to rotations of the octahedra is due to a competition between repulsive forces in the CuO_2 plane and the long-range Coulomb forces determining the Madelung energy.

2.3 $Nd_{2-x}Ce_xCuO_4$ Compounds

The crystal structure of the Nd_2CuO_4- based superconducting compounds with electron conductivity is close to that of the lanthanum compounds. It is described by the same space group I4/mmm but with the displaced oxygen ions O2 from their apex positions to sites on the faces of the tetragonal cell. In Fig. 2.10 [2.28] the tetragonal cell for the compounds La(Sr) (the T phase), Nd(Ce) (the T' phase), and mixed compounds Nd(Sr, Ce) (the T* phase) are represented for comparison.

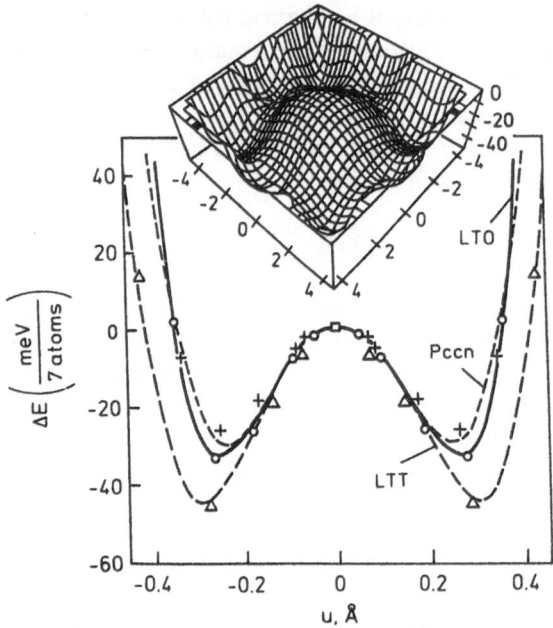

Fig. 2.9. Variation of the ground state energy of the La_2CuO_4 crystal under a rotation of the octahedra in the orthorhombic (LTO, LTOII, Pccn) and low-temperature tetragonal (LTT) phases as a function of the apex oxygen displacement $u(O3)$(Å) [2.18]

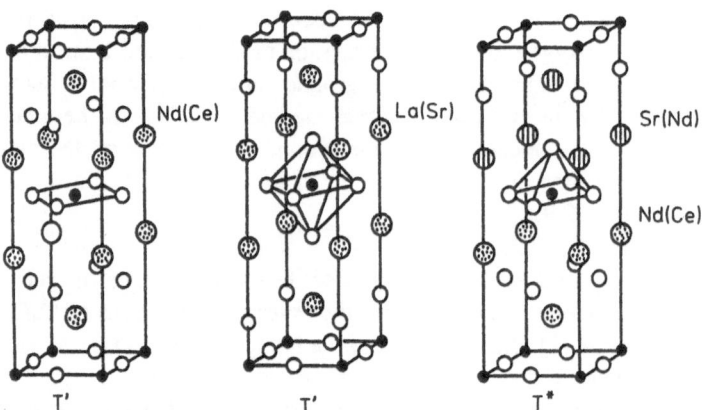

Fig. 2.10. The tetragonal unit cells of the T, T', and T^* phases $(Nd–Ce–Sr)_2CuO_4$ [2.28]

    In the T* phase, the apex oxygens are maintained only in the layer Nd–Sr, while in the layer Nd–Ce the oxygen ions are shifted to the faces. This reconstruction of the lattice in the T' phase brings about a corresponding variation in its parameters compared to the T phase. The length of the tetragonal axis increases by about 4 %, $a_t = 3.94$ Å while the c axis decreases by 8 %, $c_t = 12.1$ Å. The lattice parameters in the T* phase assume average values intermediate between those in the T and T' phases. The variation in packing of the O2 ions in the T' and T* phases compared to the T phase can be related to the difference in the sizes of the La and Nd ions.

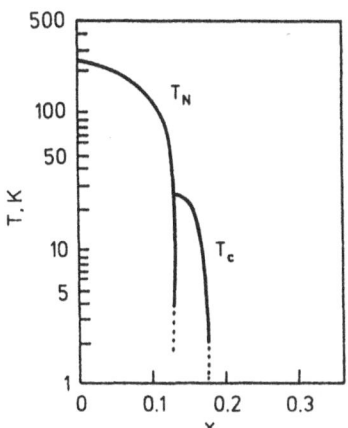

A primitive cell of the T* phase contains two formula unit and coincides with the tetragonal cell in Fig. 2.10 (P4/mmm − D_{4h}^{17}).

The phase diagram for the $Nd_{2-x}Ce_xCuO_4$ compound is shown in Fig. 2.11 [5.88]. It is qualitatively similar to the phase diagram of the LMCO compounds (Fig. 2.4). At $x = 0$ the compound displays an antiferromagnetic insulating phase ($T_N \simeq 240$ K), which is destroyed only by doping at much higher concentration of Ce ions ($x \simeq 0.12$) than in the LMCO compounds ($x \simeq 0.02$). The superconducting phase appears only in the vicinity of the antiferromagnetic phase. It exists, however, in a narrower interval of concentrations $0.12 \leq x \leq 0.18$ than in the LMCO compounds and has a lower transition temperature, $T_c \leq 24$ K. In the T′ and T* phases, no structural transitions related to displacements of O1 ions in modes of the tilting type have been observed. This can be related to the absence of complete CuO_6 octahedra in these compounds. The absence of the apex oxygen O2 in the T′ phase manifests itself in a number of the electronic properties of Nd—Ce compounds, for example, in the fact that T_c is independent of pressure (see Sect. 5.2). The role of the apex oxygen in copper-oxide layered compounds can be traced most completely in the $YBa_2Cu_3O_{7-y}$ compounds where one observes a considerable correlation between the electronic properties of the CuO_2 plane and the structural parameters of the apex oxygen.

2.4 Y—Ba—Cu—O-Based Compounds

A vast literature is devoted to the study of the compound $YBa_2Cu_3O_{7-y}$ (YBCO or 1−2−3) and its various modifications (see [2.1]). It was the first high-temperature superconductor found whose transition temperature exceeds the boiling point of nitrogen (see Chap. 1). The physical properties of this compound can be changed over a wide range without any significant changes in structure by varying the oxygen content. A whole class of compounds with similar physical properties can be obtained by replacing Y by rare-earth ions R = La, ... (see Table 1.1). As with the layered copper-oxide compounds, YBCO allows a certain modification of its structure through variation of the coupling between copper-oxygen layers. In particular, compounds with two chains $YBa_2Cu_4O_8$ $(1 - 2 - 4)$ or $Y_2Ba_4Cu_7O_{15}$ $(2 - 4 - 7)$, and compounds of type $Pb_2(Y-Ca)$ $Sr_2Cu_3O_{8+y}$ $(2 - 1 - 2 - 3)$ have been synthesized via the modification of a layer of $Cu-O$ chains into a more complicated structure Pb_2CuO_2. All these features of YBCO have brought about a huge interest in the study of its structure and other properties.

2.4.1 The Structure of $YBa_2Cu_3O_{7-y}$

The original compound $YBa_2Cu_3O_{7-y}$ is observed in two modifications. The first is the orthorhombic phase Pmmm (D_{2h}^1). The second is the tetragonal phase P4/mmm (D_{4h}^1) whose elementary cells with one formula unit are shown in Fig. 2.12. The main structural parameters in the orthorhombic phase at room temperature (in Å) are the following. The lattice constants are $a = 3.828$, $b = 3.888$, $c = 11.65$. The lengths of the bonds are Cu1 – O1 = 1.94, Cu1 – O4 = 1.94, Cu2 – O2 = 1.92, Cu2 – O3 = 1.96, Cu2 – O4 = 2.3. The length of the four Cu – O bonds for the four oxygen ions nearest to the copper both in the plane Cu2 – O2, O3 and in the chains Cu1 – O1, O4 are approximately the same and correspond to the lengths of bonds in the CuO_2 plane for LMCO compounds. The distance Cu2 – O4 in the CuO_5 pyramid is already smaller than the corresponding distance Cu – O2 along the z axis in LMCO. This may play a certain role in increasing T_c in YBCO compounds as compared to LMCO. The distance Cu2 – O4 (as the lattice constant c) varies strongly, as the oxygen content decreases under the transition into the tetragonal phase. This point will be discussed below (see Fig. 2.15). The rest of the structural parameters are close to those in the orthorhombic phase.

The orthorhombic phase is observed at low temperatures for an oxygen content $x = 7 - y \geq 6.4$. The tetragonal phase is observed at temperatures $T \geq 500°$ C when a decrease of oxygen content takes place together with its disordering in the Cu1 – O1 plane (Fig. 2.12b). It is seen that YBCO has a typical layered perovskite-like structure with two CuO_2 planes separated by a layer of Y ions which are bounded by the layers Ba – O4, Cu1 – O1, and Ba – O4. The oxygen O2 and O3 are strongly coupled with Cu2 in the CuO_2 planes, unlike the weakly coupled oxygen O1 in the Cu1 – O1. chains. Upon heating above 500° C, the latter oxygen readily leaves the sample. This enables the oxygen content to be smoothly varied from $x = 7$ $(y = 0)$ to $x = 6$ $(y = 1)$, when all oxygen of type O1 has been derived out of the compound. In the latter case, the tetragonal phase always occurs. At intermediate values of x, the structure of the compound depends on the way in which oxygen is removed [2.29]. Quenching from the high-temperature tetragonal phase at $x \leq 6.5$, preserves this tetragonal phase with disordered O1 positions in the Cu1 – O1. If the samples are prepared by the lower-temperature Zr-gettered annealing technique, the orthorhombic phase can be maintained up to $x \simeq 6.2$. In this case, several modifications of the orthorhombic phase occur. Besides an O I phase shown in Fig. 2.12 a when all the Cu1 – O1 chains are filled and $x = 7$, a phase O II can occur when alternate chains in the Cu1 – O1 plane

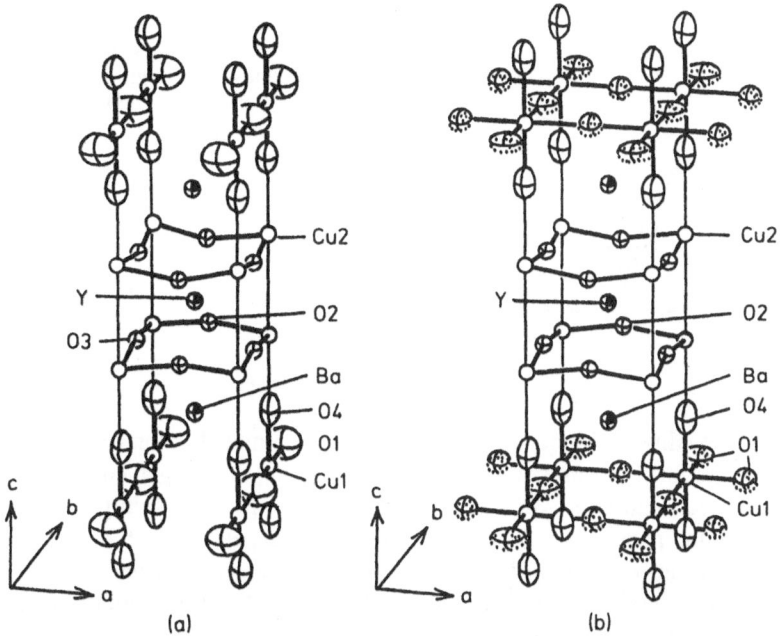

Y

O3

Cu2

O2

Ba

O4

O1

Cu1

c

b

a

(a)

Y

Cu2

O2

Ba

O4

O1

Cu1

c

b

a

(b)

Fig. 2.12. The structure of $YBa_2Cu_3O_{7-y}$ in the orthorhombic (**a**) and tetragonal (**b**) phases [2.8]

turn out to be empty at $x = 6.5$. More complicated phases at intermediate values of x are also observed. For example, at $x \simeq 6.35$ a phase $2\sqrt{2}a \times 2\sqrt{2}a$ also occurs [2.30] when half–filled chains alternate with chains that are one–quarter filled. Theoretical calculations of the phase diagram in the $x - T$ plane predict the occurrence of a number of more complicated phases with a periodic filling of chains, and with the formation of chains of a finite length, etc. (see, for example, [2.31]). Experimental studies of these phases are in progress.

The physical properties of $YBa_2Cu_3O_{7-y}$−based compounds depend considerably on oxygen content. The metallic phase with high $T_c \simeq 90\,K$ at $y = 0$ transforms into the semiconducting phase at $y \simeq 0.6$. In the latter phase, long-range antiferromagnetic order appears with a maximum Néel temperature $T_N \simeq 500\,K$ at $y = 1$. The way in which T_c depends on $x = 7 - y$ is determined by the type of sample preparation. In Fig. 2.13, the curve $T_c(x)$ is plotted. The dots correspond to high-temperature quenching and the crosses to the lower-temperature Zr-gettering (the solid line) [2.29]. In the latter case, two plateaux are observed at $T_c = 90\,K$ and $T_c = 60\,K$; these can be related to the aforementioned two orthorhombic phases OI and OII. Thus, the short-range order in the Cu1 − O1 chain has an essential effect on the electronic properties of the superconductor. This points to a local nature of the doping of the conducting CuO_2 plane due to charge (hole) transfer from the Cu1 − O1 chains [2.32, 33].

The dependence of T_c and the lattice parameters of the YBCO compound on oxygen content has been studied in greater detail in [2.32] by the lower-temperature

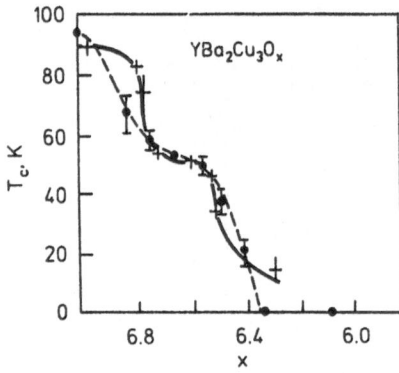

Fig. 2.13. The dependence of T_c on oxygen content x in $YBa_2Cu_3O_x$ for samples with oxygen removed by high-temperature quenching (*dots*) and lower-temperature Zr-gettering (*solid line*) [2.29]

Zr-gettering annealing technique. This technique has permitted a number of low-temperature equilibrium phases to be obtained by varying the oxygen content and has revealed a correlation between changes in electronic properties, in particular, T_c, and structural parameters. Figure 2.14 plots the dependence of the lattice parameters a, b, and c in the orthorhombic phase on oxygen content $x = 7 - y$ [2.32]. At $x \simeq 6.4$, under the transition from the orthorhombic (O) to tetragonal (T) phase, a considerable increase in the c axis lattice constant is observed. The length of the Cu2–O4 bond (O4 is the apex oxygen in the CuO_5 pyramid) undergoes equally strong variation under a slight variation in bond lengths in the plane Cu2–O2, O3 (Fig. 2.15 [2.32]).

In the work [2.32], in order to find the correlation between the change in structural parameters and electronic properties, a sum of valence bonds

$$V = \sum \exp[(R_0 - R_i)/B_i]$$

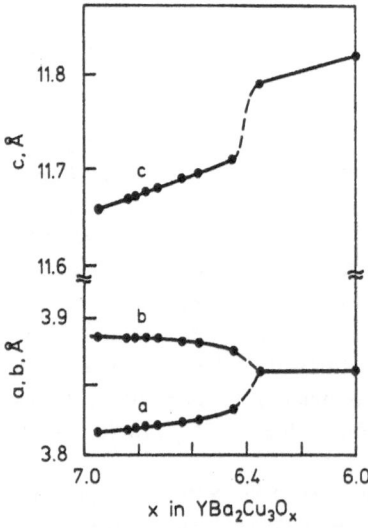

Fig. 2.14. The dependence of the lattice constants of $YBa_2Cu_3O_x$ on oxygen content x under low-temperature annealing (440° C) [2.32]

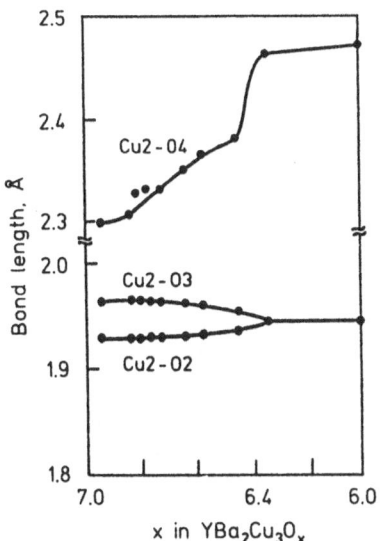

Fig. 2.15. The dependence of the copper-oxygen bond length in the plane (Cu2–O2, O3) and for the apex oxygen (Cu2–O4) [2.32]

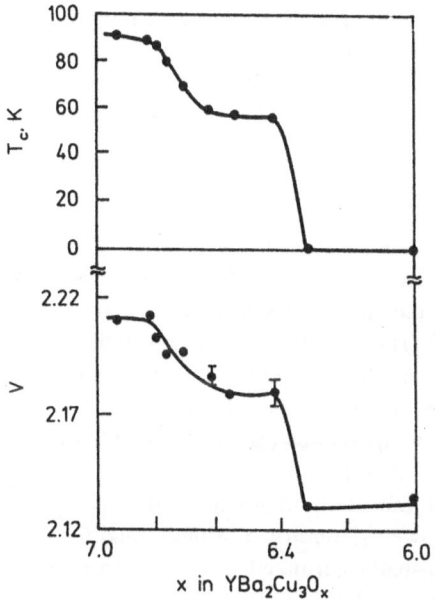

Fig. 2.16. The dependences of T_c and the effective copper valence in the plane on oxygen content [2.32]

has been calculated for various ions specified by by the constants R_i and B_i. The sum determines the effective valence of a given atom. The effective valence of copper ions in Cu1 chains has turned out to be linear in x. It varies from $V = 2.5$ at $x = 7$ to $V \simeq 1.3$ at $x = 6$. The effective valence of Cu2 undergoes a downwards jump at the transition from the orthorhombic to tetragonal phase which is accompanied by a correspondingly sharp variation in the length of the Cu2–O4 bond. In Fig. 2.16 this dependence is shown together with the dependence $T_c(x)$.

The identical behavior of these dependences indicates a correlation of T_c with the value of the effective charge in the CuO_2 plane. The decrease of T_c from 90 K to 60 K is due to the transfer of (negative) charge of about 0.03 e from the chains to the plane. The disappearance of superconductivity at $x = 6.45$ is connected with a further transfer of charge of about 0.05 e to the plane. As the oxygen content further decreases to $x = 6$, the effective charge of the copper in the plane remains constant; this should be taken into account when discussing the x−dependence of the Néel temperature of antiferromagnetic ordering in this region of oxygen concentration (Sect. 3.2).

Thus, detailed structural studies of $YBa_2Cu_3O_x$ compounds have unambiguously demonstrated the local nature of the charge transfer from CuO chains to CuO_2 planes and revealed drastic changes in the electronic properties of the system, including superconductivity, related to the transfer. In [2.32] it is noted that for samples with the same oxygen content, the transition temperature T_c can vary considerably depending on the oxygen ordering in the Cu1–O1 chains.

2.4.2 Modifications of the YBCO Structure

In synthesizing single crystals of YBCO in the orthorhombic phase, poly-domain samples with a twin plane of the type (110) usually appear. In such twinned crystals, the anisotropy of physical properties in the (a, b) plane of the orthorhombic lattice cannot be studied. In this respect, the synthesis of untwinned single crystals $YBa_2Cu_4O_8$ $(1 - 2 - 4)$ was of a great interest [2.1]. In Fig. 2.17 [2.34], the structures of three compounds of the $Y_2Ba_4Cu_{6+n}O_{14+n}$ family are shown for comparison.

The original $1 - 2 - 3$ structure corresponds to $n = 0$. The structure $1 - 2 - 4$ corresponds to $n = 2$. At $n = 1$ an intermediate structure $2-4-7$ occurs. The most important feature of new modifications is the appearance of double chains instead of the single O1–Cu1 chains in $1 - 2 - 3$. The oxygen coordination in the chains increases from two (Cu1–O1–Cu1) to three which stabilizes the entire structure. Due to this strengthening of the oxygen binding in the chains, the $1 - 2 - 4$ compound can be heated to much higher temperatures (of order 800° C) without any significant loss of oxygen. In the $1 - 2 - 4$ and $2 - 4 - 7$ compounds the temperature of the superconducting phase transition attains the values $T_c = 80$ K and $T_c = 87$ K, respectively [2.29]. The lower T_c observed in these modifications compared to $T_c = 91$ K in $1 - 2 - 3$ is usually attributed to the decrease of the effective charge of copper in the CuO_2 plane [2.32]. The synthesis of untwinned $1 - 2 - 4$ and $2 - 4 - 7$ crystals has enabled a number of interesting studies of their physical properties to be carried out (see, for example, [2.34]).

There exist several modifications of the YBCO structure obtained by means of transformations of the three layers Ba−O4, Cu1−O1, and Ba−O4 into a more complicated structure with the conservation of the bilayer CuO_2- R − CuO_2. One such modification has the general formula Pb_2 $ASr_2Cu_3O_8$ with $T_c = 70$ K [2.36] where A = Y, R, Ca. In this structure, the main $1 - 2 - 3$ unit $ASr_2Cu_2O_6$ is preserved upon the replacement of Ba by Sr. The connection between these units

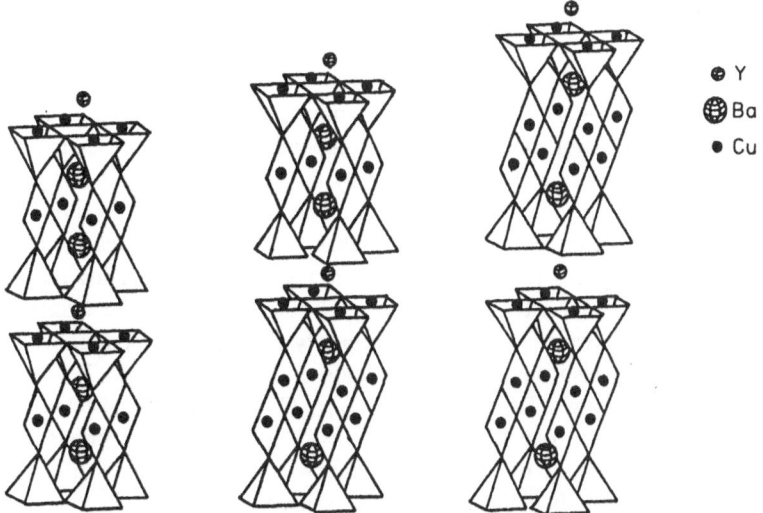

Fig. 2.17. The crystal structure of the $Y_2Ba_4Cu_{6+n}O_{14+n}$ compounds at $n = 0, 1, 2$ [2.34]

is no longer provided by Cu – O chains. It is now due to a more complicated structure Pb_2CuO_2 which consists of two PbO layers separated by a layer of copper with two-fold coordination. The structure of this compound is described by an orthorhombic elementary cell with the parameters $a = 5.40\,\text{Å}$, $b = 5.43\,\text{Å}$, and $c = 15.8\,\text{Å}$ (Cmmm) [2.36]. The small distortion of the tetragonal cell is accounted for by an ordering of oxygen in the PbO plane in non-centrosymmetric positions. In quenching from the high temperature phase at 500° C, the tetragonal structure (P4 /mmm) is observed [2.1].

A modification of the $1 - 2 - 3$ structure which does not contain any Cu – O chains is investigated in [2.37]. This is the compound $RSr_2GaCu_2O_7$ where the main element of the $1 - 2 - 3$ structure, i.e, the CuO_2–R–Cu_2O_7 bilaer, is preserved. However, instead of the copper chains, the bilayers are now bounded by GaO_4 octahedra which also form chains. This considerable change in the structure brings about quite different electronic properties. Having the stoichiometric composition with respect to oxygen O_7, this compound possesses semiconducting properties. Substitution of the trivalent ions R = Y, Yb, Er by divalent ones, e.g., Ca, which can be performed only under high oxygen pressure results in metallic properties and superconductivity at $T_c = 40$–$50\,\text{K}$ in multiphase samples [2.5].

2.5 Bi−Ca−Sr−Cu−O and Tl−Ca−Ba−Cu−O Compounds

The discovery of the compounds $Bi_2CaSr_2Cu_2O_8$ (Bi/2 − 1 − 2 − 2) [2.38] and $Tl_2CaBa_2Cu_2O_8$ (Tl/2−1−2−2) [2.39] with superconducting transition temperatures T_c above $100\,K$ was of major interest. These compounds can have various numbers of copper-oxygen planes and are described by the general formula $A_2Ca_{n-1}B_2Cu_nO_{2n+4}$ where A = Bi(Tl) and B = Sr(Ba). The transition temperature T_c depends on the number of the copper-oxygen planes and takes the values 10, 85, and 110 K for Bi compounds and 85, 105, and 125 K for Tl compounds for n = 1, 2, 3, respectively [2.1]. The synthesis of a Tl compound with the number of planes equal to $n = 4$, which occurs as an impurity phase in compounds with a smaller value of n, is also possible. Compounds of Tl with a single Tl − O layer with the general formula $TlCa_{n-1}Ba_2Cu_nO_{2n+3}$ (Tl/1...), where the number of the copper-oxygen layers reaches the value $n = 5$, have also been obtained. Figure 2.18 shows the structure of Tl compounds for $n = 1, 2, 3$ [2.40]. The ideal structure of the Bi compounds coincides with that shown in Fig. 2.18b.

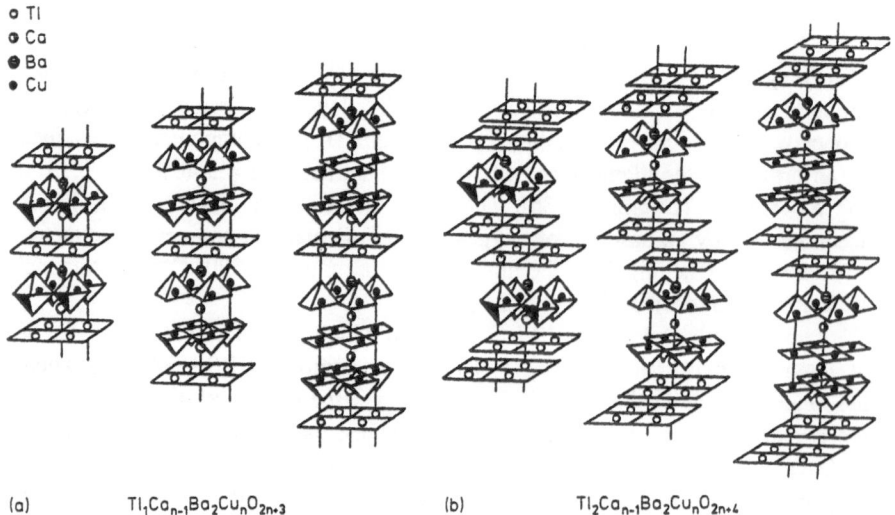

Fig. 2.18. The crystal structure of Tl−Ca−Ba−Cu−O compounds, Tl/1 ... (**a**) and Tl/2 ... (**b**), containing $n = 1, 2, 3$ copper planes [2.40]

The simplest pseudo-tetragonal unit cell (I4 /mmm − D_{4h}^{17}) of these compounds has the dimensions $3.9 \times 3.9\,\text{Å}$ in the basal plane. The lattice constant along the c axis depends on the number of copper-oxide planes n. In the compounds (Tl/2) and Bi it is c = 23.2, 29.4, 36 (Å) and c = 24.4, 30.8, 37.1 (Å) for $n = 1, 2, 3$, respectively.

The actual structure of the Bi compounds has orthorhombic distortions. The pseudo-rhombic elementary cell (Fmmm − D_{2h}^{23}) has the dimensions $a_0 \simeq b_0 \simeq a_t\sqrt{2} \simeq 5.4\,\text{Å}$ $c_0 \simeq c_t$. Moreover, a modulation of the oxygen displacements in the Bi − O layer along the b axis is observed that results in the increasing of the cell dimension to $b_0 = 5a_0$. The average Cu − O distances in the plane have a typical values 1.9 Å. The Cu − O distances along the c axis in the CuO_5 pyramids is 2.6 Å which is greater than the corresponding Cu2 − O4 distance of about 2.3 Å in the YBCO compound. The weak coupling between Bi − O layers due to the large distance $\sim 3\,\text{Å}$ between them is typical for Bi compounds. This feature makes crystals of Bi compounds look like mica.

The structure of Tl compounds undergoes much less distortion of the ideal tetragonal lattice due to the relatively strong coupling between Tl − O layers. The distance between them is $\sim 2\,\text{Å}$ while the lattice constant is smaller than in Bi compounds. The length of the Cu − O bonds in the CuO_2 planes is 1.92 Å, while the distance copper − apex oxygen along the c axis reaches 2.7 Å, which is much greater than the corresponding distance in YBCO.

More detailed studies of the structure of Tl compounds have shown, however, a considerable statistical disorder in the Tl − O layers [2.41]. The ions Tl and O3 (oxygen in the Tl − O) layer have been found to be displaced from there centrosymmetric positions in the layer in such a way that they form Tl − O chains with a shorter bond. This structure, however possesses only short-range order so that it maintains the tetragonal symmetry on average [2.42]. The CuO_2 planes also undergo short-range distortions. In the planes, a distortion of the oxygen atoms along the tetragonal c axis is observed near the temperature of the superconducting transition [2.43]. This correlation between displacements of oxygen atoms in the CuO_2 layer and the transition into the superconducting phase indicates to a strong coupling of the charge carriers (holes) with lattice vibrations.

It should be noted that considerable disorder in the arrangement of oxygen atoms in buffer layers connecting the conducting CuO_2 planes is a feature of all the layered copper-oxide superconductors [2.41]. In the LMCO compounds, such a disordering of the O2 in La − O2 layers is due to the structural phase transition to the orthorhombic phase (see Fig. 2.8). In YBCO compounds $(1 - 2 - 3)$, one observes a disordering of O1 along the Cu1 − O1 chain over two positions shifted with respect to the chain axis by $\pm 0.1\,\text{Å}$ perpendicular to it along the a axis. This gives rise to anomalously large thermal factors for O1 (see Fig. 2.12a). In compounds of Bi and Tl, a disordering of oxygen over four positions near the centrosymmetric oxygen positions in Bi − O and Tl − O planes is also observed [2.41]. The existence of several equivalent positions for oxygen atoms brings about strongly anharmonic vibrations of these atoms with relatively low frequencies. Such anharmonic phonons can significantly enhance the effective attraction between electrons (holes) and thereby increasing T_c (see Sects. 6.3, 7.4).

We now emphasize once more the most important results obtained in the study of the crystal structure of oxide superconductors.

1. All the copper-oxide superconductors have structure of a block nature [2.44]. The main unit determining the metallic and superconducting properties of a compound is the CuO_2 plane which is a square lattice formed by copper ions bound to each other through oxygen ions. Each copper ion can additionally be coordinated to oxygen ions in the apex positions of an octahedron; this gives rise to the possible coordination of copper ions equal to four, five, and six. The effective charge of the CuO_2 complex is determined by buffer blocks binding them in a crystal structure. In the infinite layer structure $A_{1-y}CuO_2$ the buffer blocks A = Sr, Ca, Nd have no oxygen ions [2.6, 7].

The structure of the $BaBiO_3-$ based three-dimensional oxide superconductors can be considered as a limiting case of the block structure consisting of a single unit with a three-dimensional net of oxygen bonds.

2. The crystal structure of oxide compounds does not completely determine their superconducting properties which crucially depend on structural defects and short-range order [2.44, 45]. The arrangement and local ordering of doping atoms in a lattice affect the superconducting properties of the system due to the local nature of charge transfer from the buffer layer to the CuO_2 plane.

3. The crystal lattice of the oxide compounds reveals structural instability which is typical for perovskite-like compounds. Both structural transitions with a change in long-range order and a short-range ordering, mainly in the oxygen sublattice, are observed.

3. Antiferromagnetism
in High-Temperature Superconductors

One of the universal features of copper-oxide compounds is the antiferromagnetic ordering of the copper spins in the CuO_2 planes. In stoichiometric compounds, the copper ions are in the state Cu^{2+} and have one hole with spin $S = 1/2$ in the $3d$ shell. A strong superexchange interaction (via oxygen ions) between hole spins at copper sites gives rise to a three-dimensional long-range antiferromagnetic order with relatively high Néel temperatures $T_N = 300 - 500$ K. Although the long-range order disappears in the metallic and superconducting phases, strong dynamical spin fluctuations with a wide spectrum of excitations are observed even at temperatures above 100 K. This fact has led to a number of hypotheses on possible electron pairing in copper-oxide compounds via magnetic degrees of freedom (Sect. 7.3). The study of the antiferromagnetic properties of high-temperature superconductors is thus important for checking hypotheses on magnetic mechanisms of superconductivity. The interaction of copper spins in a plane is of two-dimensional nature. Their small value, $S = 1/2$, is the reason behind the important quantum fluctuations. In this respect, besides exploring the interplay of antiferromagnetism and superconductivity, the study of quantum two-dimensional antiferromagnets is of great interest in its own right.

The first indications of the existence of antiferromagnetism in copper-oxide compounds were obtained on the basis of macroscopic measurements of susceptibility. However, a detailed study of both magnetic structure and spin correlations in the metallic phase became possible only with the aid of neutron scattering. Since these experiments required large single crystals, most of the results have been obtained for the compounds La_2CuO_4 and $YBa_2Cu_3O_{6+x}$ which can be synthesized as large crystals of due quality. In the present chapter, the magnetic properties of these compounds, namely, magnetic structure, spin correlations, and magnetic excitations, are discussed. Spin dynamics in the superconducting phase is also responsible for variation in the spin-lattice relaxation rates of nuclear spins. The main results obtained by the NMR method are discussed at the end of the chapter.

3.1 Antiferromagnetism in La$_2$CuO$_4$ Compounds

3.1.1 Magnetic Structure

Stoichiometric La$_2$CuO$_4$ is an antiferromagnet (AF) with the Néel temperature $T_N \simeq 300$ K. In Fig. 2.5 one can see a phase diagram of La$_{2-x}$M$_x$CuO$_4$. It shows that the antiferromagnetic state occurs in the orthorhombic phase with the lattice constants $a \simeq c \simeq a_t\sqrt{2}$ and $b \simeq c_t$. Figure 3.1 shows a unit cell of La$_2$CuO$_4$ in the orthorhombic phase with axes $b > a > c$ where the direction of spins in the antiferromagnetic phase is shown [3.1]. The spins $S = 1/2$ on the Cu^{2+} ions are directed along the orthorhombic axis c, and the antiferromagnetic modulation is along the a axis with the wave vector $Q_{AF} = (1, 0, 0)$. The value of the magnetic moment at a copper site is $\mu = (0.5 \pm 0.15)\mu_B$ [3.1]. For the copper ion Cu^{2+} with spin $S = 1/2$, the magnetic moment should be equal to $\mu = gS\mu_B = 1.14\mu_B$. The smaller observed value of magnetic moment may be due to quantum spin fluctuations and an influence of the covalent Cu$-$O bond. In the case of a two-dimensional Heisenberg magnet with spin $S = 1/2$, the spin fluctuations reduce the magnetic moment to 0.62 of its static value, i.e. to $0.68\mu_B$ for Cu^{2+}.

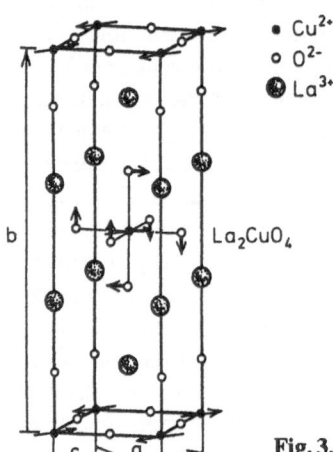

Fig. 3.1. Magnetic structure in the orthorhombic phase of La$_2$CuO$_4$ [3.1]

   An interesting feature of the antiferromagnetic ordering of spins in La$_2$CuO$_4$ is the occurrence of a weak ferromagnetic moment in the CuO$_2$ planes which is directed perpendicular to the plane and has opposite directions in neighboring planes [3.2]. The ferromagnetic moment has a small value of $\mu_1 = 2 \cdot 10^{-3}\mu_B$ per copper atom. It results from the displacement of copper spins from the (a,c) plane under rotation of the spins by a small angle ($\simeq 0.17°$) due to the antisymmetric exchange interaction. In an external field along the axis b, a spin reorientation transition may occur. In this transition the weak ferromagnetic moment in those planes where it is directed against the field changes its direction to be along the field. It results from the rotation of the antiferromagnetic moments by 180° in the (a, c) plane so that the wave vector of the antiferromagnetic structure changes its direction from [100] (along the a axis) to [001] (along the c axis). This is easily seen from the magnetic structure shown in Fig. 3.1 with the spin of a copper atom of opposite orientation in the center of a cell. In Ref. [3.3], the disappearance of the magnetic Bragg peaks from the (100) plane with increasing external magnetic field H and appearance of new Bragg peaks (201) at $H > H_{cr}(T)$ have been directly observed. Here $H_{cr}(T)$ is a critical magnetic field which depends on temperature and $H_{cr} \simeq 5$ Tesla as $T \to 0$.

In order to describe the magnetic phase diagram of La$_2$CuO$_4$, a Heisenberg model with antisymmetric exchange (the Dzyaloshinsky–Moria interaction) has been introduced in [3.2]

$$H = \sum J_{ij}^{\alpha\beta} S_i^\alpha S_j^\beta , \qquad (3.1)$$

where $J_{ij}^{\alpha\beta} = \{J^{aa}, J^{bb}, J^{cc}, J^{bc} = -J^{cb}\}$ is the exchange interaction for the nearest–neighbour sites (i, j). The antisymmetric exchange gives rise to a spin tilted by an angle $\theta \simeq J^{bc}/J_{nn}$, where $J_{nn} = (1/3)(J^{aa} + J^{bb} + J^{cc})$ within each layer. For small values of the anisotropy, the spin lies in the (b, c) plane deviating by an angle θ from the c axis. An inter–plane interaction brings about an antiferromagnetic ordering of the magnetic moments of each layer along the b axis. The parameters of the model obtained from the comparison of the theoretical calculation with experimental data are [3.1]

$$J_{nn} \simeq 116 \,\text{meV}, \quad J^{bc} \simeq 0.55 \,\text{meV}. \qquad (3.2)$$

The anisotropy of the interaction is small, $\Delta_{bc} = J^{bb} - J^{cc} \simeq 0.004$ meV, $\Delta_{ac} \simeq 10^{-4}$ meV. The average value of the inter–plane coupling is $J_\perp \simeq 0.002$ meV. The values of the parameters show that the system of copper spins in La$_2$CuO$_4$ is sufficiently well described by the two-dimensional Heisenberg model with a very small anisotropy.

3.1.2 A Phenomenological Theory of Magnetic Phase Transitions in La$_2$CuO$_4$

Studies of the antiferromagnetic phase transition in La$_2$CuO$_4$ and in the related compounds La$_2$NiO$_4$ and La$_2$CoO$_4$ (see [3.1]) show that the magnetic ordering of spins is symmetrically connected with structural transitions. In view of this, we shall consider a phenomenological theory of magnetic phase transitions taking into account the theory of structural transitions discussed in Sect. 2.2.2 [3.4].

As is shown in sect. 2.2.2, the structural transitions from the tetragonal (HTT) to orthorhombic (LTO) and low–temperature tetragonal (LTT) phases are specified by a two-component order parameter (C_1, C_2) related to the irreducible representation X_3^+ on the two-arm star of the wave vector $k_x(1, 2)$ (2.2). The order parameter (OP) describes the condensation of the soft rotational mode (2.1) $R_{1,2} = R_x \mp R_y$ for the wave vectors $k_x(1, 2)$ respectively. The antiferromagnetic phase transition in La_2CuO_4 is specified by a two-component order parameter (S_1, S_2) related to the freezing−in of spin fluctuations at copper sites

$$S_1[k_x(1)] \propto (S_x - S_y), \quad S_2[k_x(2)] \propto (S_x + S_y), \tag{3.3}$$

which also have symmetry X_3. The weak ferromagnetic moment in the CuO_2 planes is specified by a secondary order parameter, namely, the spin component S_z at the copper sites with the wave vector $k_z = \pi(0, 0, 1/\tau_z)$ in the notation of Sect. 2.2.2. (We note that the notation for the HTT and LTO phases used in this chapter does not coincide with that in [3.1].)

A symmetry analysis of the spin-phonon interaction in the tetragonal phase, which is chosen as the basic phase in describing the sequence of structural and magnetic phase transitions, enables one to write down a complete expansion for the free energy in the order parameters [3.4]

$$F = F_c + F_\varepsilon + F_{c\varepsilon} + F_s + F_{s\varepsilon} + F_{sc}. \tag{3.4}$$

Here, the first three terms are the lattice part of the free energy. They are determined by expansions (2.3−5). A magnetic term F_s and magneto− deformational part $F_{s\varepsilon}$ have the same form as expressions (2.3) and (2.5) with the substitution $C_i \rightarrow S_i$ if one also takes into account the secondary order parameter S_z. For example,

$$F_s = \frac{1}{2}r_N(S_1^2 + S_2^2) + \frac{1}{2}u_N S_1^2 S_2^2 + \frac{1}{4}v_N(S_1^4 + S_2^4) + \frac{1}{2}r_z S_z^2, \tag{3.5}$$

where $r_N = a_N(T - T_N)$ and r_z does not depend on temperature. The interaction of the magnetic and structural order parameters is described by the expansion

$$\begin{aligned} F_{sc} = &(\alpha_s(S_1^2 + S_2^2) + \beta_s S_z^2)(C_1^2 + C_2^2) \\ &+ \gamma_s(S_1^2 - S_2^2)(C_1^2 - C_2^2) + \lambda(C_1 S_2 S_z - C_2 S_1 S_z), \end{aligned} \tag{3.6}$$

where the mixed invariant of the third order may occur due to the fact that the sum of wave vectors $k_x(1) + k_x(2) + k_z$ is equal to the reciprocal lattice vector b_1 (see (2.2)). This invariant is directly related to the antisymmetric interaction in the microscopic model (3.1), $J^{bc} \propto \lambda C_{1,2}$, which is zero in the HTT phase at $C_1 = C_2 = 0$.

The analysis of the complete expansion of the free energy (3.4) enables one to study the phase diagram in the space of order parameters of the structural and magnetic phase transitions. In particular, in the LTO phase at $T < T_N < T_0$, the solution $C_1 \neq 0$, $S_2 \neq 0$, $S_z \neq 0$ at $C_2 = S_1 = 0$ is possible for one domain. A similar solution under the interchange of indices 1 and 2 is possible for the second domain. This solution describes a non-collinear antiferromagnetic structure in La_2CuO_4, where the direction of the spins S_2 (S_1) is perpendicular to the wave vector $k_x(2)$ ($k_x(1)$) (see (3.3)) and the axis of rotation of the octahedra R_1 (R_2) (see (2.1)). The weak ferromagnetic moment perpendicular to the plane $S_z \sim \lambda C_1 S_2$ ($\lambda C_2 S_1$) points in opposite directions for neighboring CuO_2 planes, since it is determined by the wave vector $k_z = \pi(0, 0, 1/\tau_z)$.

The magnetic ordering of spins in La$_2$NiO$_4$ and La$_2$CoO$_4$ is related to another irreducible representation X_5 whose pseudo−vector basis functions are transformed as spin components

$$S_1'[\boldsymbol{k}_x(1)] \propto (S_x + S_y), \quad S_2'[\boldsymbol{k}_x(2)] \propto (S_x - S_y). \tag{3.3a}$$

In this case, the direction of the spins S_1' (S_2') coincides with that of the wave vector $\boldsymbol{k}_x(1)$ ($\boldsymbol{k}_x(2)$). The expansion of the free energy in (S_1', S_2') has the same form as (3.5) and (3.6) except for the invariant of third order in (3.6) which becomes

$$F_3 = \lambda(C_1 S_1' M_z - C_2 S_2' M_z),$$

where M_z is a ferromagnetic moment along the tetragonal axis z. The collinear antiferromagnetic structure in La$_2$NiO$_4$ and La$_2$CoO$_4$ in the orthorhombic phase is described by the solution $S_2' \neq 0$, $S_1' = M_z = 0$ for the domain $C_1 \neq 0$, $C_2 = 0$ and a similar solution (under permutation of indices 1 and 2) for the second domain.

The magnetic structure of La$_2$CuO$_4$ in an external magnetic field $H_z > H_{cr}$ due to a spin-reorientation transition is also described by the basis functions (3.3a). Under the spin-reorientation transition in the orthorhombic phase, the antiferromagnetic structure with order parameter $S_2 \neq 0$, $S_z \neq 0$ for domain $C_1 \neq 0$ transforms into a structure with the order parameters $S_1' \neq 0$, $M_z \neq 0$ when the energy of the external field $M_z H_z$ becomes larger than the energy difference $F(S_i') - F(S_i)$. The antiferromagnetic vector $\boldsymbol{k}_x(2)$ for the parameter S_2 (3.3) transforms into the antiferromagnetic vector $\boldsymbol{k}_x(1)$ for the parameter S_1' (3.3a), while the weak ferromagnetic moment in the plane $S_z \propto (\lambda/r_z)C_1 S_2$ transforms into the homogeneous moment $M_z \propto (\lambda/r_z)C_1 S_1'$.

The above phenomenological theory also allows one to analyze possible magnetic structures in the LTT phase P4$_2$/ncm which is observed in La$_2$CoO$_4$ [3.1, 4].

3.1.3 Spin Dynamics in La$_{2-x}$M$_x$CuO$_4$

The temperature of the antiferromagnetic phase transition turns out to be very sensitive to the concentration of divalent impurities M = (Ba, Sr) which replace the trivalent La ions and to the concentration of oxygen vacancies. The phase diagram in Fig. 2.5 shows that, even at a concentration $x = 0.02$, the long-range antiferromagnetic order already disappears. Only a spin-glass phase, i.e., the phase of frozen spins at copper sites, remains in the region of low temperatures. This phase has been observed most distinctly in μSR experiments. Figure 3.2 shows the concentration dependence of the Néel temperature $T_N(x)$, $T_c(x)$ and the average local magnetic field $\langle | B_\mu | \rangle$ at a temperature of 35 mK measured by the μSR method [3.5].

The dependence implies that the long-range antiferromagnetic order disappears at $x \simeq 0.02$, while the static local magnetic field remains over a wide range of concentrations x including the superconducting phase. At the same time, Fig. 3.2

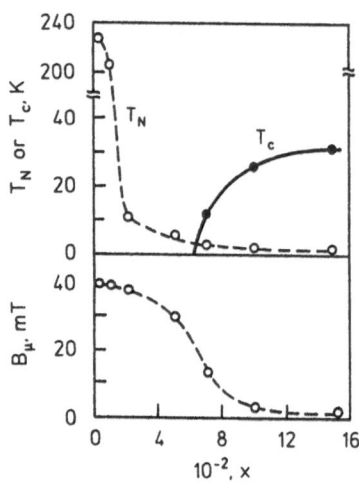

Fig. 3.2. Concentration dependence of $T_N(x)$, $T_c(x)$ and the average local field $\langle\langle| B_\mu^2 |\rangle\rangle^{1/2}$ measured by the μ SR method in $La_{2-x}M_xCuO_4$ [3.5]

clearly shows anticorrelation of the intensities of magnetic fluctuations $\propto \langle| B_\mu |^2\rangle$ and $T_c(x)$. As x increases, the former falls off rapidly, while T_c increases.

The fast destruction of the long-range antiferromagnetic order when holes appear in the CuO_2 plane can be partly accounted for by a frustration mechanism [3.6]. Experimental studies of the electronic structure of copper-oxide compounds (Sect. 5.4) show that, under doping, the holes appear at oxygen ions. A strong exchange interaction occurs between the hole spin σ and two neighboring copper spins S_1 and S_2:

$$H = -J_\sigma\sigma(S_1 + S_2).$$

Independent of the sign of J_σ, this coupling gives rise to an effective ferromagnetic interaction of the spins S_1 and S_2 competing with the antiferromagnetic exchange of spins due to J_{nn} in (3.1). A considerable contribution to the destruction of the long-range order is also made by the motion of the hole on antiferromagnetic background. The kinetic energy of the hole is much larger than the exchange interaction J_{nn} (Sect. 7.1). Therefore, delocalization of the hole accompanied by the destruction of the long-range antiferromagnetic order may turn out to be energetically more favorable.

In the same time, experimental studies by means of quasi−elastic and inelastic neutron scattering in $La_{2-x}M_xCuO_4$ show the existence of strong antiferromagnetic correlations of short-range order of copper spins in the region $T > T_N(x)$. In these experiments, a spin pair correlation function has been measured. Its Fourier transform determines the cross−section of inelastic magnetic scattering of neutrons

$$\frac{d^2\sigma}{d\Omega\,dE} \propto \sum_\alpha (1 - \tilde{Q}_\alpha^2)S^{\alpha\alpha}(Q,\omega). \tag{3.7}$$

Here $Q = k - k_0$, $\tilde{Q}_\alpha = Q_\alpha/| Q |$, and $\omega = E - E_0$ are respectively transferred momentum and energy of a neutron, while $k_0(k)$ and $E_0(E)$ are the incoming (outgoing) neutron momentum and energy. The Fourier transform of the pair spin correlation function

$$S^{\alpha\alpha}(Q,\omega) = \frac{1}{2\pi N} \int_{-\infty}^{\infty} dt\, e^{i\omega t} \langle S^{\alpha}(-Q,0)S^{\alpha}(Q,t)\rangle \tag{3.8}$$

determines a fluctuation of the spin density

$$S^{\alpha}(Q,t) = \sum_{n} e^{iQn} S_{n}^{\alpha}(t)\,, \tag{3.9}$$

where the summation is performed over all the lattice sites n.

For a system with long-range magnetic order, for example with antiferromagnetic ordering with a wave vector Q_{AF}, the expectation value of the spin in (3.9) differs from zero:

$$\langle S^{\alpha}(Q,t)\rangle = N\Delta(Q - Q_{AF})S_{\alpha}\,. \tag{3.10}$$

This gives rise to magnetic Bragg peaks in the cross−section (3.7)

$$\frac{d\sigma}{d\Omega\, dE} \propto N\Delta(Q - Q_{AF})\delta(\omega)\sum_{\alpha}(1 - \tilde{Q}_{\alpha})^{2}S_{\alpha}^{2}\,, \tag{3.11}$$

where the function $\Delta(k)$ is equal to unity if k is zero or a reciprocal lattice vector, and equal to zero otherwise. By measuring the dependence of the cross section intensity on the direction of Q_{α} one can determine the direction of spin ordering.

When integrated over all scattering energies (at fixed Q), the function (3.8) determines a spin correlation function

$$S^{\alpha\alpha}(Q) = \int d\omega S^{\alpha\alpha}(Q,\omega) = \frac{1}{N}\sum_{n,m} e^{iQ(n-m)}\langle S_{n}^{\alpha}S_{m}^{\alpha}\rangle\,. \tag{3.12}$$

In the paramagnetic phase, at $T > T_{N}$, the spin correlations decay exponentially, as the spin separation r increases

$$\langle S^{\alpha}(0)S^{\alpha}(r)\rangle \propto \frac{1}{r}e^{-\kappa r}\,. \tag{3.13}$$

In this case, the scattering cross−section is specified by the function

$$S^{\alpha\alpha}(Q) \propto \frac{1}{q^{2} + \kappa^{2}}\,. \tag{3.14}$$

Its measurement at various values of the scattering vector $q = Q - Q_{AF}$ enables the correlation length $\xi = 1/\kappa$ to be determined. It is convenient to perform the measurement of the integral scattering intensity (3.12) with the aid of a two-axes spectrometer, where only the direction of outgoing neutrons is fixed. This has the effect of integrating over the scattering energy ω.

In view of the quasi-two-dimensional nature of the spin interaction in the model (3.1) for La$_2$CuO$_4$, a quasi-two-dimensional nature of spin correlations should be expected in this compound. In this case, the diffuse scattering intensity described by the quantity $S^{\alpha\alpha}(Q)$ (3.12) should have the form of rods connecting the reciprocal lattice sites in the plane (a, c) in the direction perpendicular to the plane. In the neutron experiments [3.1], a special scattering geometry was chosen, where the momentum of scattered neutrons k is parallel to the reciprocal lattice vector b^* : $k = (0, \zeta', 0)$ (here a^*, b^*, and c^* are the reciprocal lattice vectors in the orthorhombic phase). In this case, for a fixed momentum vector of the incoming neutrons $k_0 = (\nu, \zeta, 0)$ in the plane (a^*, b^*) of reciprocal space, the momentum transfer $Q = k - k_0 = -(\nu, \zeta - \zeta', 0)$ would have a constant value ν for the component $q_\parallel$ in the plane (a^*, c^*). In this experiment, integrating over all the momenta of scattered neutrons k, it is therefore possible to determine the correlation function $S^{\alpha\alpha}(q_\parallel)$ (3.12) describing the two-dimensional spin correlations in the CuO_2 plane.

We shall now discuss the results of neutron experiments [3.1] where a detailed study of the diffuse magnetic scattering was carried out. Measurement of the function $S^{\alpha\alpha}(q_\parallel)$ (3.12) has shown that, in La_2CuO_4, scattering from spin fluctuations is observed far above the Néel temperature. The correlation length $\xi_\parallel$ for magnetic spin correlations in the plane varies from 40 Å at 500 K to 400 Å as the temperature tends to T_N.

The integral intensity of the diffuse scattering rises slightly as the temperature decreases and approaches T_N. However, below the temperature of magnetic ordering, the intensity gradually goes down and tends to zero as $T \to 0$. At the same time, the intensity of the magnetic Bragg peak (100) increases as usual and approaches its maximum value at $T = 0$. Thus, a transformation of the diffuse scattering into the Bragg peak takes place. This is observed in traditional two-dimensional antiferromagnets, for example, in K_2NiF_4. In La_2CuO_4, however, this transformation is smooth, while in K_2NiF_4 it is rather sharp occurring over a small temperature interval of about $\propto 2\%T_N$. This reflects the fundamental difference in the nature of three-dimensional ordering in these planar antiferromagnets. In K_2NiF_4, which is an Ising type magnet, the transformation to long-range order is essentially of two-dimensional nature. In La_2CuO_4, three-dimensional long-range order is formed due to an inter–plane interaction. In isomorphic magnets La_2NiO_4 and La_2CoO_4 which possess atomic spins $S = 1$ and $S = 3/2$ respectively, the phase transition to long-range order is of the Ising type and is similar to that in K_2NiF_4. However, only the La_2CuO_4 compound reveals special properties. While possessing distinct quasi-two-dimensional fluctuations, at the phase transition it behaves like a Heisenberg antiferromagnet with $S = 1/2$.

Experiments with a three-axes spectrometer where the energy of the scattered neutrons is also measured, have enabled the dynamics of spin fluctuations to be investigated. Unlike the low–energy dynamics of spin fluctuations in the vicinity of the phase transition in normal three-dimensional magnets, spin excitations in La_2CuO_4 have turned out to have high energy at temperatures well above T_N. The dispersion of the excitations is of a two-dimensional character, $\omega(q) = vq_\parallel$, i.e., it does not depend on the component of the wave vector perpendicular to the CuO_2 plane. The velocity of the spin excitations turns out to be extremely high: at $T = 300$ K it exceeds 0.6 eV· Å. This value agrees with data obtained for spin-wave excitations from the evaluation of the in−plane exchange integral J_{nn} (3.2) and also with data from a two-magnon Raman scattering [3.7].

In Ref. [3.8], the spin dynamics of La$_2$CuO$_4$ in the region of high-energy spin excitations up to 76 meV has been investigated. The results of the experiment turned out to fit well into the two-dimensional spin-wave theory with dispersion of spin waves equal to $v \simeq 0.9$ eV· Å. This quantity remains practically constant, as the temperature increases from 18 to 300 K.

Studies of the temperature dependence of the density of spin excitations have shown that the intensity of inelastic peaks does not depend on T in the interval 200–300 K (in the paramagnetic phase) and that it falls in the interval 150–5 K following the factor $n(\omega) + 1$ of the Bose distribution function $n(\omega)$ [3.1]. This implies that the spin excitations of the system at $T < T_N$ are the usual spin waves in the Néel antiferromagnetic state.

We see that the $S = 1/2$ Heisenberg antiferromagnet La$_2$CuO$_4$ reveals unusual properties also in paramagnetic region of temperatures. They consist in the existence of two-dimensional spin correlations at distances of order 200 Å and in relatively high-energy spin excitations. Such a state has been called a quantum spin liquid (QSL) [3.1]. The term liquid reflects the fact that the structural factor, i.e., the quantity $S^{\alpha\alpha}(Q)$, is of a purely dynamical nature.

Quantum effects play a very important role in this spin liquid behaviour. One of their manifestations is a considerable decrease of the correlation length compared to the classical two-dimensional Heisenberg magnet. Furthermore, in classical three- and two-dimensional systems a slowing down of fluctuations is observed as soon as the correlation length becomes large. In contrast to this, nothing similar occurs in La$_2$CuO$_4$ and the fluctuations remain of a high-energy nature. In this compound, the velocity of spin excitations exceeds the velocity of sound by an order of magnitude.

Let us now proceed to doped compounds La$_{2-x}$M$_x$CuO$_4$. Figure 2.5 shows that Sr−doped compounds undergo a metal−insulator phase transition at $x \simeq 0.05$. In the region $x \leq 0.05$ and $T \leq 100$ K, a typical hopping conductivity is observed where $\ln\sigma \propto - (T_0/T)^{1/4}$. The carriers turn out to be localized due to disorder. At $x > 0.05$, the carriers (holes) are delocalized so that the compound becomes a metal and superconductor. As we have already seen, T_N depends drastically on the concentration x of a divalent doping metal. In the metallic phase, in the absence of long-range magnetic order, doped compounds can also be in the state of quantum spin liquid. In this case, the high-energy spin fluctuations could result in a pairing interaction of electrons. This idea has motivated researchers of high-temperature superconductivity to study spin fluctuations in doped systems in greater detail.

Experiments performed on a two-axes spectrometer have shown that the correlation length of spin fluctuations drastically falls as the dopant concentration increases (Fig. 3.3). A solid line in Fig. 3.3 describes the function $3.8/\sqrt{x}$ (Å), which determines the average distance between O$^-$ holes in the CuO$_2$ planes due to Sr doping. The good agreement obtained between the correlation length ξ in these experiments and the average distance between holes shows that holes really

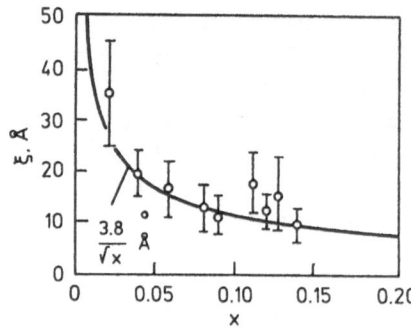

Fig. 3.3. The dependence of the magnetic correlation length ξ on x in $La_{2-x}Sr_xCuO_4$ [3.1]

do destroy the magnetic state in the system of spins at Cu^{2+} ions. A correlation length much smaller than the average distance between holes has been obtained in [3.8], $\xi \simeq 7.5\,\text{Å}$ at $x = 0.1$. It is clear that, in order to explain the dependences $T_N(x)$ and $\xi(x)$, a consistent microscopic theory and more thorough experiments with homogeneous samples are required.

The decrease in the correlation length with increasing x should correlate to the variation of T_N in the magnetically ordered phase. In fact, the temperature of the three-dimensional phase transition in the quasi-two-dimensional model can be written in terms of a correlation length $\xi_\parallel$ of two-dimensional spin fluctuations expressed in units of the average distance between neighbors

$$kT_N \simeq J_\perp \xi_\parallel^2(T_N). \tag{3.15}$$

The measurement of the intensities of peaks of spin correlations $S^{\alpha\alpha}(Q)$ integrated over transfer momentum has shown that these quantities which determine the local magnetic moment at a Cu^{2+} ion are independent of dopant concentration. This leads to an important general conclusion that, in doped compounds, holes only affect the correlation of spins at copper ions while they do not change the value of the atomic magnetic moment of the copper due to its local nature.

Spin fluctuations in strongly doped samples are mainly high-energy at room temperatures. However, even at 350 K, there is a good fraction of fluctuations with low energy $E > 0.5\,\text{meV}$. These low–energy fluctuations prove to be three-dimensional. In [3.8], a considerable decrease in spin rigidity under doping has been found, although the description of inelastic paramagnetic scattering, particularly in the region of low energies, encounters a number of difficulties within the theory of spin waves. As the temperature decreases, the fraction of the low–energy component increases. The μSR experiments show that all the spins are frozen at temperatures below about 4 K [3.5].

Inelastic neutron scattering experiments reveal also an incommensurate anti-ferromagnetic structure in doped crystals of $La_{2-x}Sr_xCuO_4$ [3.1, 9]. The maxima of inelastic magnetic scattering appear at the incommensurate AF wave vectors $Q_{AF}^* = (1, q_y), (q_x, 1)$ where $q_{x,y} = (1 \pm \delta)$, $\delta \simeq 0.24$ for $La_{1.86}Sr_{0.14}CuO_4$ [3.9].

In $La_{2-x}Sr_xCuO_4$, superconductivity thus occurs in the presence of a slowly fluctuating spin liquid. The origin of such a peculiar magnetic state is the holes introduced by the dopants.

3.2 Antiferromagnetism in YBa$_2$Cu$_3$O$_{6+x}$ Compounds

3.2.1 The Magnetic Phase Diagram

In the insulating phase of YBa$_2$Cu$_3$O$_{6+x}$, i.e., at $x < 0.4$, one observes antiferro-magnetic ordering of magnetic moments at Cu2 sites in CuO$_2$ planes. It is quite similar to the antiferromagnetic transition in La$_2$CuO$_4$. First reliable indications of the antiferromagnetic phase transition have been obtained with the aid of magnetic neutron scattering [3.10, 11]. Further investigations of large single crystals have yield rather detailed information on both the static and dynamic characteristics of the spin subsystem and antiferromagnetic correlations (see [3.1, 12 − 14]).

The antiferromagnetic ordering in the tetragonal phase of YBa$_2$Cu$_3$O$_{6+x}$ is described by the wave vector $Q_{AF} = (1/2, 1/2, l)$, which corresponds to a magnetic unit cell with the parameters $(a\sqrt{2}, a\sqrt{2}, c)$ where a and c are the lattice constants of the tetragonal unit cell of Fig. 2.12b. The observation of magnetic Bragg peaks $(1/2, 1/2, l)$ with integral values l unambiguously shows that the magnetic moments at Cu2 ions lie in the basis plane as shown in Fig. 3.4. The absence of scattering with $l = 0$ proves that the magnetic moments in the bilayer Cu − Y − Cu are also antiferromagnetically ordered and that there is no magnetic moment on the Cu1 ions. The value of the magnetic moment on the Cu2 ions at low temperatures in the region $x < 0.2$ is $\mu = 0.64\mu_B$. This value corresponds to the magnetic moment in La$_2$CuO$_4$. Due to quantum fluctuations, the value of magnetic moment in YBCO is also smaller than its static value.

As the oxygen concentration x in the Cu1− O1 chains increases, the values of the ordered magnetic moment μ and of the Néel temperature T_N decrease as shown in Fig. 3.5 so that $\mu(x) \propto T_N(x)$. The value of the local moment on the Cu2 ions, like in La$_{2-x}$Sr$_x$CuO$_4$, remains constant. This indicates a localization of holes at Cu2 ions in the $3d^9$ state. The dependences $\mu(x)$ and $T_N(x)$ can be subdivided into three regions [3.12]. At $x < 0.20$ these quantities are constants. Then, they start to gradually decrease up to $x \simeq 0.35$, and afterwards they rapidly tend to zero

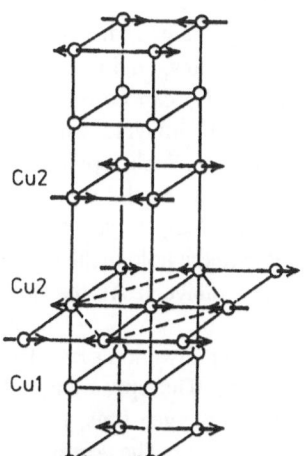

Cu2

Cu2

Cu1

Fig. 3.4. The antiferromagnetic structure of YBa$_2$Cu$_3$O$_{6+x}$ [3.11]

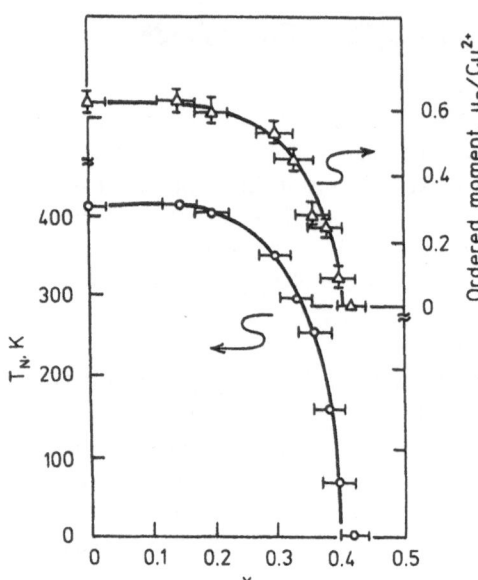

Fig. 3.5. The average magnetic moment μ and the Néel temperature T_N in as a function of oxygen concentration [3.12]

as x tends to a critical oxygen concentration $x \rightarrow x_c = 0.41$, at which the long-range antiferromagnetic order completely vanishes. The study of diffuse magnetic scattering in the CuO_2 plane permits one to determine the dependence of the magnetic correlation length $\xi(x)$ and to evaluate the maximum hole concentration in the plane which behaves like $n \propto (1/\xi^2)$. The maximum value at $x_c = 0.41$ is equal to $n_c \simeq 0.02$, as in the case of $La_{2-x}Sr_xCuO_4$, when the long-range antiferromagnetic order disappears (see Fig. 2.5).

Thus the specific variation of $T_N(x)$ in YBCO is accounted for by a more complicated mechanism of charge transfer from $Cu1 - O1$ chains to CuO_2 planes, as the oxygen concentration in $YBa_2Cu_3O_{6+x}$ increases. As was noted in Sect. 2.4, the effective valence of copper Cu2 in the planes is almost constant up to an oxygen concentration $x = 0.4$ (Fig. 2.16). In this region, only charge redistribution at Cu1 ions ($Cu^{+1} \rightarrow Cu^{+2}$) and the formation of holes at oxygen ions O1 in the chains take place. In this region, a gradual decrease of T_N can be expected due to weakening of the coupling between $Cu2- Y - Cu2$ bilayers with the same strong antiferromagnetic correlations within each bilayer. Such a picture is supported by the observed decrease of the Bragg peak intensity in the region of low temperatures $T < T_r$, where T_r is the so-called reentrant temperature. Below it the antiferromagnetic long-range order is suppressed due to the hole localization and a decrease of the antiferromagnetic coupling between bilayers. As the hole concentration increases $x \rightarrow x_c$, the temperature T_r tends to zero $T_r \rightarrow 0$. In the region $x \simeq 0.4$, the number of holes in the bilayer rapidly increases. The valence of the Cu2 copper sharply increases (see Fig. 2.16). The appearance of a large number of holes in the layer $Cu2- O2- O3$ at the oxygen ions, like in LSCO, suppresses two-dimensional antiferromagnetic correlations. The correlation length $\xi_\parallel$ decreases (Fig. 3.3) and long-range antiferromagnetic order disappears. Measurement of the critical exponent of magnetization of the antiferromagnetic sublattices $L \propto (T - T_N)^\beta$ shows that, in the region $x < 0.2$, it is equal to $\beta = 0.25$, which corresponds to the quasi-two-dimensional XY model [3.12].

3.2.2 Spin Dynamics in $YBa_2Cu_3O_{6+x}$

The most complete data on spin dynamics and antiferromagnetic short-range order correlations have been obtained with the aid of magnetic inelastic neutron scattering [3.12 − 14]. Spin waves in the region of the pure antiferromagnetic state at $x < 0.2$ and in the region of low hole concentration $x < 0.4$ have been investigated. The spectrum of antiferromagnetic spin fluctuations in the metallic phase at various hole concentrations including the superconducting phase with $T_c \simeq 40 − 90\,K$ has also been studied.

The spectrum of spin waves in the pure antiferromagnetic phase can be described with the aid of a model Hamiltonian for $S = 1/2$ copper spins in the CuO_2 plane [3.12]

$$H = -\sum_{ij} \left\{ J_{ij}^\parallel S_i^z S_j^z + J_{ij}^\perp (S_i^x S_j^x + S_i^y S_j^y) \right\},$$ (3.16)

The difference $\Delta J = J^\perp - J^\parallel$ determines the anisotropy of coupling for the in−plane $J^\perp$ and out−of−plane $J^\parallel$ spin components. The most important coupling constants are those of a superexchange interaction via oxygen ions for two neighboring spins on the Cu2 ions in the plane $2J$, the exchange interaction of spins in the bilayer $2J_b$, and the exchange interaction of $2J'$ spins on Cu2 ions via the Cu1− O1 chains. Due to existence of two spins in each antiferromagnetic unit cell, i.e., bilayer Cu2− Y − Cu2, acoustic and optical antiferromagnetic spin waves are possible. The velocity of acoustic spin waves turns out to be very large $v = 4\sqrt{2}JSa = 1\,eV\cdot Å$. After taking into account quantum corrections, the corresponding spin coupling Cu2− Cu2 in the plane is $2J \simeq 0.15\,eV$. The gap in the excitation spectrum for longitudinal components determines the value of the anisotropy $\Delta J = 10^{-4}J$ which turns out to be a small quantity, as in the case of La_2CuO_4. The coupling between spins in the bilayer which determines the gap in the optical spectrum of spin waves $\Delta E = 8JS\sqrt{J_b/J}$ is evaluated by the quantity $J_b \simeq 10^{-2}J$. Finally, the coupling between bilayers J' which determines the dispersion of the acoustic mode along the z axis is estimated by the quantity $J' = 10^{-5}J$. The values of these parameters show that the spin dynamics in $YBa_2Cu_3O_{6+x}$ should be determined by the XY Heisenberg model for spins $S = 1/2$ with very small anisotropy and strong two-dimensional fluctuations, as in La_2CuO_4.

Like in LSCO, the appearance of holes in the Cu2− (O2, O3) layers drastically changes the spin dynamics. The correlation length ξ goes down, the damping of spin waves in the plane increases and their velocity decreases. For example, according to [3.12], at $x = 0.37$, we have $\xi = 7.5a$ ($n_h = 0.018$), $v = 0.45\,eV\cdot Å$. At the critical concentration, when the long-range antiferromagnetic order disappears, the spin rigidity vanishes $v \rightarrow 0$.

The most interesting results have been obtained for the transition to the metallic and superconducting phases at $x > 0.41$. The increase of hole concentration in the CuO_2 layer and their delocalization at the transition to the metallic phase drastically decrease the magnetic correlation length to $\xi \simeq 2.2a$ ($x = 0.45$), which weakly depends on temperature. However, like in LSCO compounds, intensive dynamic antiferromagnetic fluctuations are maintained for wave vectors near rods of two-dimensional antiferromagnetic diffusion scattering at $Q = (1/2, 1/2, l)$. The investigation with the better resolution in $YBa_2Cu_3O_{6.6}$ ($T_c = 53$ K) [3.14] suggests the incommensurate structure, as in LSCO−compounds [3.9].

The spectrum of spin fluctuations is distributed over a wide energy range from zero up to $40 - 50$ meV. The maximum energy $\hbar\omega_{max}$ increases with hole concentration. At $x = 0.45$ (0.51), it is $\hbar\omega_{max} = 8$ (22) meV. In the region of low energies, $E < E_G$, the intensity of fluctuations falls sharply as temperature decreases. This can be described as the appearance of a gap in the spectrum of spin fluctuations at some temperature T_s close or above the temperature T_c of the superconducting phase transition. For $x = 0.45(0.51)$ and $T_c = 37(47)$ K, we have $E_G = 3(4)$ meV. In Fig. 3.6, the dependence of the imaginary part of the dynamical spin susceptibility $\mathrm{Im}\,\chi(Q, \omega)$ is shown at $x = 0.69$ and $T_c = 59$ K. It is related to the dynamical structure factor $S(Q, \omega)$ (3.8) by the equation

$$S(Q, \omega) = \frac{1}{\pi} \left[1 - \exp\left(-\frac{\hbar\omega}{kT}\right) \right]^{-1} \mathrm{Im}\,\chi(Q, \omega). \qquad (3.17)$$

In the inset of Fig. 3.6 the temperature dependence of the spectrum at $\hbar\omega = 8$ meV is clearly seen. The main figure shows the energy dependence of the fluctuation spectrum at various temperatures. The value of the gap in the spectrum of spin fluctuations $E_G = 16$ meV is comparable with $3kT_c$, although it is smaller than the value of the superconducting gap in the plane $2\Delta_{ab} \simeq 6kT_c$ (see Sect. 5.6), like in samples with $x = 0.45 - 0.51$ where $E_G \simeq kT_c$. Nevertheless, the appearance of the gap E_G at $T \simeq T_c$ points to a certain relation between antiferromagnetic spin fluctuations at Cu2 ions and the formation of superconducting Cooper pairs in the electron liquid. The existence of a magnetic gap has been supported in recent experiments on neutron scattering in single−crystals, in particular, in samples with $x = 0.92$ at $T_c = 91$ K [3.13].

The relationship between these two subsystems, namely, the spin quantum liquid, i.e. the system of spins at copper sites in the CuO_2 plane with strong an-tiferromagnetic exchange coupling, and the strongly correlated system of holes at copper and oxygen sites in the same CuO_2 plane, is not yet quite clear. NMR stud-ies show that a picture of a single one−component spin liquid made of strongly bound singlet states of two holes at copper and neighboring oxygen sites looks most plausible. The data on the rate of spin−lattice relaxation obtained in these experiments agree with the density of spin fluctuations measured by neutron ex-periments (3.17). NMR studies will be discussed in greater detail in Sect. 3.3.

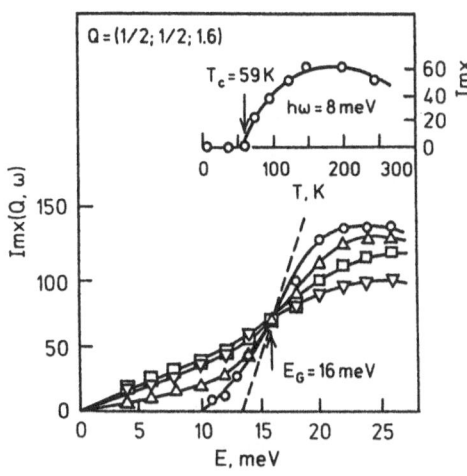

Fig. 3.6. The spectrum of spin fluctuations $\mathrm{Im}\,\chi(Q,\omega)$ in $YBa_2Cu_3O_{6+x}$ at $x = 0.59$ and $T_c = 59$ K [3.12]

3.2.3 Antiferromagnetism of Rare-Earth Ions in REBa₂Cu₃O₆₊ₓ

Immediately after the discovery of the YBCO system, it was found that the replacement of Y by trivalent rare-earth (RE) ions RE = Nd, Sm, Eu, Gd, Dy, Ho, Er, Tm, and Yb preserves the crystal structure and only slightly affects T_c despite the large RE atomic magnetic moment [3.15]. This phenomenon indicates a weak coupling of the $4f$ electrons of the RE ions with holes in adjacent CuO_2 planes, although the distance RE − (O2, O3) does not exceed 2.4 Å. The magnetic subsystem of $4f$ electrons can therefore be considered as uncoupled to strong antiferromagnetic correlations of copper spins in CuO_2 planes. Indeed, for a very low temperature , an AF ordering of RE ions was observed. The Néel temperatures for the rare-earth ions RE = Yb, Nd, Er, Dy, and Gd turned out to be $T_{NR}(K) = 0.35, 0.5, 0.5,$ 1.0, and 2.2 respectively (see, e.g., [3.16]). The AF ordering temperatures of the RE sublattice is so low because the localized moments of $4f$ shells are weakly coupled to electrons on the Fermi surface, and T_{NR} are determined primarily by a weak dipole−dipole interaction. An increase of T_{NR} for Er, Dy and Gd ions may be related only to an increase of their magnetic moments $\mu = 4.9, 7.2,$ and $7.4\mu_B$, respectively.

Neutron experiments demonstrate that the magnetic moments of RE ions form a simple AF structure with unit cell dimensions doubled in all three directions. For substances with Dy, Gd and Nd the magnetic moments are oriented along the c axis, while for Er compounds one observes a chain AF structure with the moments along the a (or b) axis [3.17]. It is important to note that T_{NR} does not depend on the oxygen content x which again confirms a weak coupling of $4f$ electrons to the holes in CuO_2 planes.

Among REBa₂Cu₃O₆₊ₓ compounds, the most detailed study of magnetic structure by means of neutron scattering has been performed for a compound with Nd as the RE ion (see, for example, [3.17, 18]). In this compound, the existence of two Néel temperatures has been found $T_{N1} > T_{N2}$. At the smaller temperature T_{N2}, one observes the disappearance of magnetic Bragg peaks (1/2, 1/2, l) with integral values of l and the appearance of new ones with $l = 1/2$ and 3/2. This implies a doubling of the period of the magnetic lattice along the c axis. This has been interpreted by the authors as ordering of magnetic moments with $\mu \simeq 0.46\mu_B$ at Cu1 ions in the chains. The antiferromagnetic coupling of spins in the bilayer $CuO_2 - RE - CuO_2$ is maintained, while the coupling between bilayers becomes ferromagnetic due to antiferromagnetic ordering of the Cu1 copper spins in the chains and in adjacent CuO_2 planes. As the oxygen content x increases, both T_{N1} and T_{N2} decrease. So the unusual behavior of the spins at Cu1 ions can presumably be related to a chemical inhomogeneity of the sample [3.1]. The ordered magnetic moment at Nd ions at $T < T_{NR} = 0.5\,\mathrm{K}$ is only $\mu = 0.38\mu_B$ and should slightly effect spin ordering at copper ions.

A much higher temperature $T_{NR} = 17\,\mathrm{K}$ of antiferromagnetic ordering of magnetic moments of RE ions has been found in compounds with Pr [3.16]. Taking into account the small value of the ordered magnetic moment $\mu = 0.24\mu_B$ and the large value of the Néel temperature T_{NR}, the authors have assumed a strong hybridization of $4f$ electrons of Pr with electrons at the Fermi surface. A large value of the Sommerfeld constant γ in the low−temperature electronic specific heat also indicates a strong coupling of localized $4f$ electrons of Pr with holes in the CuO_2 plane. The value of the specific heat is comparable to that for systems of heavy fermions. Unlike in the another compounds with RE ions, the replacement of Y by Pr also suppresses superconductivity. Further investigations of $Y_{1-x-y}Ca_yPr_xBa_2Cu_3O_7$ compounds [5.18] have shown that the suppression of superconductivity is due to two factors. First, the substitution of Pr^{4+} for Y^{3+} decreases the number of holes in the CuO_2 plane and transforms the compound to the insulating state, as $x \rightarrow 1$. Second, the magnetic moment of the $4f$ electrons of Pr gives rise to magnetic scattering and causes depairing of the Cooper pairs, as do paramagnetic impurities in conventional superconductors. Thus, the properties of $PrBa_2Cu_3O_{7+x}$ differ drastically from those of other compounds with RE ions due to much smaller localization of $4f$ electrons in the Pr ion which, after Ce, has the smallest charge in the group of $4f$ elements. The role of Pr substitution for Y in YBCO compounds is discussed in some details in Sect. 5.2 (see Fig. 5.5).

3.3 Spin Dynamics Studied by the NMR Method

Nuclear magnetic resonance (NMR) and nuclear quadrupole resonance (NQR) studies have played an important role in understanding the nature of low−energy spin and electron excitations in copper-oxide superconductors. These methods yield both static (the Knight shift) and dynamic (the nuclear−spin−lattice relaxation rates) characteristics for a given ion and its nearest neighbors. The most complete data have been obtained by NMR and NQR for the nuclei ^{63}Cu, ^{65}Cu, ^{17}O, and ^{89}Y in YBCO compounds (see reviews [3.19, 20]). By studying samples with various oxygen content, one can investigate the variation in electron and spin characteristics in these compounds in the transition from the antiferromagnetic insulating phase to the metallic state and superconducting phase.

Let us now discuss the main physical parameters measured by the NMR method (see, for example, [3.19]). Under a dc magnetic field H_0, the frequency of NMR in a transverse radio–frequency (rf) field is determined by the Zeeman splitting energy

$$\hbar\omega_0 = \hbar\gamma_n H_0 (1 + K), \tag{3.18}$$

where γ_n is the nuclear gyromagnetic ratio and K is the total shift of the NMR frequency due to the interaction of the nuclear moment with electrons. The total shift is represented as a sum $K = \sigma + K^L + K^S$ where σ is a diamagnetic contribution due to inner electron shells, K^L is an orbital Van Vleck contribution, and K^S is actually the Knight shift in metals due to spin paramagnetism of the conduction electrons. The diamagnetic term is usually small and together with the orbital shift K^L (for s–electrons, $K^L = 0$) determines a chemical shift independent of temperature. The paramagnetic term K^S is due to the finite density of s–electron states at the nucleus and is described by a contact, in general, anisotropic hyperfine coupling

$$H_{hf} = \sum_{\alpha,i} A_{\alpha\alpha} I_i^\alpha S_i^\alpha, \tag{3.19}$$

Here $A_{\alpha\alpha}$ is the constant of the hyperfine interaction of nuclear spin I_i with electron spin S_i at lattice site i. The value of the Knight shift is determined by the static paramagnetic susceptibility $\chi_0^{\alpha\alpha}$

$$K_\alpha^S = A_{\alpha\alpha} \frac{\chi_0^{\alpha\alpha}}{\hbar^2 \gamma \gamma_n}. \tag{3.20}$$

For a non-interacting electron gas it is isotropic and given by

$$\chi_0^{\alpha\alpha} = (\hbar\gamma)^2 \frac{1}{2} N(0) = 2\mu_B^2 N(0), \tag{3.21}$$

where γ is the gyromagnetic ratio for electron $\hbar\gamma = 2\mu_B$ and $\mu_B = e\hbar/2mc$ is the Bohr magneton, and $N(0)$ is the density of electron states per atom per spin direction at the Fermi surface.

The width of the NMR line is determined by the rate of longitudinal relaxation $1/T_1$ of magnetization $M_0(t) \propto \exp(-t/T_1)$. In metals the relaxation rate is mainly determined by interaction of nuclear spins with conduction electrons and may be written as [3.19]

$$\frac{1}{T_1} = \frac{kT}{\hbar^2 \omega_0} \frac{1}{(\hbar\gamma)^2 N} \sum_q A_\perp^2 \operatorname{Im} \chi^{+-}(q, \omega_0), \tag{3.22}$$

where $A_\perp$ is the component transverse with respect to $M_0 \propto H_0$ of the hyperfine interaction tensor in (3.19). The dynamical spin susceptibility for a circularly polarized rf field $\chi^{+-}(q, \omega_0)$ is related to the dynamical form−factor (3.8) by the equality (3.17) where, in the paramagnetic phase for the isotropic case, one has

$$\chi^{+-}(q, \omega_0) = 2\chi^{\alpha\alpha}(q, \omega_0).$$

For a non-interacting electron gas we have

$$\chi^{+-}(q, \omega) = \frac{(\hbar\gamma)}{N} \sum_p \frac{f(\varepsilon_{p+q}) - f(\varepsilon_p)}{\hbar\omega - \varepsilon_{p+q} + \varepsilon_p}, \tag{3.23}$$

where

$$f(\varepsilon_p) = (\exp(\varepsilon_p - \mu)/kT + 1)^{-1}$$

is the Fermi distribution, ε_p is the electron energy with momentum p, and μ is the chemical potential. The static paramagnetic susceptibility introduced in the expression (3.21) is related to the expression (3.23) in the following way

$$\chi_0^{\alpha\alpha} = (1/2) \operatorname{Re} \chi^{+-}(q \to 0, \omega = 0).$$

Since in the NMR experiments the inequality $\hbar\omega_0 \ll kT$ usually holds, the integration in (3.23) can be performed in the limit $\hbar\omega_0 \to 0$. Using (3.23), for a non-interacting electron gas, we obtain

$$\frac{1}{T_1} = \frac{\pi k T A_\perp^2}{\hbar} \int_\infty^\infty d\varepsilon\, N^2(\varepsilon) \left(-\frac{df}{d\varepsilon}\right) = \frac{\pi k T A_\perp^2}{\hbar} N^2(0). \tag{3.24}$$

If we express the constant of hyperfine interaction in terms of the Knight shift, we get the Korringa relation

$$\frac{1}{T_1 T (K_\perp^S)^2} = \frac{4\pi k}{\hbar} \left(\frac{\gamma_n}{\gamma}\right) \Lambda. \tag{3.25}$$

The coefficient Λ is introduced to take into account corrections due to interactions in the electron gas. For non-interacting electrons, we have $\Lambda = 1$. Apart from universal constants, the Korringa relation contains only quantities which can be directly measured in experiments such as the Knight shift K^S and the spin-lattice relaxation time T_1. It is therefore convenient for processing experimental data. A deviation from the Korringa law is usually related to additional mechanisms of relaxation of nuclear spins.

At the transition to the superconducting state, electrons form Cooper pairs and a gap occurs at the Fermi surface. In the case of singlet pairing, the electrons with opposite spins in a Cooper pair do not contribute to the paramagnetic susceptibility. As $T \to 0$, the Knight shift (3.20) vanishes $K_\alpha^S(T = 0) = 0$. A fast decrease in the Knight shift at $T < T_c$ indicates singlet electron pairing. In the superconducting phase due to the formation of a gap Δ at the Fermi surface, the density of states varies: $N_s(|\varepsilon| < \Delta) = 0$ while $N_s(|\varepsilon| \geq \Delta) \simeq N_0(0)(|\varepsilon| / \sqrt{\varepsilon^2 - \Delta^2})$. (The finite life-time effects of quasi-particles are neglected here). In view of (3.24), the appearance of a peak in the density of states at $|\varepsilon| \gtrsim \Delta$ gives rise to a peak in the

velocity of spin-lattice relaxation at $T \leq T_c$. This is referred to as the Hebel –
Slichter peak [3.19]. As the temperature T further decreases, the relaxation rate
falls exponentially $(1/T_1) \propto \exp(-\Delta/kT)$. The above results apply to the case of
an s–wave pairing (orbital moment of a pair is equal to zero, $l = 0$), when the
gap is nonzero everywhere at the Fermi surface. In the case of a d–wave pairing
($l = 2$), the energy gap vanishes along certain directions in k–space and the spin-
lattice relaxation rate does not increase at $T \leq T_c$. There is no Hebel–Slichter
peak and, instead of the exponential decay, a power–law decay of the Knight shift
and spin-lattice relaxation rate should be observed.

3.3.1 The Knight Shift

As was discussed in Sect. 3.2, in copper-oxide compounds superconductivity occurs in the metallic phase against the background of the spin quantum liquid
characterized by strong dynamic antiferromagnetic correlations. A considerable
contribution to understanding the nature of antiferromagnetic spin fluctuations has
been made by studying the Knight shift and spin-lattice relaxation rate in the compounds $YBa_2Cu_3O_{7-y}$ and $La_{2-x}Sr_xCuO_4$ (see [3.21, 22] and references therein).
In view of variety of local symmetries of the nuclei of copper, oxygen, and yttrium
in these crystals, antiferromagnetic fluctuations of copper spins contribute differ-
ently to the measured quantities. In this way one can study the interplay of holes
at copper and oxygen sites in the CuO_2 plane and clarify whether the holes have
independent spin degrees of freedom, i.e., whether they form a two-component
spin liquid, or the hybridization between them is so strong that they should be
treated as a one–component spin liquid.

Let us consider the results of measurements of the Knight shift at oriented
(along the c axis) powders of $YBa_2Cu_3O_{7-y}$ at $y \simeq 0$, $T_c = 90\,K$ and $y = 0.37$,
$T_c = 62\,K$ as discussed in [3.23]. Figure 3.7a, shows the temperature dependence
of the total Knight shift at ^{63}Cu nuclei for copper ions in the Cu2 plane for external
field $H \parallel c(K_c)$ and $H \perp c(K_{ab})$. The solid line represents these quantities for
a sample with $y = 0$ according to data of [3.24], while the dots correspond to
$y = 0.37$. In Fig. 3.7b one can see the temperature dependence of the Knight shift
at ^{17}O for oxygen ions O2 and O3 for components along $(K_{\parallel})$ and perpendicular
to $(K_{\perp})$ the Cu – O bond and along the c axis (K_c). The solid line represents the
results for $K_{\perp}$ for a sample with $y = 0$ according to data of [3.25].

First of all, we see quite different temperature dependences of the Knight shift
$^{63}K_{ab}$ and $^{17}K_{\perp}$ for samples with maximal $y = 0$ and decreased $y = 0.37$ oxygen content which is observed in many experiments also for other nuclei (see
[3.19, 20]). In the first case (the solid line) the dependence $K(T)$ is typical for
conventional superconductors. The Knight shift, i.e., the K^S contribution, is constant in the normal state and rapidly decreases in the superconducting phase. In
samples with smaller oxygen content $y = 0.37$ a smooth variation of $K(T)$ is observed over the entire range of temperature without any remarkable change at T_c.
In view of the general expression for the paramagnetic contribution to the Knight
shift (3.20), the variation of $K(T)$ can be related to the temperature dependence

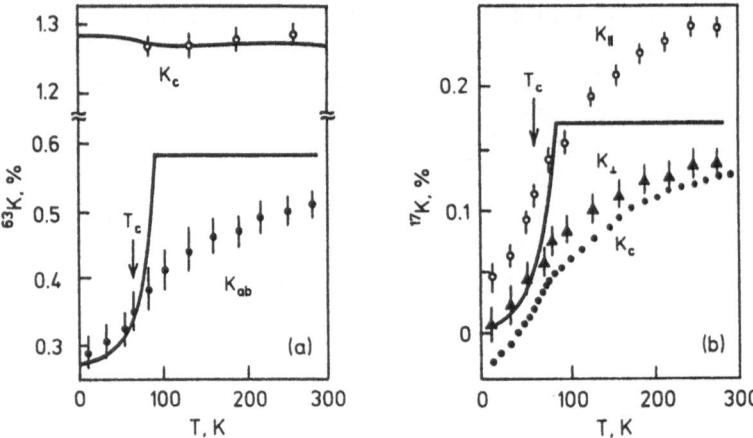

Fig. 3.7a, b. The temperature dependence of the Knight shifts: (**a**) for Cu2 ions for $H \parallel c$ (K_c) and $H \perp c$ (K_{ab}) and (**b**) for oxygen ions O2 and O3 for an external field along ($K_{\parallel}$) and perpendicular ($K_{\perp}$) to the Cu–O bond and along the c axis (K_c), for $YBa_2Cu_3O_{7-y}$ (dots and solid line correspond to $y = 0.37$ and $y = 0$ for $K_{\perp}$, respectively) [3.23]

of the static paramagnetic susceptibility $\chi_0(T)$. For samples with $y = 0$ the susceptibility is of a metallic nature, where the main contribution is made by the paramagnetic Pauli susceptibility which is independent of temperature; see (3.21). The sharp decrease in $K(T)$ and, therefore, in $\chi_0(T)$ at $T < T_c$ unambiguously indicates the formation of singlet Cooper pairs (with the zero total spin, $S = 0$). At lower oxygen content when $y = 0.37$, a more important role to be played by antiferromagnetic fluctuations and, related to them, the temperature dependence of the susceptibility $\chi_0(T)$.

Assuming that in the superconducting phase $K^S(T \to 0) \to 0$, we can estimate orbital terms K^L. According to measurements reported in [3.24], performed on single–crystals of YBCO at $y = 0$, the orbital $^{63}K_\alpha^L$ and spin $^{63}K_\alpha^S$ terms in the normal phase for the Cu2 nucleus are equal to: $K_a^L = K_b^L = 0.28$, $K_c^L = 1.3$, $K_a^S = K_b^S = 0.3$, $K_c^S = -0.02$ (%) (for comparison, see Fig. 3.7a). Measurements made in the same work of the Knight shift for the nucleus Cu1, i.e., for the chains, have shown that the paramagnetic term $^{63}K_\alpha^S$ is approximately isotropic $K_a^S = 0.24$, $K_b^S = 0.29$, $K_c^S = 0.32$ and is accompanied by the same anisotropy of the orbital term $K_a^L = 1.08$, $K_b^L \simeq K_c^S = 0.27$, as for the Cu2 nucleus.

In order to account for the considerable anisotropy of the Knight shift for Cu2, ($K_c^S \simeq 0$, $K_{ab}^S \simeq 0.3$), the nature of the hyperfine interaction of nuclear spins with electrons should be considered in greater detail. According to [3.26], the Hamiltonian of the hyperfine interaction in copper-oxide superconductors is largely determined by the polarization of the s–electrons of the corresponding atoms due to their hybridization with holes in the $3d(x^2 - y^2)$ states at planar copper sites. The hyperfine interaction for copper nuclei consists of two parts: the first is an anisotropic interaction due to a spin-orbit coupling of $3d$ holes and a dipole interaction at the same copper site. The second is a transferred hyperfine interaction due to polarization of $4s$ copper shells by $3d$ spins at neighboring copper sites. As a result, the Hamiltonian of the hyperfine interaction for copper nuclei can be written in the form [3.27]

$$H_{hf} = \sum_{i\alpha} A_{\alpha\alpha} I_i^\alpha S_i^\alpha + \sum_{ij\alpha} B I_i^\alpha S_j^\alpha , \qquad (3.26)$$

where the transferred hyperfine interaction with an isotropic coupling constant B is determined by the contribution of the four nearest to i copper sites j in the plane. In view of Eq. (3.20), for the Knight shift, we obtain

$$^{63}K_\alpha^S = \frac{A_\alpha + 4B}{^{63}\gamma_n \gamma \hbar^2} \chi_0 , \qquad (3.27)$$

where $\alpha = c$ or $\alpha = ab$ for an external magnetic field along or perpendicular to the axis c of the YBCO crystal, respectively. Since $K_c^S = 0$ regardless of the oxygen content y (Fig. 3.7a), the condition $A + 4B = 0$ should hold. This mutual compensation of the direct and transferred hyperfine interactions for planar copper ions is also observed in other compounds such as $La_{2-x}Sr_xCuO_4$ [3.22].

The hyperfine interaction for oxygen nuclei in the CuO_2 plane (O2, O3) is determined by two contributions, namely, by the spin density for the $2p$ orbitals and the induced polarization of $2s$ orbitals due to their coupling with the spins of the $3d$ copper orbitals at neighboring sites. The hyperfine field from $2p$ orbitals gives rise to a anisotropic dipole interaction, which determines the axial contribution to the Knight shift

$$^{17}K_{ax}^S = \frac{1}{3}(K_{\parallel} - K_\perp) = \frac{A_p}{^{17}\gamma_n \gamma \hbar^2} \chi_0 . \qquad (3.28)$$

As in (3.26), the induced hyperfine interaction is isotropic and, after the summation over the two nearest copper sites, determines the isotropic part of the Knight shift

$$^{17}K_{iso}^S = \frac{1}{3}(K_{\parallel} + K_\perp + K_c) = \frac{2C}{^{17}\gamma_n \gamma \hbar^2} \chi_0 , \qquad (3.29)$$

where C is the constant of the transferred hyperfine interaction for planar oxygen ions. Since the anisotropic part $^{17}K_{ax}^S > 0$ for samples with $y = 0$ [3.25] and with $y = 0.37$ [3.23], i.e. $K_{\parallel} > K_\perp$, the spin density at ions O2 and O3 should be concentrated at $2p\sigma$ orbitals along the $Cu - O$ bond. The study of the Knight shift for oxygen nuclei at O4 sites (the bridge oxygen, see Fig. 2.12) also indicates the existence of spin density at $2p$ oxygen orbitals [3.25].

In writing equations (3.27–29), we have implicitly assumed that the static susceptibility at copper and oxygen sites is described by a single paramagnetic susceptibility $\chi_0(T)$, i.e., we have accepted the model of a one−component spin liquid. The most important result of the work [3.23] is an experimental proof of the validity of this model. Figure 3.8 shows the temperature dependence of the Knight shift $^{63}K_{ab}$ related to the susceptibility at copper sites χ_0^d and $^{17}K_\alpha$ related to the susceptibility at oxygen sites χ_0^p for a sample with $y = 0.37$. The dependence

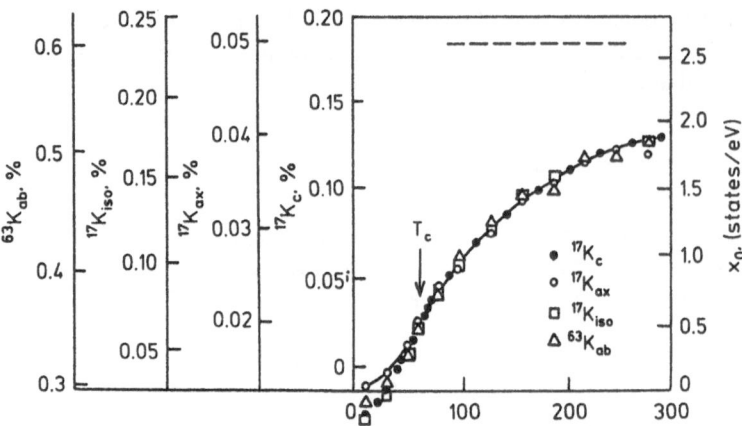

Fig. 3.8. Reduced temperature dependence of the Knight shift for YBa$_2$Cu$_3$O$_{7-y}$ at $y = 0.37$ (*dots*) and $y = 0$ (*dashed line*). On the right axis, the values of static susceptibility $\chi_0(T)/\mu_B^2$ are shown [3.21, 23]

fits a single temperature−dependent function $\chi_0(T)$ (in units χ_0/μ_B^2 on the right axis) under a corresponding choice of scale (to take into account the constants of the hyperfine interaction) and shift of the origin (due to the orbital terms). To within the experimental accuracy, for a sample with $y = 0$, one obtains for the susceptibility in the normal phase $\chi_0/\mu_B^2 = 2N(0) \simeq 2.6\,\mathrm{eV}^{-1}$ [3.21] shown by the dashed line.

Measurements of the Knight shift for the nuclei ^{89}Y maintain the picture of a one−component spin liquid. The study of the temperature dependence of the Knight shift ^{89}K performed in [3.28, 29] at various oxygen content $0 \le y \le 0.6$ has shown that its temperature−dependent part is proportional to a macroscopic spin susceptibility

$$^{89}K_{\mathrm{iso}}^S = \frac{8D}{^{89}\gamma_n\gamma\hbar^2}\chi_0, \tag{3.30}$$

where D is the constant of the transferred hyperfine interaction. The latter is related to the interaction of the ^{89}Y nucleus with the eight nearest oxygen sites, i.e. with the spin density of the O $2p$ orbitals, while the macroscopic susceptibility χ_0 is on the whole determined by $3d$ spins at Cu2 sites. On the other hand, the relation (3.30) is valid as y varies over a wide range. This shows that the degree of hybridization Cu$3d-$ O$2p$ does not depend on the concentration of doped holes. There are thus no independent spin susceptibilities for copper ($3d$ holes) and oxygen ($2p$ holes) sites. The system of holes in the CuO$_2$ planes should be described by a one−component spin quantum liquid.

3.3.2 Spin–Lattice Relaxation

The very first experiments [3.19] on copper-oxide superconductors showed an anomalous nuclear relaxation at the copper sites in CuO_2 planes which deviates from the Korringa law. Figure 3.9 shows the temperature dependence of the spin-lattice relaxation rate $1/T_1$ in field $\boldsymbol{H} \parallel c$ for ^{63}Cu.

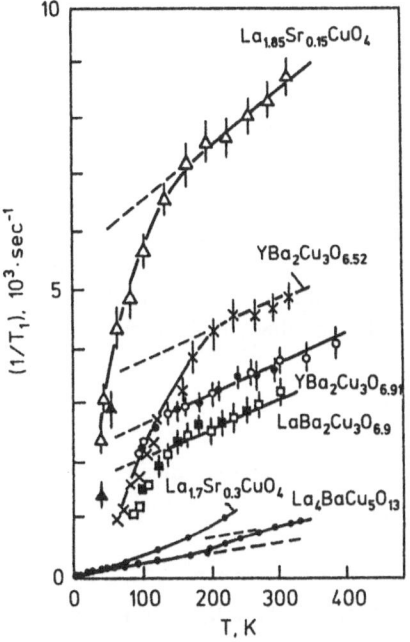

Fig. 3.9. The temperature dependence of the spin-lattice relaxation rate for ^{63}Cu nuclei [3.30]

For superconducting compounds the relaxation rate turns out to be several times higher than that in copper-oxide compounds with normal metallic conductivity which are shown at bottom of the figure. However, at a certain temperature $T_s \leq 150\,K$ much higher than T_c, the relaxation rate falls. In [3.30], it has been noted that, at $T > T_s$, the temperature dependence of the relaxation rate can be represented as the sum of two terms $(1/T_1) = aT + b$. The first term behaves the same as in normal metals in agreement with the Korringa law. The second term has been attributed to the contribution of antiferromagnetic fluctuations of spins at copper sites. The authors of [3.30] have accounted for the decrease of the second term at $T < T_s$ via the appearance of a gap in the spectrum of antiferromagnetic fluctuations. As will be shown below, the temperature dependence of the relaxation rate can also be described by taking into consideration the decrease in the correlation length $\xi(T)$ of antiferromagnetic fluctuations with increasing temperature, see (3.40),

$$b \propto T\xi^2(T) \propto T/(T + T_s).$$

Further investigations have shown that the temperature dependence of the spin-lattice relaxation rate strongly depends on doping, and that the simple Korringa law does not usually hold due to the rather complicated temperature dependence of the susceptibility. For O2, O3, and Y nuclei, however, one can obtain the relation

$$T_1 \, T \, K^S = \text{const}.$$

This relation maintains the model of one−component spin liquid. Relaxation at Cu2 nuclei has a more complicated temperature dependence due to the above mentioned auxiliary contribution of antiferromagnetic spin fluctuations. Let us discuss these results in greater detail.

Figure 3.10a shows the temperature dependence of the function $(T_1 T)^{-1}$ for ^{17}O at sites O2 (O3) in a field $H \parallel c$ in YBCO. In Fig. 3.10b, the same dependence is shown for ^{63}Cu at Cu2 sites. The circles and dots correspond to the results for $y = 0$ [3.31] and $y = 0.37$, respectively. If we compare the temperature dependences of the relaxation rate and the Knight shift for oxygen nuclei (Figs. 3.10a and 3.7b), we find that, for the sample with $y = 0$ the Korringa law (3.25) holds with constant $\Lambda = 1.4$. In view of the temperature dependence of the static susceptibility shown in Fig. 3.8, the Korringa law is no longer valid for the sample with $y = 0.37$. However, one has instead the relation

$$^{17}T_1 \, T \, ^{17}K_\perp = \text{const}.$$

The same relation applies to ^{89}Y nuclei over a wide range of values $0 < y < 0.6$ [3.28, 29]

$$^{89}T_1 \, T \, ^{89}K = \text{const}.$$

At the same time, there is no simple relation between the relaxation rate and the Knight shift for ^{63}Cu nuclei. An attempt to compare these quantities on the basis of the Korringa law (3.25) for the sample with $y = 0$ yields $\Lambda = 11$ at $T = 100 \, \text{K}$ [3.31]. The results obtained indicate the existence of an additional mechanism for relaxation at Cu2 nuclei compared to O2 (O3) and Y nuclei.

The different behavior of the relaxation rate at Cu2 sites and at O2 (O3), Y can be accounted for if a non-local nature of the hyperfine interaction in copper-oxide superconductors is taken into consideration as it has been done by Millis-Monien-Pines (MMP) [3.27].

According to the equation (3.26), the hyperfine interaction contains not only a contact contribution but also a transferred hyperfine interaction induced by $3d$ spins at neighboring Cu2 sites. As a result, in the calculation of the spin-lattice relaxation rate in the relation (3.22), one needs to use the wave-vector-dependent coupling constants

$$^{63}A_\alpha(q) = A_\alpha + 2B(\cos q_x a + \cos q_y a), \tag{3.31a}$$

$$^{17}A_\alpha(q) = 2C \cos(q_\alpha a/2), \tag{3.31b}$$

$$^{89}A_\alpha(q) = 8D \cos(q_x a/2) \cos(q_y a/2) \cos(q_z c/2), \tag{3.31c}$$

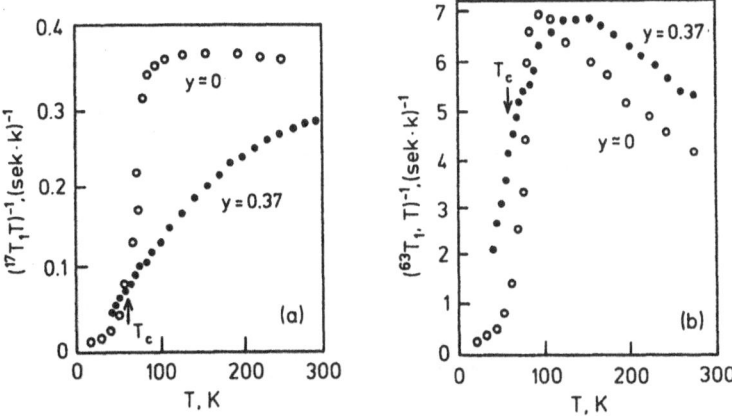

Fig. 3.10. The temperature dependence $(T_1 T)^{-1}$ for sites O2 (O3) (a) and for sites Cu2 (b) in a field $H \parallel c$ in YBCO for $y = 0$ (*circles*) [3.31] and $y = 0.37$ (*dots*) [3.23]

instead of the local coupling constant $A_\perp$ [3.27, 29, 30]. The Knight shifts are determined by the values of these functions at $q \to 0$ and by the uniform static susceptibility $\chi(q \to 0, \omega = 0)$. At the same time, antiferromagnetic spin fluctuations give the largest contribution to the susceptibility at $q_x = q_y = \pi/a$ when the spins at neighboring Cu2 sites have opposite directions. Since sites O2 (O3) and Y are arranged in a symmetric way with respect to Cu2 sites, their net contribution to antiferromagnetic fluctuations is equal to zero $^{17}A_\alpha(Q) = \, ^{89}A_\alpha(Q) = 0$ where $Q = (\pi/a)(1, 1)$. Thus, antiferromagnetic fluctuations of $3d$ spins are filtered out for the sites O2 (O3) and Y but give a finite contribution for Cu2 sites $^{63}A_\alpha(Q) = A_\alpha - 4B$.

In order to describe the contribution of antiferromagnetic spin fluctuation to the of spin-lattice relaxation rate (3.22), Millis et al. [3.27] proposed a phenomenological model for dynamic spin susceptibility consisting of two terms

$$\chi(q, \omega) = \chi_{QP}(q, \omega) + \chi_{AF}(q, \omega). \tag{3.32}$$

The quasi−particle term

$$\chi_{QP}(q, \omega) = \bar{\chi}_0 \frac{1}{1 - i\omega\pi/\hbar\Gamma}. \tag{3.33}$$

is determined by two parameters, namely, the static susceptibility χ_0 and the specific electronic energy $\hbar\Gamma/\pi$ which is of order of the Fermi energy E_F. A term due to antiferromagnetic fluctuations

$$\chi_{AF}(q, \omega) = \chi_Q \frac{1}{1 + \xi^2(Q - q)^2 - i\omega/\omega_{sf}}. \tag{3.34}$$

is described by the static susceptibility χ_Q for wave vector Q and the typical energy of antiferromagnetic fluctuation $\hbar\omega_{sf}$. According to [3.27], these parameters are related to the quasi−particle parameters by the relations

$$\chi_Q = \bar{\chi}_0(\xi/\xi_0)^2, \quad \omega_{sf} = (\hbar\Gamma/\pi)(\xi_0/\xi)^2, \tag{3.35}$$

where $\xi(T)$ is the correlation length of antiferromagnetic spin fluctuations. It is assumed that $(\xi/\xi_0)^2 \gg 1$, and therefore $\chi_Q \gg \chi_0$ and $\Gamma \gg \omega_{sf}$.

The Knight shift is determined by the homogeneous static susceptibility

$$\chi_0 = \chi(0,0) = \bar{\chi}_0[1 + \sqrt{\beta}/2\pi^2], \tag{3.36}$$

where the parameter β is given by $\beta = (a/\xi_0)^4 \simeq 10$ [3.21]. The contribution of antiferromagnetic fluctuations is small here of the order of 16%. The temperature dependence of all the Knight shifts is determined by a single function which is the quasi–particle static susceptibility $\bar{\chi}_0(T)$.

The temperature dependence of the spin-lattice relaxation rate (3.22) is determined by the function

$$S(\boldsymbol{q}, \omega \to 0) = \lim_{\omega \to 0} \frac{kT}{\hbar\omega} \operatorname{Im} \chi(\boldsymbol{q}, \omega + i\delta)$$

$$= \pi \frac{kT}{\hbar\Gamma} \bar{\chi}_0(T) \left[1 + \beta \left(\frac{(\xi(T)/a)^2}{1 + \xi^2(T)(\boldsymbol{Q} - \boldsymbol{q})^2} \right)^2 \right]. \tag{3.37}$$

The contribution of antiferromagnetic fluctuations is determined here by the second term. Its contribution is small at $q = 0$ being of order $\beta/4\pi^4 \simeq 0.03$. However it becomes the major term at $q = Q$ of the order $\beta(\xi/a)^2 \gg 1$. Therefore, the spin-lattice relaxation rate at Cu2 sites depends essentially on antiferromagnetic spin fluctuations, since according to (3.31a), it contains contributions from the whole region of q. In view of (3.31a), the integration over q in (3.22) yields [3.21]

$$^{63}T_1^{-1} \propto B^2 \frac{kT}{\hbar^2\Gamma} \frac{\bar{\chi}_0(T)}{\mu_B^2} \left[1 + 1.67 \frac{\beta}{\pi^2} \left(\frac{\xi(T)}{a} \right)^2 + \cdots \right], \tag{3.38}$$

where only the leading term in $\xi(T)$ is written explicitly. According to (3.31b, c), the region $q \simeq Q$ does not contribute to the relaxation rate at sites O2 (O3) and Y. It therefore does not contain any essential dependence on antiferromagnetic fluctuations. This accounts for the very different temperature behavior of the relaxation times at the nuclei Cu2 and O2 (O3) or Y.

The calculation of the Knight shifts (3.27, 29, 30) on the basis of the representation (3.36) enables the temperature dependence $\bar{\chi}_0(T)$ to be determined and the constants of the hyperfine interaction in (3.31) to be evaluated. The remaining three parameters of the model, i.e., $(\xi(T)/a)$, Γ and $\beta = (\xi_0/a)^4$, can be determined by comparing calculations with experimental data for the relaxation rates for Cu2, O2, and Y nuclei. In particular, the temperature dependence of the correlation length $(\xi(T)/a)$ can be obtained from the ratio of the relaxation times

$$^{17}T_1/^{63}T_1 \propto \frac{B^2}{C^2} \left[1 + 1.7 \frac{\beta}{\pi^2} \left(\frac{\xi(T)}{a} \right)^2 + \cdots \right], \tag{3.39}$$

where only the leading term is written. Figure 3.11a shows the temperature dependence of the ratio for YBCO at $y = 0$ (circles) according to data of [3.25]

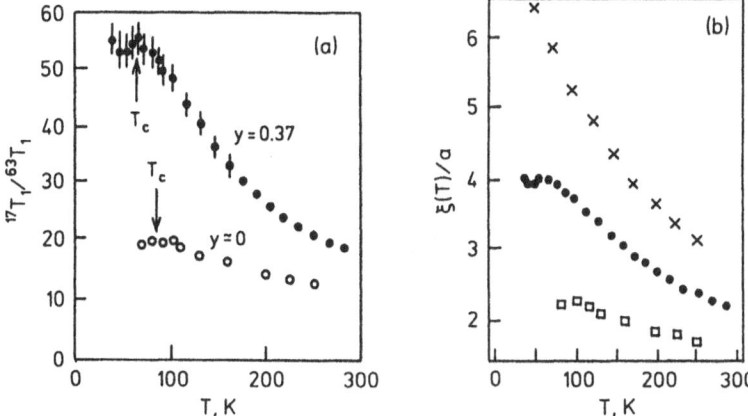

Fig. 3.11. The temperature dependence of the ratio of relaxation rates of the nuclei ^{63}Cu and ^{17}O for YBCO at $y = 0$ (*circles*) and $y = 0.37$ (*dots*) [3.23] (**a**) and the temperature dependence of the correlation length $\xi(T)/a$ of antiferromagnetic fluctuations for YBCO at $y = 0$ (*squares*) and $y = 0.37$ (*dots*) and LSCO at $x = 0.15$ (*crosses*) [3.22] (**b**)

and $y = 0.37$ (dots) [3.23]. Figure 3.11b gives the temperature dependence $\xi(T)/a$ calculated on the basis of the MMP model [3.22] for YBCO (squares and dots correspond to $y = 0$ and $y = 0.37$, respectively) and LSCO (crosses represent $x = 0.15$). These data imply that, in the temperature region $T \geq 130$ K, the dependence $\xi(T)$ can be represented in the form

$$\left(\frac{\xi(T)}{a}\right)^2 = \text{const} \left| \frac{T_s}{T + T_s} \right|, \tag{3.40}$$

where T_s determines the scale of variation of the correlation length with temperature. For YBCO at $y = 0$, it is $T_s \simeq 120$ K. However, it becomes negative for samples with lower concentration of O $2p$ holes. For YBCO at $y = 0.37$ and for LSCO at $x = 0.15$, we have $T_s = -30$ K and $T_s = -38$ K, respectively. The occurrence of a negative effective Néel temperature $T_s < 0$ indicates that the system of spins is close to antiferromagnetic instability. According to Refs. [3.21, 22], as the temperature decreases $T \to T_c$, superconducting correlations suppress the increase of the antiferromagnetic correlation length $\xi(T)$ and the system of spins remains a virtual antiferromagnet.

Thus, according to the MMP theory, the rather complicated temperature dependence of the spin-lattice relaxation rate (3.38) at Cu2 sites is accounted for by an interplay of the temperature dependences of the static susceptibility $\chi_0(T)$ shown in Fig. 3.8 and the correlation length $\xi(T)$. In particular, the maximum of the $1/T_1T$ curve for ^{63}Cu at $y = 0.37$ (Fig. 3.10b) in the region $T = 150$ K is due to an increase in $\xi(T)$ and a decrease of $\chi_0(T)$, with decreasing temperature. It requires no assumption about the occurrence of a gap in the spectrum of antiferromagnetic fluctuations which was found in neutron experiments (Fig. 3.6). A detailed comparison of experimental data with calculations on the basis of the

MMP theory performed in [3.32] also shows that, taking into account a weak temperature dependence of the parameter $\hbar\Gamma = 0.4 - 0.5\,\text{eV}$, the model parameter $\beta = (\xi_0/a) = 10$ can be considered as a constant. In this case, the spin-lattice relaxation rates at O2 (O3) nuclei and Y are determined only by the temperature-dependent static susceptibility so that they satisfy the relation $TT_1 K^S = \text{const.}$ [3.29].

On the basis of the MMP model, the transverse relaxation rate for copper nuclei in CuO$_2$ planes may also be computed. The transverse relaxation is determined by the spin–spin interaction of nuclei which is proportional to $\text{Re}\,\chi(q,\omega = 0)$ in (3.32). Such calculations are performed in [3.33] without additional fitting parameters of the MMP model and they give good agreement with experiment. This agreement also supports the MMP model. The explanation of the temperature dependence of the static susceptibility $\bar{\chi}_0(T)$ in the MMP model and the calculation of the temperature dependence $\xi(T)$ are the most complicated issues from the point of view of the microscopic theory. Some theoretical models proposed to describe the system of Cu $3d$ and O $2p$ holes will be considered in Sect. 7.1.3.

But it should be pointed out that strong temperature dependence for the correlation length $\xi(T)$ (3.40) was not proved in the neutron scattering experiments [3.9, 12, 13, 34]. At the same time a considerable suppression of AF spin fluctuations at low frequencies below $T_s \gtrsim T_c$ (and spin gap formation in YBCO) was observed. It permits one to give an alternative interpretation of the suppression of the spin-lattice relaxation rate below T_s [3.28].

While on the subject of spin–lattice relaxation at $T < T_c$, we note two very important facts. The first is the absence of a Hebel–Slichter peak at $T < T_c$ for ^{17}O nuclei in YBCO at $y = 0$ (Fig. 3.10a). The second is a faster decrease in the relaxation rate at $T < T_c$ than in the conventional superconductors described on the basis of the BCS theory. As has already been noted, the contribution of antiferromagnetic fluctuations at O2 (O3) sites does not play any significant role. The anomalous behavior of the relaxation rate at $T < T_c$ has led some authors to conclude a d–wave nature of pairing [3.37, 38].

However, taking into account the final lifetime effects of quasi–particles at $T < T_c$, these two results can actually be explained in the frame of the usual theory of s–wave pairing (see, for example, [3.35]). At the same time, the description of experimental results on the spin-lattice relaxation rate at $T < T_c$ in the frame of the MMP theory encounters a number of difficulties and requires some additional mechanisms of relaxation besides antiferromagnetic fluctuations, as e.g. due to interaction with the flux lattice. One such problem is how to explain the temperature–dependent anisotropy $(T_1)_\parallel/(T_1)_\perp$ of the relaxation at Cu2 nuclei for $T < T_c$, as found in [3.36].

To summarize the study of magnetic properties of copper-oxide compounds we point out the following most important results.

1. In the insulating phase of the parent compounds, one observes long-range antiferromagnetic order with a sufficiently high Néel temperature $T_N \simeq 300 - 500\,\text{K}$. The antiferromagnetic state is due to a strong exchange interaction $J \propto$

0.13 eV in the CuO_2 planes between $S = 1/2$ spins of $3d$ holes at copper sites with effective magnetic moment $\mu \simeq 0.6\mu_B$.

2. Under doping, the long-range antiferromagnetic order is destroyed. This happens at a concentration of doped holes equal to $n_h \simeq 0.02$. In $Nd_{2-x}Ce_xCuO_4$ the corresponding electron concentration is $n_e \simeq 0.15$. Strong two-dimensional antiferromagnetic correlations are maintained, however, in the superconducting phase. The value of the effective magnetic moment of Cu $3d$ holes does not vary with the doping, while the correlation length of the antiferromagnetic fluctuations decreases to $1 - 2$ interatomic distances in the superconducting phase.

3. The spectrum of magnetic excitations in the insulating phase is determined by antiferromagnetic spin waves with a high propagation velocity in the plane $v \simeq 1\,eV \cdot Å$. Under doping, the velocity rapidly decreases. In the metallic phase, antiferromagnetic spin fluctuations with a wide spectrum of excitations are observed. The characteristic excitation energy of $20 - 30\,meV$ increases with concentration of doped holes while the intensity of fluctuations decreases. Inelastic neutron scattering indicates a gap in the spectrum of antiferromagnetic fluctuations at temperatures $T_s \gtrsim T_c$.

4. Measurement of the Knight shift in the normal metallic phase supports the model of a one−component spin liquid with temperature−dependent paramagnetic susceptibility. Studies of the spin-lattice relaxation rate show the existence of antiferromagnetic spin fluctuations at copper sites under a strong hybridization of Cu $3d$ and O $2p$ holes.

5. In the superconducting phase, the decrease of the Knight shift indicates singlet pairing at T_c. The absence of the Hebel−Slichter peak and the rapid decrease in the spin-lattice relaxation rate at $T \leq T_c$ suggest either d−wave pairing or gapless s-wave pairing with a short lifetime of quasi−particles in the strong−coupling limit.

4. Thermodynamic Properties
of High-Temperature Superconductors

A study of the thermodynamic properties of high-temperature superconductors enables a number of macroscopic parameters of these compounds to be determined. It also yields certain boundary conditions that must be satisfied by microscopic theories. The measurement of thermodynamic quantities such as critical magnetic fields and related critical currents is also important from the point of view of applications.

In order to describe the thermodynamic properties of superconductors near the phase transition, the phenomenological Ginsburg–Landau theory is usually used. Its generalization to the anisotropic case is considered in the next section. The measurement of the specific heat $C(T)$ and its jump ΔC under the transition from the normal into the superconducting state allows one to evaluate one of the most important parameters of superconductors: the Sommerfeld constant γ in the low-temperature electronic specific heat $C_e = \gamma T$. The results of these measurements for LSCO and YBCO compounds are discussed in Sect. 4.2. Experimental data for the critical magnetic fields H_{c1} and H_{c2} and an estimate of the related parameters – the correlation length $\xi(T)$ and penetration depth of magnetic field $\lambda(T)$ are considered in Sect. 4.3.

4.1 The Anisotropic Ginsburg–Landau Model

In the phenomenological Ginsburg–Landau theory, the superconducting state is described by the complex scalar order parameter

$$\Psi(r) = |\Psi(r)| \exp[i\Phi(r)] . \tag{4.1}$$

The modulus of the order parameter is usually normalized to the concentration of superconducting electron pairs $(n_s/2)$

$$| \Psi(r) |^2 = n_s/2 ,$$

while the phase of the order parameter is related to the superconducting current. Thus, in the Ginsburg–Landau theory, the superconducting phase is described by a two-component $(n = 2)$ order parameter.

The equilibrium properties of a superconductor are determined by a functional of the free energy which depends on the order parameter and external magnetic field (see [4.1]). Anisotropy is usually taken into account by the introduction of an anisotropic effective mass of superconducting pairs. In this approximation, we arrived at the following expression for the free-energy functional (see, for example, [4.2])

$$F(\Psi) = F_{no} + \int dV \left\{ \frac{B^2}{8\pi} + a|\Psi|^2 + \frac{1}{2}b|\Psi|^4 \right.$$
$$\left. + \sum_{\alpha}(4m)_{\alpha}^{-1}|(-i\hbar\nabla_{\alpha} - \frac{2e}{c}A_{\alpha})\Psi|^2 \right\}, \tag{4.2}$$

where F_{no} is the free energy in the normal phase, B = curl A is the magnetic induction, $2m_{\alpha}^{-1}$ are the principal values of the tensor of inverse masses for a superconducting pair of electrons with charge 2e. Near the superconducting phase transition with temperature T_c it is usually assumed that the parameter $a = \alpha(T - T_c) = \alpha T_c \tau$, where $\tau = (T/T_c - 1)$ and the parameter b and the effective mass m_{α} do not depend on temperature.

Equilibrium values of the order parameters and the superconducting current in an external field $B \neq 0$ are determined by the minimum of functional (4.2) under its variation with respect to $\Psi(r)$ and $A(r)$

$$\frac{1}{4m_{\alpha}}(-i\hbar\nabla_{\alpha} - \frac{2e}{c}A_{\alpha})^2\Psi + a\Psi + b|\Psi|^2\Psi = 0, \tag{4.3}$$

$$j_{\alpha} = -\frac{ie\hbar}{2m_{\alpha}}(\Psi^*\nabla_{\alpha}\Psi - \Psi\nabla_{\alpha}\Psi^*) - \frac{2e^2}{m_{\alpha}c}|\Psi|^2 A_{\alpha} . \tag{4.4}$$

In the absence of a magnetic field, the homogeneous equilibrium value of the order parameter is

$$|\Psi_0|^2 = \frac{n_s}{2} = -\frac{a}{b} = \frac{\alpha}{b}(T_c - T), \quad T < T_c . \tag{4.5}$$

Under the transition from the superconducting to normal state, the specific heat undergoes a jump

$$\Delta C = C_s - C_n = \alpha^2 T_c / b . \tag{4.6}$$

The thermodynamic critical field $H_c(T)$, which is determined from the condition that the free energy should take the same value in the normal and superconducting phase, is determined by the relation

$$H_{c0} = \left(\frac{4\pi a^2}{b}\right)^{1/2} = \left(\frac{4\pi\alpha^2}{b}\right)^{1/2}(T_c - T) . \tag{4.7}$$

The jump in the specific heat (4.6) is related to the derivative of the critical field by the Rutgers formula

$$\frac{\Delta C}{T_c} = \left[\frac{1}{4\pi}\left(\frac{dH_c}{dT}\right)_{T_c}\right]^2 . \tag{4.8}$$

The penetration depth of an external magnetic field is determined by the corresponding screening current. In the anisotropic case, the relations between current, external field, and the penetration depth are of a complicated nature. Figure 4.1 shows these quantities for the case of external field $H \parallel c$ and $H \parallel a$, where a, b, and c are the principal symmetry axes of the crystal [4.3]. Since the symmetry of copper-oxide superconductors is close to tetragonal, we consider below the case of axial symmetry $m_a = m_b = m_\perp$, $m_c = m_\parallel$ where the symbols $\parallel$ and $\perp$ denote the orientations parallel and perpendicular to the symmetry axis c. Figure 4.1 shows that, for an external field along the crystal axis c, the penetration depths $\lambda_{a,b}^c$ along axes a and b are determined by currents in the basis plane $j_b \approx j_a$ (4.4), and therefore $\lambda_a^c \approx \lambda_b^c$. For an external field $H \perp c$, the penetration depths $\lambda_b^a = \lambda_a^b$ in plane ab are determined by the current j_c along axis c, while the penetration depth λ_c^a along axis c is determined by the current j_b in plane ab. Since the currents in the plane ab for fields of the same magnitudes along axes c and a are equal $j_b^c = j_b^a$, we have $\lambda_c^a = \lambda_a^c$. If axis c is a "rigid" direction for screening currents in a superconductor, i.e., $j_c \ll j_{ab}$ (for $m_c \gg m_{ab}$), for the penetration depth we obtain $\lambda_{ab}^c \ll \lambda_b^a$.

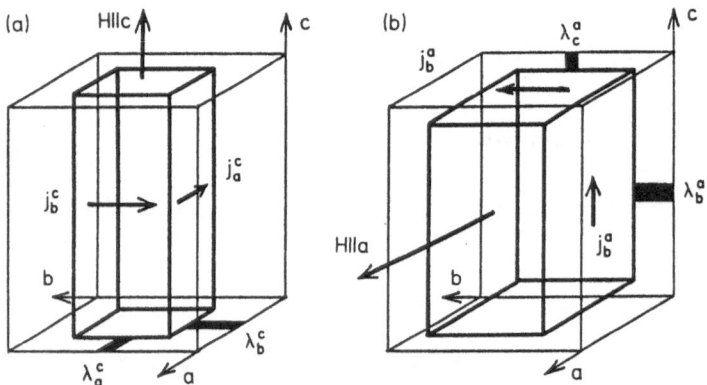

Fig. 4.1. The penetration lengths λ_α^i and the screening currents for the magnetic fields H^i oriented along the crystal axes $i = c$ **(a)** and $i = a$ **(b)** [4.3]

In order to calculate the penetration depth of the external field and superconducting current in the Ginsburg–Landau theory, it is necessary to apply the equation for the current (4.4) and the Maxwell equation $j = (c/4\pi)\,\mathrm{curl}\,B$. Let us consider, for example, an external field H parallel to axis c, H_c, and calculate the current j_a and magnetic induction B_c in the plane ac (see Fig. 4.1). The solution of this set of equations gives the following dependence

$$j_a(y) = j_a(0)\exp(-y/\lambda_a)\,, \quad B_c(y) = B_c(0)\exp(-y/\lambda_a)\,, \tag{4.9}$$

where the coordinate y starts on the surface of the superconductor and goes along the b axis. The screening of the magnetic induction by the current $j_a(y)$, which is perpendicular to the induction, is determined by the penetration depth

$$\lambda_\alpha^2(T) = \frac{m_\alpha c^2}{4\pi e^2 n_s} = \frac{m_\alpha c^2 b}{8\pi e^2 |a|} = \frac{\lambda_\alpha^2(0)}{|\tau|} . \tag{4.10}$$

Thus, the penetration depth λ_a is related to the corresponding screening current j_a. In the anisotropic case, it can be represented in the form of a tensor with principal values $\lambda_\alpha = \lambda(m_\alpha/m)^{1/2}$, where $m = (m_a m_b m_c)^{1/3}$, $\lambda = (mc^2/4\pi e^2 n_s)^{1/2}$. The penetration depths shown in Fig. 4.1 are related to the quantity (4.10) by the equalities $\lambda_a = \lambda_b^c$, $\lambda_b = \lambda_a^c$, $\lambda_c = \lambda_b^a$.

Another special length in the Ginsburg–Landau theory is a correlation length $\xi(T)$. It determines a specific distance, within which the order parameter is coherent

$$\langle \Psi(r)\Psi(0)\rangle \propto \frac{1}{r}\exp(-r/\xi) .$$

where the average $\langle ... \rangle$ is calculated on the basis of the Ginsburg–Landau functional. In the anisotropic case (4.2), the correlation length is determined by the relation

$$\xi_\alpha(T) = \left(\frac{\hbar^2}{2m_\alpha |a|}\right)^{1/2} = \frac{\xi_\alpha(0)}{|\tau|^{1/2}} . \tag{4.11}$$

The behavior of superconductors in an external magnetic field is determined by the Ginsburg–Landau parameter $\kappa = \lambda/\xi$. In the case of superconductors of second kind ($\kappa > 1/\sqrt{2}$) there are two critical fields $H_{c1} < H_{c0} < H_{c2}$. High-temperature superconductors with small correlation length $\xi \ll \lambda$ belong to this particular class of superconductors. The upper critical field $H_{c2}(T)$ determines the stability limit of the superconducting phase. In the anisotropic case it is equal to

$$H_{c2}^c = \frac{\phi_0}{2\pi\xi_{ab}^2} = \kappa_c\sqrt{2}H_{c0} , \qquad H_{c2}^{ab} = \frac{\phi_0}{2\pi\xi_{ab}\xi_c} = \kappa_{ab}\sqrt{2}H_{c0} , \tag{4.12}$$

where $\phi_0 = hc/2e = \pi\hbar c/e = 2\cdot 10^{-7}$ Gs·cm^2 is the quantum of magnetic flux. Due to the definition (4.7) for the thermodynamic critical field

$$H_{c0} = \frac{\phi_0}{2\pi\sqrt{2}}\frac{1}{\xi_\alpha\lambda_\alpha} , \tag{4.13}$$

relations (4.12) yield the following expressions for the anisotropic Ginsburg–Landau parameters

$$\kappa_c = \frac{\lambda_{ab}}{\xi_{ab}} = m_{ab}\frac{c\sqrt{b}}{e\hbar 2\pi} , \qquad \kappa_{ab} = \left(\frac{\lambda_{ab}\lambda_c}{\xi_{ab}\xi_c}\right)^{1/2} = \left(\frac{m_c}{m_{ab}}\right)^{1/2}\kappa_c . \tag{4.14}$$

The lower critical field $H_{c1}(T)$ determines the stability limit of the homogeneous Meissner superconducting phase, which does not contain magnetic vortices. The formation of vortices at $H > H_{c1}$ becomes profitable when the energy of a vortex becomes smaller than the energy of the external field $H_{c1}\phi_0/4\pi$. In the limit $\kappa \gg 1$, calculations for the anisotropic case yield (see, for example, [4.4])

$$H_{c1}^c = \frac{\phi_0}{4\pi\lambda_{ab}^2}\ln\kappa_c = \frac{\ln\kappa_c}{\kappa_c\sqrt{2}}H_{c0}\,,$$

$$H_{c1}^{ab} = \frac{\phi_0}{4\pi\lambda_{ab}\lambda_c}\ln\kappa_{ab} = \frac{\ln\kappa_{ab}}{\kappa_{ab}\sqrt{2}}H_{c0}\,. \tag{4.15}$$

The temperature dependence of all critical fields (4.12) and (4.15) is determined by the function $H_{c0}(T)$. According to equation (4.7), it vanishes following a linear law as $T \to T_c$. We therefore obtain $H_{ci}^\alpha \propto (T_c - T)$. The measurement of the upper critical field (4.12) enables the correlation length $\xi_\alpha(T)$ to be estimated. The parameter κ_α can be found from the relation

$$\frac{H_{c1}^\alpha}{H_{c2}^\alpha} = \frac{\ln\kappa_\alpha}{2\kappa_\alpha^2}\,. \tag{4.16}$$

The penetration length can be determined by equations (4.14). The results of measurements of the critical fields in high-temperature superconductors are discussed, for example, in the review [4.5] and are to be considered in the present book in Sect. 4.3.

The description of superconductors on the basis of the phenomenological Ginsburg–Landau theory in the approximation of anisotropic mass (4.2) is reasonable only when the correlation length (4.11) satisfies the relations $\xi_\alpha(T) \gg d$ where d is a specific interatomic spacing of the order of the lattice constant. In high-temperature superconductors this relation may be violated due to a small correlation length and strong anisotropy. For example, in compounds of bismuth and thallium, we have $r = \xi_c(0)/d \ll 1$. In this case, the Ginsburg–Landau functional should be written for the order parameter $\psi_n(x, y)$ in layer n. The Josephson coupling between adjacent layers $(n, n + 1)$ should also be taken into account (see [4.2]). Then, a more complicated picture of the penetration of magnetic field into a superconductor at $H > H_{c1}$ and a more complicated temperature dependence of critical fields arise [4.2, 5].

Another possible violation of the applicability of the Ginsburg–Landau theory is connected with critical fluctuations at $T \to T_c$. As with any mean-field theory, the Ginsburg–Landau theory fails in the region of temperatures, where the critical fluctuations of the order parameter become larger than its equilibrium value. This region of the critical fluctuations is estimated by the dimensionless temperature (see, for example, [4.2]):

$$\tau_{G3} = \left(\frac{1}{8\pi^2\xi_{ab}^2(0)\xi_c(0)\Delta C}\right)^2 \simeq \left(\frac{T_c}{\varepsilon_F}\right)^4 \frac{m_c}{m_{ab}}\,, \tag{4.17}$$

where ΔC is the specific heat jump (4.6) and ε_F is the Fermi energy. In the temperature region $\tau > \tau_{G3}$, Gaussian fluctuation corrections of order $(|\tau|/\tau_G)^{1/2}$ to the classical temperature dependences may occur. For quasi-two-dimensional superconductors, the Gaussian fluctuations may be observed in the temperature region $\tau > \tau_{G2}$ where

$$\tau_{G2} = \frac{1}{4\pi d \xi_{ab}^2(0)\Delta C} \simeq \frac{T_c}{\varepsilon_F}. \tag{4.17a}$$

In the usual quasi-isotropic superconductors, the fluctuation region τ_{G3} is extremely small due to large value of the correlation length. The mean-field approximation then applies to the whole region of temperatures which can be attained experimentally. In high-temperature superconductors at $\xi \sim d$, the region of critical fluctuations is much broader. Thus, evaluations performed in [4.2] show that in YBCO compounds $\tau_{G3} \simeq 3 \cdot 10^{-3}$, while in strongly anisotropic compounds of Tl and Bi two-dimensional fluctuations of the order parameter can occur at $\tau < \tau_{G2} \simeq 10^{-2}$. The effect of the fluctuations on the jump of the specific heat in YBCO is discussed in the next section.

4.2 Specific Heat

The study of the specific heat makes it possible to investigate bulk properties of solids, primarily, integral characteristics of the excitation spectrum of electronic, phononic and magnetic degrees of freedom. In particular, the observation of the specific-heat jump occuring at the superconducting phase transition in oxide superconductors has confirmed the bulk nature of high-temperature superconductivity. The specific heat of YBCO and LMCO compounds has been best studied, while that of Bi and Tl compounds has been investigated less. A detailed review of these studies is given in [4.6]. Here, we shall only discuss the results, that are most important for understanding the nature of high-temperature superconductivity.

The study of the specific heat and, in particular, its electronic part in high-temperature superconductors encounters a number of difficulties. First of all, due to the high value of the critical temperature T_c, the phononic contribution to the specific heat is large. Against its background, it is difficult to single out the electronic specific heat. For example, Fig. 4.2 shows the temperature dependences of the specific heat and the ratio C/T (left scale) for YBCO compounds [4.6]. Here, the jump of the specific heat $\Delta C(T_c)$ at the superconducting transition characterizes the electronic contribution C_e, Fig. 4.3. The jump of the specific heat observed experimentally in other high-temperature superconductors turns out to be even smaller. In order to obtain any quantitative results, high measurement accuracy is therefore required.

The specific heat of YBCO compounds at room temperature is far from its maximum value in the classical limit, $3Nk$ per gram-atom. It has been estimated that it constitutes only 85 % of the maximum value. Consequently, in the spectrum of lattice oscillations there are high-frequency modes which are actually observed in inelastic neutron scattering (see Chap. 6).

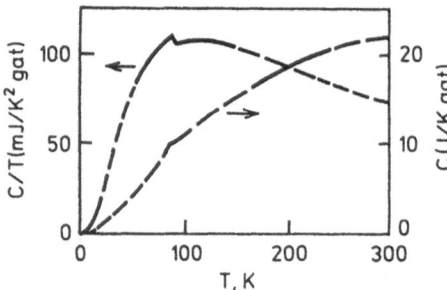

Fig. 4.2. The temperature dependences of the specific heat (*right-hand scale*) and the ratio C/T (*left-hand scale*) for YBCO compounds [4.6]

The high sensitivity of the electronic contribution to structural defects such as impurities, atomic disorder, etc. is another important feature which makes it difficult to obtain quantitative results. This is due to the small value of the correlation length ξ which is comparable to interatomic spacings and therefore sensitive to short-range disorder. This disorder occurs in all high-temperature superconductors, since they are simply many-component solid solutions. Moreover, when the synthesizing compounds such as YBCO, it is difficult to avoid the appearance of impurity phases. In some of them such as $Y_2Cu_2O_5$ or $BaCuO_{2+x}$, the low-temperature specific heat is $10 - 100$ times as large as the electronic part in YBCO compounds [4.6].

As was noted in Chap. 1, the density of electronic states $N(0)$ at the Fermi surface is one of the most important parameters for a superconductor. It is connected with the Sommerfeld constant γ in the low-temperature electronic specific heat by the relation

$$\gamma = \frac{\pi^2}{3} k^2 2N(0)(1 + \lambda) \equiv \frac{\pi^2}{3} k^2 \nu(0). \tag{4.18}$$

Here $N(0)$ is the density of electronic states per atom and per spin direction, $\nu(0) = 2N(0)(1 + \lambda)$ is an effective density of states per atom which takes into account the renormalization of the band density of states $N(0)$ due to an electron–phonon coupling λ. In the region above the Debye temperature $T \geq \theta_D$, the renormalization can be neglected, $\lambda \to 0$, so that $\nu(0) = 2N(0)$. The specific heat is usually measured in the units of J/mol or J/g-atom, where a mole of a substance contains $N_A = 6.02 \times 10^{23}$ formula units and a gram-atom contains N_A atoms. For $YBa_3Cu_3O_7$, one mole is equal to 13 g·atom. In order to determine the density of states $\nu(0)$ according to (4.18) we proceed from the specific heat and use the following relation:

$$\nu(0) \left(\frac{\text{states}}{\text{eV} \cdot \text{atom}} \right) = 0.414\gamma \left(\frac{\text{mJ}}{\text{K}^2 \cdot \text{g-at}} \right). \tag{4.19}$$

In discussing the specific heat of copper-oxide superconductors, we shall also take into account that its electronic part is related to free carriers in the CuO_2 plane. It is therefore convenient to refer it to a Cu atom or a Cu mole in (4.19).

Numerous measurements of the temperature dependence of the specific heat in a wide temperature interval and a special procedure for subtracting the phononic contribution have given the following values for the constant γ in YBCO compounds [4.6]:

$$\gamma = (20 - 35) \left(\frac{mJ}{K^2 \cdot mol} \right) = (1.5 - 2.7) \left(\frac{mJ}{K^2 \cdot g - at} \right) . \tag{4.20}$$

If we refer the density of states to a Cu atom (in YBCO compounds, there are three Cu atoms per formula unit), according to (4.19), we obtain $\nu(0) = 2.8 - 4.8$ (state/eV· Cu atom). If we compare this value with the results of the band-structure calculations $2N(0) \simeq 1 - 2$ (state/eV·Cu atom) (Sect. 5.3), we can deduce a strong electron–phonon renormalization $\lambda = 1 - 2$. In Sect. 5.3, it is mentioned, however, that the one-electron approximation is difficult to prove in copper-oxide compounds. The accuracy in the measurements of the electron contribution to the specific heat in $La_{2-x}M_xCuO_4$ compounds is much lower. However, the values $\gamma = 8 - 10$ mJ/K²·mol obtained in these measurements agree with the values (4.20) in YBCO compounds referred to a Cu mole.

In a number of experiments, the paramagnetic susceptibility $\chi(T)$ has been studied in addition to the measurement of the electronic contribution to the specific heat and the evaluation of γ (4.20). The paramagnetic susceptibility contains several contributions which we discussed in Sect. 3.3. One of them – a paramagnetic Pauli contribution – is directly related to the density of states (cf. Eq. (3.21))

$$\chi^S = 2\mu_B^2 N(0)S, \tag{4.21}$$

where $S = [1 - IN(0)]^{-1}$ is the Stoner factor which describes the increase of the susceptibility due to an exchange interaction I in the electron gas. Since it is not clear how to estimate other contributions to the susceptibility (diamagnetic, orbital, or antiferromagnetic fluctuation), a direct determination of the contribution χ^S turns out to be insufficiently accurate. We now mention data for the spin susceptibility for YBCO compounds at $y = 0$ shown in Fig. 3.8. These data, $\chi^S/\mu_B^2 \simeq 2.7$ state/eV· Cu atom, agree rather well with the value of the total density of states obtained from (4.20), provided $S \simeq 1$ and $\lambda \ll 1$. On the other hand, the comparison with the band-structure calculations gives the estimate $S = 1.5 - 2$ at $\lambda = 1 - 2$.

Another possible way of determining the Sommerfeld constant (4.18) consists in the measurement of the specific-heat jump at the superconducting phase transition (4.6)

$$\frac{\Delta C}{C_n} = \frac{\Delta C}{\gamma T_c} = A, \tag{4.22}$$

where $A = 1.43$ is a universal constant in the weak-coupling limit of the BCS theory. In the strong-coupling limit, the constant A can attain rather large values, $A \sim 10$ (for strong coupling via high-frequency phonons) and values smaller than one, $A < 1$ (for strong coupling via low-frequency modes) [4.7]. Therefore, in order to determine the constant γ from (4.22), certain assumptions concerning the

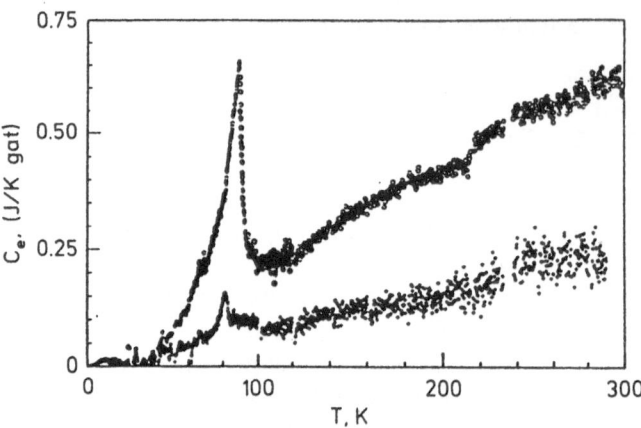

Fig. 4.3. The electronic contribution to the specific heat in YBCO−123 compounds (*upper curve*) and YBCO−124 compounds (*lower curve*) [4.8]

nature of the superconducting transition are required. On the other hand, the latter information can be obtained if the parameter A is determined from (4.22) when the jump $\Delta C/T_c$ and the constant γ are measured independently.

For example, let us consider the electronic contribution to the specific heat near T_c in YBCO−123 and YBCO−124 compounds shown in Fig. 4.3 [4.8]. The comparison of the data for these two closely related compounds is of interest in its own right. For $T > T_c$, the contribution to the specific heat linear in temperature determines the electronic specific heat (which constitutes 1–2 % of the total specific heat at 100 K). In Ref. [4.8] the electronic specific heat is estimated by the constants

$$\gamma(\mathrm{mJ/K^2 \cdot mol}) \simeq 12 \ (\mathrm{YBCO} - 124) - 28 \ (\mathrm{YBCO} - 123). \tag{4.23}$$

The corresponding specific-heat jumps at T_c are equal to

$$\Delta C_p/T_c(\mathrm{mJ/K^2 \cdot mol}) \simeq 28 \ (\mathrm{YBCO} - 124) - 67 \ (\mathrm{YBCO} - 123). \tag{4.24}$$

A much smaller value of the constant γ and the specific-heat jump in YBCO−124 compounds indicates a smaller density of electronic states, $\nu(0) \simeq 1.24$ (state/eV · Cu atom). At the same time, the transition temperature in 124-compounds, $T_c = 82$ K is only 13 % lower than that in 123-compounds, $T_c = 90$ K. The ratio (4.22) however takes similar values, $A = \Delta C/\gamma T_c \simeq 2.3 - 2.4$ for both compounds. A much larger value of A compared to $A_{BCS} = 1.43$ points to the strong coupling.

Rather large jumps of the specific heat (4.24) obtained in Ref. [4.8] are apparently related to a high degree of homogeneity of the samples. Smaller values are usually observed. In [4.9], for example, the following results are reported:

$$\Delta C_p/T_c(\mathrm{mJ/K^2 mol}) = 15 \ (\mathrm{YBCO} - 124) - 55 \ (\mathrm{YBCO} - 123). \tag{4.25}$$

In Ref. [4.9], due to a new method of synthesis, a higher transition temperature, $T_c \simeq 88$ K, has been obtained in calorimetric measurements in 124-compound.

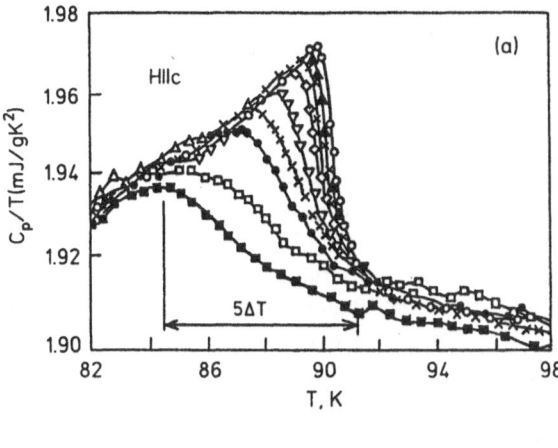

Fig. 4.4. The dependence of $C_p(H,T)/T$ for a single-crystal sample of $YBa_2Cu_3O_{7-y}$ in a magnetic field: (**a**) along the c axis and (**b**) in the ab plane for magnetic fields from $H = 0$ up to $H = 7$ Tesla (for details see [4.10])

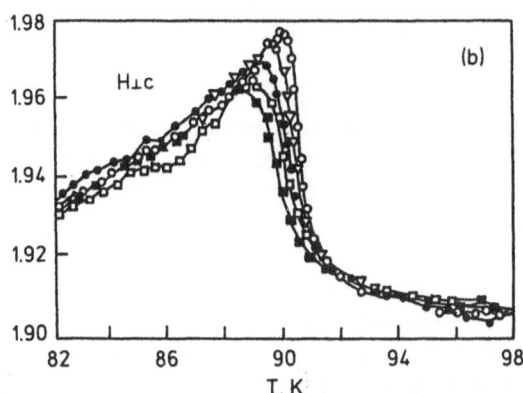

(The resistivity vanishes at $T_c \simeq 81$ K). In doped Ca samples, $YBa_{2-x}Ca_xCu_4O_8$ at $x \simeq 0.2$ an increase of the specific heat jump up to 22 mJ/K· mol and an increase of T_c up to 90 K have been observed.

A remarkable peak in $\Delta C(T)$ at T_c and a tail of this dependence at $T > T_c$ seen in Fig. 4.3 are related to the contribution of critical fluctuations of the order parameter at the superconducting transition. In Ref. [4.10] the contribution of critical fluctuations to the specific heat has been studied in greater detail in monodomain samples of YBCO–123 in magnetic fields. The results of these measurements are shown in Fig. 4.4 for magnetic field parallel to the axis c, $\boldsymbol{H} \parallel \boldsymbol{c}$ (a) and perpendicular to axis c, $\boldsymbol{H} \perp \boldsymbol{c}$ (b). The dependence of the specific-heat jump on the value and direction of the magnetic field unambiguously indicates the electronic nature of this contribution to the specific heat. In this experiment, an anisotropy of the bulk superconductivity of YBCO compounds is also distinctly seen.

In zero magnetic field, the specific-heat jump can be described by the three-dimensional Ginsburg–Landau theory for the two-component ($n = 2$) order parameter when Gaussian fluctuations are taken into account, as was discussed at the end of Sect. 4.1. We note that, in earlier experiments (see, for example, [4.11]) on YBCO samples of poorer quality, the conclusion was reached that this theory does not work and that the number of components of the order parameter should be larger, $n = 4 - 6$.

In a magnetic field, the phase transition gets smeared and the maximum of the specific heat shifts to lower temperature, $T_c(H) < T_c(0)$. In Ref. [4.10] an analysis of the shape of the curve $C_p(T, H)$ has been performed on the basis of the scaling theory. The analysis has shown that, as $T \rightarrow T_c(H)$, the critical fluctuations diverge logarithmically. They cannot be treated in the Gaussian approximation. It is important that the correlation length of the fluctuations in a magnetic field is limited. This experiment has thus confirmed the existence of a critical region in the superconducting phase transition.

The specific-heat jump is smeared out not only in a magnetic field, but also when the sample is doped or under perturbations of its structure. For example, under 2% Fe doping of $YBa_2(Cu_{1-x}Fe_x)_3O_{7-y}$, the specific-heat jump falls to almost one third of its initial value. At $x = 4\%$, the jump is no longer observed. The total suppression of superconductivity, $T_c = 0$, occurs only at $x = 7\%$ [4.6]. A similar effect is produced by a decrease of the oxygen content or by doping with non-magnetic Zn. The specific-heat jump is halved at $x = 2.5\%$ and, finally, the critical temperature vanishes, $T_c \simeq 0$, at a Zn concentration $x = 7\%$ [4.6]. The reason why the specific-heat jump decreases so sharply and the phase transition gets smeared is presumably the small value of the correlation length ξ (4.11) in copper-oxide superconductors. The critical temperature is determined locally by the energy of condensation averaged over volume $(\bar{\xi})^3$. When the distance between structural defects (dopants) becomes comparable with ξ, a considerable fluctuation of the local temperature of the superconducting transition occurs and the specific-heat jump becomes smeared out. A remarkable decrease of the Meissner effect, (i.e., the decrease of the fraction of superconducting volume in a sample) under doping [4.6] is also indicative of this mechanism.

The value of the specific-heat jump in the superconducting transition with the formation of Cooper pairs should be equal to $\Delta C \simeq N_s k$ where N_s is the number of Cooper pairs. In the conventional superconductors, they constitute only a small fraction $N_s \simeq (T_c/E_F)N$ of the total number of electrons N, since $T_c/E_F \leq 10^{-3}$. If the superconducting transition is connected with the Bose condensation of the pairs that exist above T_c, we can write $\Delta C \simeq Nk$ where N is the number of these pairs, which is of the order of the total number of particles. Eq. (4.24) implies that in YBCO the specific-heat jump per mole is equal to $0.25R$, where $R = N_A k_B$. Such a high value of the jump is sometimes treated as confirmation of a mechanism of Bose condensation of bipolaron pairs (see Sect. 7.4). It should however be noted that, in bismuth and thallium based superconductors of 2212−type, the specific-heat jump is so small that it is almost unobservable even in perfect single crystals [4.6]. Therefore, the jump in YBCO compounds should presumably be attributed to a low value of the Fermi energy E_F and a three-dimensional nature of the superconducting transition. In Tl and, particularly, Bi compounds with a high degree of anisotropy due to weak coupling between CuO_2 planes,

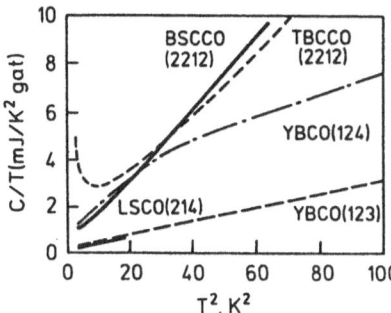

Fig. 4.5. The low–temperature specific heat of copper-oxide superconductors. The ratio C/T is plotted as a function of T^2 [4.6]

the superconducting transition is of a quasi-two-dimensional nature with a much smaller transition entropy and specific-heat jump than in the three-dimensional phase transition.

In the transition to the superconducting state with the formation of Cooper pairs, the electronic specific heat should decrease exponentially at $T < T_c$ and vanish at $T \ll T_c$. However, numerous experiments to measure the specific heat in copper-oxide superconductors at $T \ll T_c$ reveal a finite value of the specific heat which is most often characterized by the linear function $C(T \to 0) \simeq \gamma_r T$. The value of the constant γ_r – the so called residual Sommerfeld constant – can be estimated by plotting the dependence of the ratio C/T on T^2. This dependence is shown in Fig. 4.5 for a number of high-temperature superconductors [4.6]. For all the compounds, finite values of C/T, i.e.,

$$\gamma_r = \left(\frac{C}{T}\right)_{\min} \left[\frac{\text{mJ}}{\text{K}^2 \cdot \text{mol}}\right] = \begin{cases} 1.2 & (\text{LSCO}, x = 0.15), \\ 4.4 & (\text{YBCO}, y \simeq 0), \\ 4.8 & (\text{BSCCO}) \end{cases} \quad (4.26)$$

are observed for $T \to 0$. These values are obtained in the measurement of the specific heat in the interval 2–8 K on the "best" samples, where the contribution to the specific heat due to doping phases, structural defects, dopants, and other "external" sources is reduced to a minimum. In the region of even lower temperatures, $T < 1$ K, a contribution to the specific heat due to nuclear spins, $C_N \sim 1/T^2$, can occur. It can be easily taken into account since its temperature dependence is known.

Both Bose and Fermi excitations, as well as excitations of two-level systems well known in glasses and disordered systems, may be internal sources of the residual specific heat of a superconductor. For example, in the two-dimensional case, antiferromagnetic fluctuations yield $C \sim T$. In the two-level systems, we also have $C_{\text{TLS}} \sim T$. The contribution to the specific heat linear in temperature may be also due to the vanishing of the superconducting gap on certain parts of the Fermi surface (for example, under the d-wave pairing) or due to non-standard mechanisms of superconductivity (Chap. 7). For this reason, the question concerning the residual linear specific heat has attracted much attention. Numerous experiments, particularly on the measurement of the residual specific heat in strong magnetic fields, have resulted in the following conclusions [4.6]:

1. The residual specific heat can be written in the most general form as $C \sim T^\alpha$ where $\alpha \simeq 0 - 2$. The linear dependence, $\alpha = 1$, is a special case.

2. There are a few possible explanations for the linear residual specific heat, some of which may be true. At the same time, the explanations of the above values (4.26) on the basis of the model of gapless superconductivity or due to two-level excitations are doubtful.

3. The residual specific heat in the antiferromagnetic and in the superconducting phases of YBCO compounds turns out to be of the same order. Therefore, it can hardly be characterized as a phenomenon specific for the superconducting phase of copper-oxide superconductors. It is known that "linear" residual specific heat is often observed in ordinally doped metallic superconductors.

4. The existence of localized magnetic moments at Cu^{2+} ions and (or) localized electronic states with a broad spectrum of excitation energy is one of the plausible reasons for the residual specific heat. In strong magnetic fields, $B > 20\,kG$, and low temperatures, $T < 0.4\,K$, the value of the residual specific heat is close to zero, at least, $\gamma_r < 1\,mJ/K^2 \cdot mol$ for YBCO and Bi/2212 compounds. An analysis of experimental data on the residual specific heat in YBCO compounds as a function of the concentration of magnetic moments at Cu^{2+} ions is reported in Ref. [4.12]. The authors have concluded that there is no residual specific heat, $\gamma_r = 0$, in the ideal sample containing no localized magnetic moments Cu^{2+}.

Interesting results supporting these conclusions have been obtained in Ref. [4.13] where the dependence of the constant $\gamma_r(x)$ on the carrier concentration x in $La_{2-x}Sr_xCuO_4$ was studied (Fig. 4.6). In the dielectric antiferromagnetic phase at $x = 0$, the constant γ_r is close to zero. As a sample is doped and free carriers begin to form, there appears a contribution to the Sommerfeld constant γ. At the transition to the superconducting state with the formation of the gap at $x > 0.05$, this contribution goes down. At large concentrations, $x > 0.25$, superconductivity disappears and, in the normal metallic phase of LSCO compounds, γ_r takes the usual value for the Sommerfeld constant $\gamma \simeq 9\,mJ/K^2 \cdot mol$. This experiment thus shows that a linear residual specific heat with the value $\gamma_r \leq 1\,mJ/K^2 \cdot mol$ is observed as a background, no matter what the magnetic order, structure and doping are. This appears to be the most convincing experiment which demonstrates the absence of a sufficiently large residual specific heat for superconductivity of a new nature in copper-oxide compounds[1].

---

[1] Electronic specific heat measurement [4.32] of $YBa_2Cu_3O_{6+x}$ for $0.16 \leq x \leq 0.97$ confirms this result and suggests a formation of a spin and/or superconducting gap for fermion charge carriers at low temperatures.

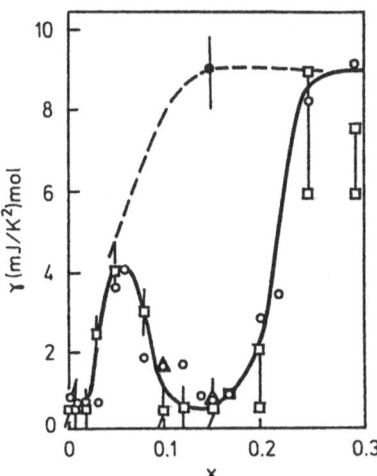

Fig. 4.6. The x-dependence of the constant γ_r in the low-temperature linear specific heat in $La_{2-x}Sr_xCuO_4$ [4.13]

4.3 Magnetic Properties

The study of macroscopic magnetic properties of copper-oxide superconductors reveals a number of novel phenomena whose interpretation is still controversial [4.14]. Even in the first magnetic measurements [4.15], an irreversibility line in the $H - T$ plane was found. It is specified by the temperature $T_c^*(H)$ below which the field-cooled susceptibility, χ_{FC}, and the zero-field-cooled susceptibility, χ_{ZFC}, take different values. The irreversibility line is described by the formula

$$(1 - T/T_c^*) = H^q , \qquad (4.27)$$

where the exponent q is close to 2/3. Such irreversible phenomena are observed in spin glasses which are viewed as metastable thermodynamical systems. In this respect, a hypothesis has been proposed [4.15] that a metastable state — the superconducting glass — arises at $T < T_c^*(H)$. This state was observed earlier in granular superconductors with weak-link structure. Later on, a number of other models were proposed to describe the phenomena of irreversible magnetization and the relaxation of the remanent moment [4.14, 16].

These unusual (in comparison with conventional superconductors) phenomena are strongly connected with a small value of the correlation length of the order parameter (4.11) and a strong anisotropy of the superconducting properties. In superconductors of the second kind at $H > H_{c1}$, a system of fluxes occurs. Their positions are fixed by the pinning centers. The pinning energies are estimated by the value $U_0 = H_c^2 \xi^3 / 8\pi$ where H_c is a thermodynamic critical field and $\xi^3 = \xi_{ab}^2 \xi_c$ is a characteristic volume. In high-temperature superconductors, due to a small value of the correlation length, this energy is small (in YBCO compounds it is equal to $U_0 \simeq 20\,meV$). At high T_c, weak flux pinning and a giant flux creep [4.14] occur. The estimation of the temperature $T_c^*(H)$, above which fluxes move almost freely (the depinning temperature), shows that it also satisfies the relation (4.27) describing the irreversibility line with $q = 2/3$ [4.14]. There also exist more complicated models: a model of collective flux pinning where the interaction of fluxes is taken into account, a model of flux-lattice melting due to thermal fluctuations at some temperature $T_c(H)$, etc. (see the review [4.17]). First measurements of critical fields in ceramic samples were additionally complicated by the presence of weak Josephson links at grain boundaries, which has made the interpretation of experimental data much more difficult. We shall now consider some experimental results on the measurement of critical fields and the penetration depth of magnetic field obtained on single crystals of high quality. The results on the critical magnetic fields in various high-temperature superconductors are discussed in greater detail in Ref. [4.5].

In Refs. [4.3, 18], the lower critical field was determined from the deviation of the magnetization $M(T)$ from the linear law $M(H) = \chi H$ which holds for $H < H_{c1}(T)$. At $H > H_{c1}$ the magnetic field of fluxes begins to penetrate into a superconductor and the susceptibility χ begins to decrease. In high-temperature superconductors, a sharp kink on the curve $M(T, H)$ which is a specific feature of conventional superconductors of the second kind, is not seen at $H_{c1}(T)$. This method does not therefore ensure sufficient accuracy for determining H_{c1} (cf. data of Table 4.1).

Table 4.1. Critical magnetic fields in single-crystals of YBCO

T_c [K]	$\left(dH_{c1}^c/dT\right)_{T_c}$ [Oe/K]	$\left(dH_{c1}^{ab}/dT\right)_{T_c}$ [Oe/K]	$H_{c1}^c(0)$ [Oe]	$H_{c1}^{ab}(0)$ [Oe]	Ref.
92			690 ± 50	120 ± 50	[4.3]
91.5			900 ± 100	250 ± 50	[4.19]
88.2	-28	-6.8	850	250	[4.20]
89	-17	-5.5	950	230	[4.21]

T_c [K]	$\left(dH_{c2}^c/dT\right)_{T_c}$ [T/K]	$\left(dH_{c2}^{ab}/dT\right)_{T_c}$ [T/K]	$H_{c2}^c(0)$ [T]	$H_{c2}^{ab}(0)$ [T]	$\xi_c(0)$ [Å]	$\xi_{ab}(0)$ [Å]	Ref.
92.5	-1.8	-10.5	122	674	3	16.4	[4.22]
93	-1.8	-10	120	670	3	16	[4.23]
91	-0.85		42			28	[4.24]

In other experiments the irreversible behavior of the magnetization at $H > H_{c1}$ due to the relaxation dynamics of fluxes is used. In Ref. [4.19], for example, the critical field H_{c1} was measured by the occurrence of irreversibility in the magnetization $dM/d\ln t \neq 0$ above H_{c1}. The analysis of data based on the flux-creep model and the critical state in the Bean model have given values of H_{c1} along the c-axis and in ab−plane (Table 4.1) and characteristic energies of flux-motion activation $U_0^{\parallel} \simeq 0.02\,\text{eV}$ and $U_0^{\perp} = 0.4\,\text{eV}$ for magnetic fields H parallel and perpendicular to the c−axis, respectively. In Ref. [4.20], the critical field H_{c1} was

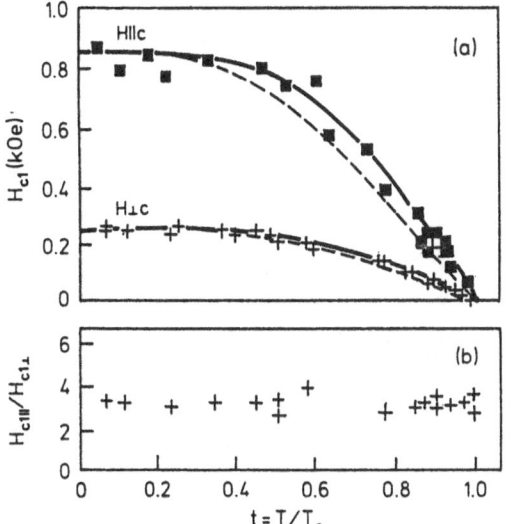

Fig. 4.7. The temperature dependence of the lower critical field H_{c1} and the anisotropy $H_{c1}^{\parallel}/H_{c1}^{\perp}$ in a single crystal of YBCO at $T_c = 88.2\,\mathrm{K}$ [4.20]

determined by measuring the penetration depth of a radio-frequency field into a sample manifesting the mixed phase at $H > H_{c1}$. The temperature dependence of H_{c1} for both directions parallel and perpendicular to the c–axis has been measured by this method. The results are plotted in Fig. 4.7. The theoretical calculations of $H_{c1}(T)$ performed on the basis of the BCS theory are shown by the solid line for $2\Delta/kT_c = 4.3$ and dashed line for $2\Delta/kT_c = 3.5$. In general, a good agreement with theory is observed for anisotropy independent of temperature $H_{c1}^{\parallel}/H_{c1}^{\perp} \simeq 3.4$ (the bottom part of Fig. 4.7). The limiting values of the critical screening current have been calculated for the single crystal of YBCO proceeding from the estimation of the flux-pinning energy. The results, $I_c^a \simeq 2.5 \cdot 10^7$ A/cm^2 for $H \parallel c$ and $I_c^c \simeq 2.3 \cdot 10^6$ A/cm^2 for $H \perp c$ are close to the data of other studies (see, for example, [4.3]).

In Ref. [4.21] an extrapolation method for determining H_{c1} from a trapped magnetic moment M_t has been used. If the magnetic field is switched on isothermally, it behaves like $M_t \sim (H - H_{c1})^2$. The value $H_{c1}(T)$ can then be found by means of the linear extrapolation $[M_t(H)]^{1/2} \to 0$. The value of $H_{c1}^{ab}(T)$ obtained in this way for a single crystal of YBCO ($T_c = 89$ K) had the usual dependence with saturation as $T \to 0$, while $H_{c1}^c(T)$ depends almost linearly on T in the entire range of temperatures. The reason for the deviation of the temperature dependence of $H_{c1}^c(T)$ from the standard one (see, for example, Fig. 4.7) observed in some experiments is not yet clear and may be of "external" nature, having nothing in common with high-temperature superconductivity.

The data summarized in Table 4.1 for the lower critical fields $H_{c1}(0)$ show that their averaged values lie in the interval $H_{c1}^c \simeq 800-900$ Oe and $H_{c1}^{ab} \simeq 230-250$ Oe. Equation (4.15) gives the ratio of effective masses in the anisotropic Ginsburg–Landau theory

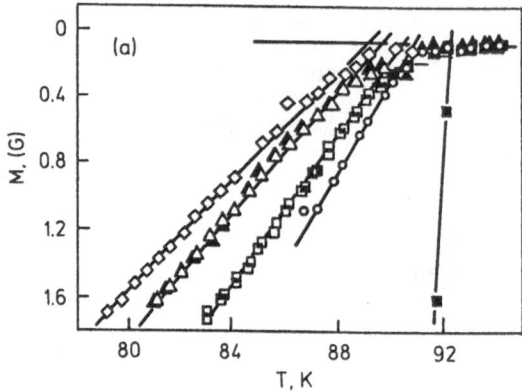

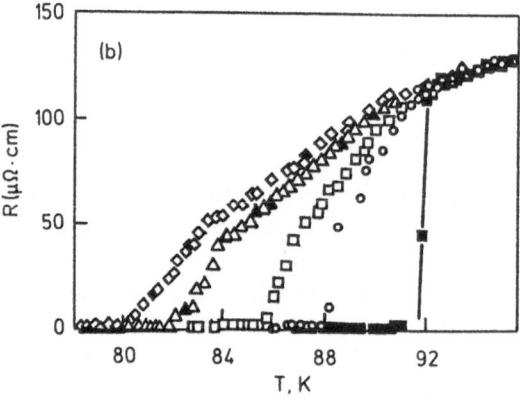

Fig. 4.8. The temperature dependence of magnetization (**a**) and resistivity (**b**) for a single crystal of YBCO for field $H \parallel c$ from $H = 0$ up to $H = 5$ Tesla (for details see [4.22]

$$\frac{m_c}{m_{ab}} = \left(\frac{H_{c1}^c \ln \kappa_{ab}}{H_{c1}^{ab} \ln \kappa_c} \right)^2 \simeq 15 . \qquad (4.28)$$

The analysis of the temperature dependence $H_{c1}(T)$ and the related penetration depth $H_{c1}(T) \sim \lambda^{-2}(T)$ (see Eq. (4.15)) show that the experimental data obtained for YBCO correspond to the "pure" limit of a superconductor with strong coupling [4.2][2].

The measurement of the upper critical field $H_{c2}(T)$ in high-temperature superconductors encounters a number of difficulties mentioned at the beginning of this section. Let us consider, for example, a resistivity method for determining the upper critical field from the measurement of the resistivity $\varrho(T, H)$ in a magnetic field H. Figure 4.8 shows the dependence of magnetization $M(T)$ (a) and resistivity $R(T)$ (b) in external fields $H \parallel c$ in a single crystal of YBCO with $T_c = 92.5$ K and 92.2 K, respectively [4.22]. As the field H increases, the magnetization curves $M(T)$ shift towards lower temperatures remaining linear in temperature. This enables the dependence $T_c(H)$ to be determined accurately enough. At the same time, as the external field is switched on, the dependence $R(T)$ is drastically smeared out

---

[2] See footnote [4] in this chapter.

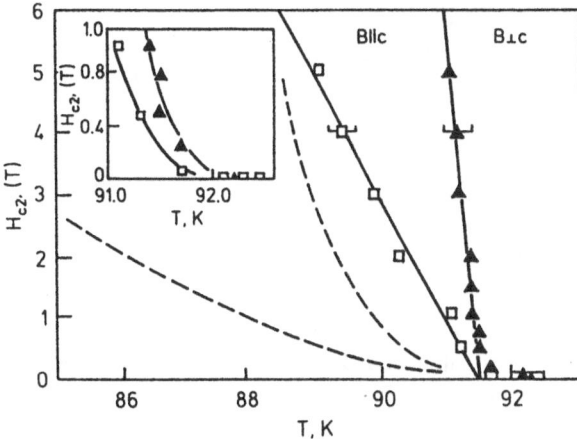

Fig. 4.9. The temperature dependence of the upper critical field $H_{c2}(T)$ for $H \parallel c$ and $H \perp c$ (*solid lines*). The dependence measured by the resistivity method ($R = 0$) is shown by the dashed line [4.22]

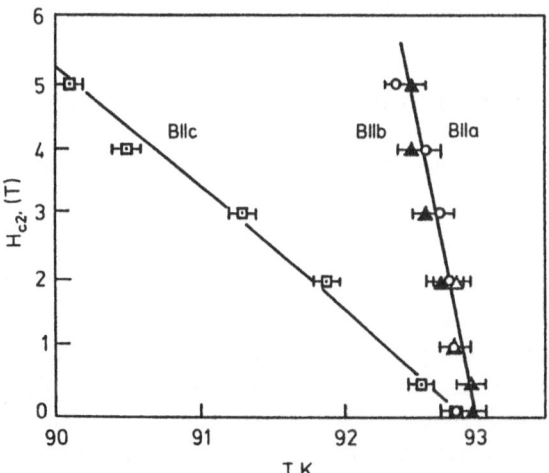

Fig. 4.10. The dependence $H_{c2}(T)$ in a monodomain crystal of YBCO with $T_c = 93$ K [4.23]

being pulled into the region of low temperatures. This prevents the temperature of the superconducting transition in a field $T_c(H)$ from being uniquely determined. The smearing of the phase transition observed in the resistivity method is specific to the copper-oxide superconductors and reflects a weak pinning of magnetic flux. The dependence $H_{c2}(T)$ obtained in this method is non-linear $H_{c2}(T) \propto (T_c - T)^{3/2}$, and as a rule corresponds to the irreversibility line (4.27) or the depinning line (flux-lattis melting) discussed above. In this resistivity method, the data for $(dH_c/dT)_{T_c}$ have greater dispersion and are smaller than those obtained by other methods.

Another method for determining the derivative $(dH_{c2}/dT)_T$ is based on the measurement of reversible magnetization $M(T)$ in the region where χ_{FC} and χ_{ZFC} coincide. The temperature dependence of $H_{c2}(T)$ near T_c measured by this method is shown in Fig. 4.9 for a polydomain [4.10] and in Fig. 4.10 for a monodomain crystal [4.23].

The region of nonlinear dependence of $H_{c2}(T)$ in low magnetic fields occurring in the polydomain sample (see the insert in Fig. 4.9) is not observed in the monodomain crystal.[3] As shown by the dashed line in Fig. 4.9, the resistivity method gives the nonlinear dependence $H_{c2}(T)$ over a wide range of temperatures $T < T_c$ and a smaller slope dH_{c2}/dT. Since the direct measurement of $H_{c2}(0)$ in high-temperature superconductors encounters remarkable difficulties, the Werthamer–Helfand–Hohenberg formula [4.5]

$$H_{c2}(0) \simeq 0.7(dH_{c2}/dT)_{T_c} T_c, \qquad (4.29)$$

obtained for the BCS model with weak coupling is often used to estimate this quantity. The estimation for $H_{c2}^{\alpha}(0)$ calculated on the basis of this formula and the corresponding correlation lengths in the anisotropic Ginsburg–Landau model (4.12) are given in Table 4.1. For the ratio of the effective masses, we obtain the value

$$\frac{m_c}{m_{ab}} = \left(\frac{H_{c2}^{ab}}{H_{c2}^{c}}\right)^2 \simeq 30, \qquad (4.30)$$

which exceeds the value (4.28) obtained from the measurement of the lower critical field.

Comparing the specific-heat jump calculated from the slope of the magnetization curve near T_c

$$\frac{\Delta C}{T_c} = \frac{1}{8\pi\kappa^2}\left(\frac{\partial H_{c2}}{\partial T}\right)_{T_c}^2, \qquad (4.31)$$

with the value obtained in calorimetric experiments can serve as an additional test of the accuracy of the measurement of $H_{c2}(T)$. The calculations according to Eq. (4.31) performed in Ref. [4.22] yield $\Delta C/T_c \simeq 42$–$30\,\mathrm{mJ/K^2 \cdot mol}$ for the parameters along the axes c and a, b, respectively. This value of the jump is in a reasonable agreement with experiment (see (4.24)).

In order to determine the parameter κ_α (4.14) in the Ginsburg–Landau theory, Eq. (4.16) which relates it to the lower and upper critical fields can be used. If we set $H_{c1}^c \simeq 800$–$900\,\mathrm{Oe}$, $H_{c1}^{ab} \simeq 230$–$250\,\mathrm{Oe}$ and $H_{c2}^c \simeq 1.2 \cdot 10^6\,\mathrm{Oe}$, $H_{c2}^{ab} \simeq 6.7 \cdot 10^6\,\mathrm{Oe}$, we obtain

$$\kappa_c \simeq 50 - 55, \quad \kappa_{ab} \simeq 275 - 290. \qquad (4.32)$$

The above values of the parameters are only estimates, because the accuracy of the measurement of $H_{c1}(0)$ and the evaluation of $H_{c2}(0)$ according to formula (4.29) is not sufficient. For example, direct measurements of $H_{c2}(0)$ in pulsed magnetic

---

[3] An anamalous temperature dependence of the resistive upper critical magnetic field was observed in overdoped single crystals of $Tl_2Ba_2CuO_6$ ($T_c \simeq 20\,K$) [4.33] and in $Bi_2Sr_2CuO_y$ thin films [4.34] which shows an upward curvature without saturation down to very low temperatures, $T/T_c \simeq 10^{-3}$.

fields by the resistivity method yield much smaller values of the upper critical fields and larger correlation length [4.24] (see Table 4.1).

Measurements of the critical fields in other high-temperature superconductors, for example, thallium- and bismuth-based superconductors reveal much more dispersion of experimental data. These data are discussed in greater detail in [4.5].

Besides the measurement of the critical fields, important information on the nature of the superconducting pairing can be obtained from the direct measurement of the penetration depth of the external field $\lambda_\alpha(T)$. According to (4.10), this is determined by the density of superconducting pairs $n_s(T)$. Therefore, its temperature behavior is directly related to the excitation spectrum of these pairs. In the case of the usual s-wave pairing, there exists a non-zero gap $\Delta(T)$ over the entire Fermi surface. The temperature dependence of the penetration depth as $T \rightarrow 0$ should be of the activation nature, i.e., $\lambda(T) \propto \exp(-\Delta(0)/T)$. For an unconventional pairing, the gap vanishes along certain lines on the Fermi surface so that a power dependence of $\lambda(T)$ occurs. In Ref. [4.25] it was shown that in superconductors with tetragonal or orthorhombic lattice symmetry a linear temperature dependence $\Delta\lambda_{ab}(T) = \lambda_{ab}(T) - \lambda_{ab}(0) \propto T$ should be observed for all types of pairing apart from the s-wave singlet pairing. The analysis of experimental data [4.26] for YBCO compounds performed in Ref. [4.25] shows that $\Delta\lambda_{ab}(T) \propto T^2$ i.e., follows neither the linear law for an unconventional pairing nor the activation law of the BCS pairing. Such an unusual behavior may be related to a strong scattering of electrons on defects or antiferromagnetic spin fluctuations which are specific for copper−oxide superconductors (see Chap. 3)[4].

Let us consider experimental results [4.27] of $\lambda_{ab}(T)$ on a single crystal of YBCO with the critical temperature 89.5 K in the region of very small external fields, $H < 1$ Oe; this excludes any magnetic irreversible phenomena. Figure 4.11 shows the quantity $\Delta\lambda_{ab}(T)$ as a function of $(T/T_c)^2$ over the entire range of temperatures $0 < T < T_c$. Although a large scatter of experimental data is observed as $T \rightarrow 0$, all the points lie between two curves obtained in the weak-coupling approximation in the pure limit of the BCS theory (dotted line) and in the two-liquid model (dashed line).

The two-liquid model gives the power dependence

$$\lambda(T)/\lambda(0) = [1 - (T/T_c)^4]^{-1/2}, \tag{4.33}$$

which rather well agrees with the strong-coupling limit in the BCS theory, as is noted in [4.28]. The dash-dotted line corresponds to the dependence $\Delta\lambda(T) \simeq T^2$ which is often observed in polycrystalline samples. This line lies well above the experimental points. By varying the fitting parameters the authors of [4.27]

---

[4] Measurements [4.35] of the microwave surface impedance on high quality crystals of YBCO show that $\lambda_{ab}(T)$ has predominantly linear temperature dependence at low temperatures. This suggests nodes in the gap function. But in crystals with impurities or defects $\lambda_{ab}(T) \propto T^2$ as observed usually in thin films [4.36]. Discovery of the paramagnetic Meissner effect (Wohlleben effect) [4.37] and inverse Josephson coupling (π - contact) in a YBCO single crystal SQUID [4.38] also point to the unconventional (d-wave) pairing.

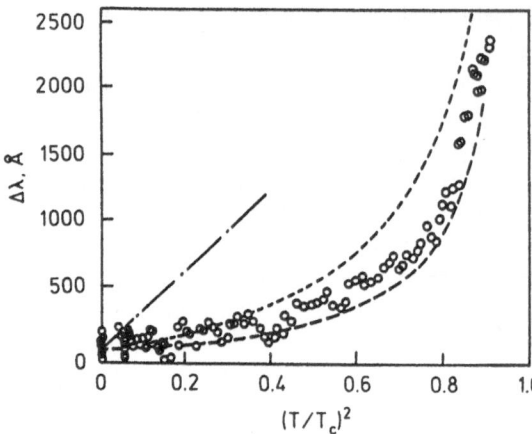

Fig. 4.11. Temperature dependence for penetration depth $\Delta\lambda_{ab}$ = $\lambda_{ab}(T) - \lambda_{ab}(0)$ in a single crystal of YBCO [4.27]

conclude that the conventional s-wave pairing with $\lambda_{ab}(0) = 1400 \pm 50$ Å occurs. If in (4.10) we set the electron density at $T = 0$ to be $n_s = n = 5 \times 10^{21}$ cm^{-3}, for $\lambda(0) = 1400$ Å we obtain a value of the effective mass equal to $m_{ab} \simeq 4m_e$. This is in a good agreement with optical measurements discussed in Sect. 5.4.3. Taking into account the ratio $\lambda_c/\lambda_{ab} = (m_c/m_{ab})^{1/2}$ and the estimate (4.28), we get $\lambda_c \simeq 5400$ Å. The results of a number of other measurements of $\lambda(T)$ near T_c by the method of reversible magnetization in single crystals of YBCO compounds [4.29] and Tl– or Bi– based compounds [4.30] are given in Table 4.2. The quantities $\lambda_{ab}(0)$ have been calculated in the weak-coupling approximation of the pure limit of the BCS theory. The data for YBCO compounds are in a rather good agreement with each other.

The specific values of $\lambda_{ab}(0)$ for the above copper-oxide superconductors reveal a certain correlation between the increase of T_c and the decrease of the penetration length. This correlation – the "Uemura plot" – has been studied in greater detail by the μSR method in [4.31]. It turned out to be $T_c \propto (1/\lambda^2(0))$ $\propto (n_s/m^*)$. Since in the frame of the BCS theory, there is no direct proportionality between T_c and the density of superconducting carriers n_s (see (1.1)), the authors of [4.31] have thus concluded that the mechanism of superconductivity in copper-oxide compounds is unconventional. An same time, the temperature dependence of $\lambda(T)$ observed in the μSR experiments [4.31] corresponds to the two-liquid model (4.32) or to the strong-coupling approximation in the BCS theory [4.28].[5]

In general, the analysis of the experimental data for the magnetic fields H_{c1}^α and H_{c2}^α confirms a sufficiently strong anisotropy of the effective masses, although the values derived, (4.28) and (4.30), differ a bit. There is also no doubt about the high values of the upper critical fields and, therefore, the small values of the correlation length ξ_α (Table 4.1) and the high values of the Ginsburg–Landau parameter κ_α

---

[5] Quite an unusual correlation between T_c and n_s/m^* in overdoped $Tl_2Ba_2CuO_{6+x}$ has been observed in μSR experiments [4.39] which points to a gapless superconductivity.

Table 4.2. Magnetic field penetration depth in copper-oxide superconductors

Compounds	T_c K	$\lambda_{ab}(0)$ Å	Ref.
	–	1425	[4.26]
$YBa_2Cu_3O_7$	89.7	1400	[4.27]
	90.5	1390	[4.29]
$YBa_2Cu_4O_8$	84.8	1980	[4.29]
$Bi_2Sr_2CaCu_2O_8$	89.5	3100	[4.30]
$Bi_{1.6}Pb_{0.4}Sr_2Ca_2Cu_3O_{10}$	104	2320	[4.30]
$Tl_2Ba_2CaCu_2O_8$	99.3	2210	[4.30]
$Tl_2Ba_2Ca_2Cu_3O_{10}$	109	1960	[4.30]

(4.32). A similar situation is observed for all the copper-oxide superconductors, although the specific values of the parameters $H_{c2}^{\alpha}(0)$, ξ_{α}, κ_{α} are dispersed, since the upper critical field is determined by extrapolation.

The temperature dependence of the magnetic field penetration depth $\lambda(T)$ at low temperatures [4.35, 36] and inverse Josephson coupling [4.37, 38] point to unconventional pairing.

5. Electronic Properties of High-T_c Superconductors

The superconducting properties of metals are essentially defined by their electronic properties in the normal state. Therefore studies of the latter are important in clarifying the mechanism of superconductivity in the new oxide superconductors. With respect to their electronic structure, these compounds may be related to the class of ionic semiconductors, in which the metallic conductivity appears upon the changes of stoichiometry. Their electronic structure is defined by a rather complicated interaction of localized and band electronic states, which is sensitive to short-range order in atomic positions. In this context, we will first discuss some crystallochemical aspects of electronic structure of oxide superconductors which are due to the individual properties of the constituent ions, and we will also consider the effect of impurity substitution on the electronic properties. The electronic structure is then discussed in greater detail on the basis of "first principles" and cluster-model electronic band calculations.

For understanding the electronic structure of oxide superconductors experimental studies are essential: X-ray and electron spectroscopy are useful in the high energy range, and optical methods are important in the domain of low-energy electronic excitations. Studies of transport properties – electrical resistivity, Hall effect, heat conductivity, etc. – provide additional information on the electronic structure of these compounds in the metallic phase. At the end of this chapter we discuss the results of measuring the superconducting gap by optical methods and on the basis of tunnelling spectroscopy.

5.1 Crystal Chemistry of Oxide Superconductors

To discuss the electronic structure of copper-oxide compounds, one can choose as a basis the ionic model, in which the state of each atom is described by a certain degree of oxidation, or formal valence z (see [5.1, 2]). In this case, the actual charge of ions such as Y^{2+}, La^{3+}, Ba^{2+}, Sr^{2+}, rare-earth ions RE=Nd^{3+}, Eu^{3+}, and others, really proves close to their formal valence in the $RBa_2Cu_3O_{7-y}$ (RBCO, R=Y, RE), $La_{2-x}M_xCuO_4$ (LMCO), and in the Bi- and Tl-based compounds. At the same time, due to strong hybridization of copper $3d$ states and oxygen $2p$ states, the charge of ions Cu^{2+} and O^{2-} may differ significantly from their formal valence, and depend on the degree of doping and the oxygen content. As a result,

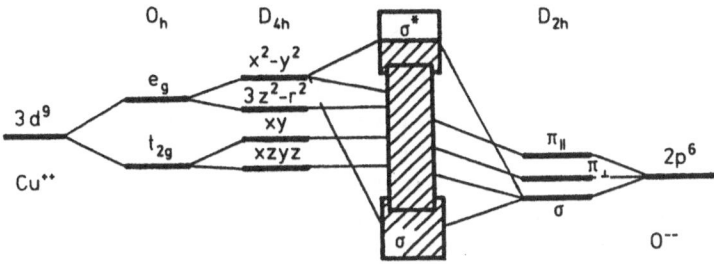

Fig. 5.1. Formation of the electronic structure of CuO_2 planes, accounting for the splitting of $3d$ and $2p$ levels in the crystal field and their covalent bonding [5.3]

there appear mixed $3d - 2p$ states in which copper ions with the formal valence $z = +2, +3$ are represented in the form

$$Cu^{2+} \rightarrow \alpha|3d^9 2p^6\rangle + \beta|3d^{10} 2p^5\rangle ,$$
$$Cu^{3+} \rightarrow \alpha_1|3d^8 2p^6\rangle + \beta_1|3d^9 2p^5\rangle .$$

(5.1)

The coefficients β and β_1 determine the degree of transfer of positive charge (hole) from the $3d$ shell of copper to the filled $2p$ shell of an oxygen ion O^{2-}. When $La_{2-x}M_xCuO_{4-y}$ is doped with divalent ions of $M=Ba^{2+}$, Sr^{2+}, Ca^{2+}, the formal valence of copper becomes $z = 2+x-2y$. In the $RBa_2Cu_3O_{7-y}$ compounds there are two nonequivalent positions of copper: in-plane Cu2 and in-chain Cu1 (see Fig. 2.12). In Sect. 2.4 we saw that under an increase of oxygen content, the formal valence of in-chain copper, Cu1, varies smoothly from Cu^{1+} ($y = 1$) to Cu^{3+} ($y = 0$), and for the in-plane copper, Cu2, the valence remains constant Cu^{2+} until a transition to the metallic phase at $y \leq 0.6$ when a transfer of charge from chains to planes with formation of the states Cu^{3+} (5.1) appears (see Fig. 2.16). An essential role in this phenomenon is played by the short-range order in the oxygen positions in the chains Cu1−O1.

In Fig. 5.1 [5.3], we outline schematically the formation of in-plane electronic structure, taking as an example La_2CuO_4. For a copper atom in a spherically symmetric field, the $3d$ levels are degenerate in energy. The value of this energy for copper is the minimum in the series of $3d$ elements, $\varepsilon^0(Cu\ 3d) = -20.14\,eV$, and it proves lower than the atomic level of oxygen $\varepsilon^0(O2p) = -14.1\,eV$ (see, e.g., [5.4]). In a crystal field of cubic symmetry O_h, for a proper octahedron CuO_6 the five $3d$ levels split into a doublet $e_g = \{d(x^2 - y^2), d(3z^2 - r^2)\}$ and a triplet $t_{2g} = \{d(xy), d(xz), d(yz)\}$. The value of this splitting is $1 - 2\,eV$ [5.2]. Upon the decrease of symmetry to tetragonal D_{4h}, a further splitting of the $3d$ levels into the states $b_{1g}\{d(x^2 - y^2)\}$, $a_{1g}\{(3z^2 - r^2)\}$, $b_{2g}\{d(xy)\}$ and $e_g = \{d(xz), d(yz)\}$ occurs. Degenerate atomic $2p$ oxygen levels $p(x)$, $p(y)$ and $p(z)$ split in the crystal field of the D_{2h} site symmetry into 3 levels: $(p\pi_\parallel)$, $(p\pi_\perp)$ and $(p\sigma)$. The π-type states correspond to the in-plane orbitals $p(x)$ or $p(y)$ ($\pi_\parallel$), or to the out-of-plane states $p(z)$ ($\pi_\perp$) which are directed perpendicular to the bond Cu−O. The σ-type states are formed by the in-plane orbitals $p(x)$ or $p(y)$, which are directed along the Cu–O bonds. These oxygen σ-type orbitals feel the strongest covalent bonding with copper orbitals $d(x^2 - y^2)$, which gives rise to broad bonding (σ) and antibonding (σ^*) bands of hybridized $pd\sigma$-states.

When discussing the structure of La$_2$CuO$_4$ in Sect. 2.2, we noted the essential deformation of the CuO$_6$ octahedron – it is stretched along the z-axis under an increase of the length Cu–O2 up to 2.4 Å as compared to the in-plane Cu–O1 distance 1.9 Å. An essential contribution to the tetragonal distortion of the octahedron for copper in the state $3d^9$ is due to the Jahn-Teller effect: a decrease of the energy of the electronic system under removal of the doublet degeneracy due to a reduction of the symmetry of the octahedron from the cubic O_h to the tetragonal D_{4h}. Indeed, in the case of singly filled levels $d(x^2 - y^2)$ and doubly filled levels $d(3z^2 - r^2)$, the electronic energy of the system decreases when they split. For Cu^{3+} or Ni^{2+}, which are described by the triplet ($S = 1$) state $3d^8$ with one electron on the level $d(x^2 - y^2)$, and one in the level $d(3z^2 - r^2)$, the Jahn-Teller effect is absent. This conclusion is supported by comparing the structurally equivalent lattices La$_2$NiO$_4$ and La$_2$CuO$_4$: stretching along the z-axis of the octahedron NiO$_6$ is twice as small as that of CuO$_6$. The typical energy of the Jahn-Teller distortion $E_{J-T} \simeq 0.5$ eV is much higher than that of the lattice vibrations, and thus one can view them as a static distortion of the structure.

It follows from the picture of the formation of the electronic structure of La$_2$CuO$_4$ sketched in Fig. 5.1 that this structure should be a metal with a half-filled antibonding $pd\sigma$ band. We arrive at the same conclusion when this scheme is applied to the electronic structure of YBa$_2$Cu$_3$O$_6$. However, these conclusions contradict experiments which demonstrate that the stoichiometric compounds LCO and YBCO are insulators with a moderately wide gap of $1-2$ eV. This discrepancy is related to the fact that, in the scheme described, one neglects the Coulomb single-site repulsion of $3d$ electrons. The typical value of the Coulomb correlation energy is $U_d \simeq 8 - 10$ eV, which is much more than the typical width of the $pd\sigma$-band $W \simeq 2$ eV [5.2] leading to a splitting of this band into one- and two-hole subbands (Cu^{2+} and Cu^{3+} in the notation of (5.1)). The possibility of the correlation splitting was first noted by *Mott* (see [5.5]) and investigated in the frame work of a simple model by *Hubbard* [5.6] and *Anderson* [5.7]. In these models it has been assumed that $U_d < \Delta$, where Δ is the energy of anion-cation charge transfer (in our case, $\Delta_{pd} = E_p - E_d$). In this Mott-Hubbard picture and in the Anderson theory of superexchange, anion states can be neglected if one considers only the narrow d-band due to the direct d-d exchange. However, the opposite situation when $U > \Delta$ is also possible, as it was first examined in [5.8, 9]. Here the insulator correlation gap is defined not by U_d but rather by the energy of charge transfer Δ. It is this case which is realized in cuprates where $\Delta_{pd} \simeq 3 - 4$ eV$ < U_d \simeq 8 - 10$ eV, in view of which one can call them insulators possessing a charge transfer gap [5.2].

After taking into account the Coulomb correlation energy, the picture of the electronic structure of copper-oxide compounds becomes similar to that shown in Fig. 5.2 [5.3]. In the insulating phase there is a filled (O2p, Cu3d) band, primarily of the O2p-type, which is the valence band, and an empty (Cu 3d, O2p) conduction band, primarily of the 3d type; these are separated by a charge transfer gap $E_g \leq$

Δ_{pd}. The conduction band, which is frequently said to be the upper Hubbard band, is separated by a large Coulomb correlation energy U_d from the two-hole (Cu^{3+}) lower Hubbard band (which is not shown in Fig. 5.2). There are several models describing the change in electronic structure of this insulating phase under doping by electrons (n) or holes (p). They are shown in Fig. 5.2. In the rigid band approximation, O2p type holes appear in the valence band, and 3d-type electrons in the conduction band. In other models it is assumed that under doping an impurity band of p- or n-type appears, or else the doping band gradually fills the gap from the Fermi level, lying close to the top of the valence band or to the bottom of the conduction band under p- or n-doping respectively. For high carrier concentrations, a wide metallic band arises, which is also predicted by the band-structure calculations (see Sect. 5.3). Experimental studies of the band-structure are discussed in Sect. 5.4.1.

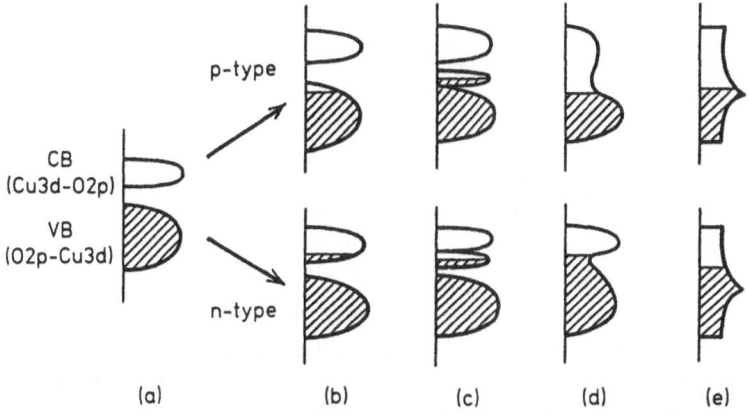

Fig. 5.2. Illustration of the electronic structure of copper-oxide superconductors accounting for the correlation splitting of the $pd\sigma$-band (**a**), and some models for its change upon p- or n-doping: rigid band (**b**), impurity band (**c**), filled gap (**d**), local density approximation (**e**) [5.3]

5.2 The Effect of Impurity Substitution

The ionic nature of bonding for Y^{3+} ions, lanthanides Ln^{3+}, alkaline ions in compounds like LMCO, RBCO, and metallic (covalent) bonding for ions of copper and oxygen, have been confirmed in numerous experiments by substituting both the ions of the first type with fixed valence, and copper ions having "mixed" valence Cu^{2+}, Cu^{3+} (see (5.1)). Results of these experiments are disscused in [5.10–13].

A rather simple situation is found in the compounds $La_{2-x}M_xCuO_{4-y}$. Substituting La^{3+} by divalent ions $M=Ba^{2+}$, Sr^{2+}, Ca^{2+} leads to the appearance of metallic conductivity, and for $0.05 < x < 0.3$ superconductivity appears. The $T_c(x)$ is plotted in Fig. 5.3a, where the black dots correspond to the appearance of the Meissner signal, and the white dots to half of its saturated value [5.13]. For

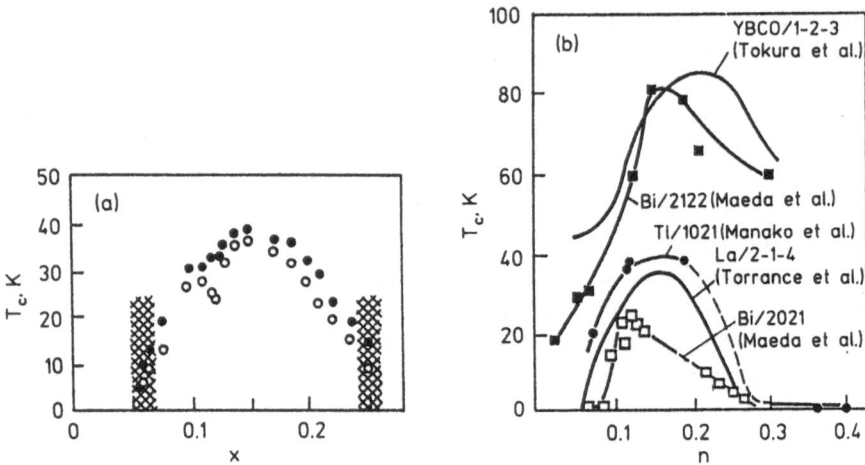

Fig. 5.3. T_c versus the concentration of Sr in $(La_{1-x}Sr_x)_2CuO_4$ (**a**) [5.13], and the general dependence $T_c(n)$ on the concentration of holes n in the CuO_2 planes for some copper-oxide superconductors (**b**) [5.15]

larger concentrations of Sr, $x > 0.15$, the oxygen vacancies can appear, $y > 0$; their formation, however, can be suppressed by performing the annealing under high oxygen pressure. In [5.14] it has been also discussed superconductivity under the transition of $La_{2-x}Sr_xCuO_4$ to the tetragonal phase for $x \geq 0.2$.

Metallic conductivity and superconductivity may also be obtained by increasing the content of oxygen above the stoichiometric value $La_2CuO_{4+\delta}$, $\delta > 0$. In this case, oxygen occupies intersital positions in layers of La – O. In general, for $La_{2-x}M_xCuO_{4-y}$ one observes the typical dependence $T_c(n)$ which is shown in Fig. 5.3b, with x substituted by $n = x - 2y$, the number of free charge carriers (holes) per cell in the CuO_2 planes.

Under the substitution of La by the trivalent rare-earth ions RE=Nd, Sm or Gd, a smooth decrease of T_c with decreasing ionic radius in the series of these ions [5.11] occurs. At the same time, the value of the magnetic moment of the RE ion has no effect on T_c, which indicates that the Cooper pairs in the CuO_2 plane are weakly coupled to the magnetic moments of ions in the La – O layers. An analogous situation is observed for electronic superconductors $Nd_{2-x}Ce_xCuO_4$ under the substitution of Nd by the rare-earth ions Pr, Sm, Eu. The decrease of T_c with decreasing radii of RE ions and primitive cell volume may be related to the decrease of the Cu–O bond length.

However, this simple picture contradicts the dependence of T_c on the external pressure. In the La – Sr compounds, T_c increases with an increase of pressure attaining a certain maximum, and even decreases at high pressures. Meanwhile, in the Nd – Ce compounds, T_c is pressure independent. This difference in the $T_c(p)$ dependence can be related to a special role of apex oxygen: in the T-phase of La – Sr compounds there are two apex oxygens in the complete CuO_6 octahedron, and

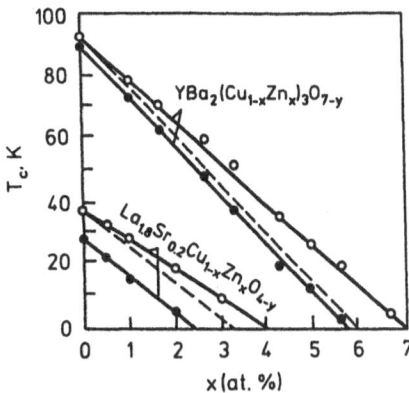

Fig. 5.4. The dependence of T_c on the concentration of Zn in $La_{1.8}Sr_{0.2}Cu_{1-x}Zn_xO_4$ and $YBa_2(Cu_{1-x}Zn_x)_3O_{7-y}$. The T_c measured with respect to the change of the $R(T)$ slope is shown by circles, and $T_c(R = 0)$ by dots

in the T'-phase of the Nd – Ce compounds apex oxygen is absent (see Fig. 2.10). This conclusion is confirmed by a rather strong dependence $T_c(p)$ [5.16] in the T^*-phase of (Nd – Sr – Ce)CuO_4, in which there is only one apex oxygen in the CuO_5 pyramid. Even a more complicated $T_c(p)$ dependence was observed in an untwinned single crystal of $YBa_2Cu_3O_7$ [5.17]. Applying uniaxial stresses along the a, b and c axes they found large and opposite in sign the pressure derivatives in the (a,b) plane and a small one for c-axis: dT_c/dp_i (K/GPa)= $-2.0(i = a)$, $+1.9(i = b)$, $-0.3(i = c)$. These results show that the small hydrostatic pressure dependence of T_c in fully oxygenated samples of YBCO is due to cancellation of a large and opposite effects in the (a, b) plane.

An isovalence substitution of copper by the ions of $3d$ metals has a much stronger effect on T_c. At a concentration of Ni, Fe $x = 5$–7 %, and at $x = 2$–3 % for Zn ions, the superconductivity in LMCO disappears (see [5.10–12]). While the destruction of superconductivity due to scattering of the Cooper pairs on magnetic impurities is well known, the suppression of T_c due to Zn impurities in CuO planes is characteristic of copper-oxide superconductors. Figure 5.4 shows the $T_c(x)$ dependence for the Zn impurity in LSCO and YBCO [5.12]. It is possible that Zn ions, which have the filled $3d$ shell ($3d^{10}$), destroy a rather complicated correlation conduction band in CuO_2 planes (which is related to the Cu−O charge transfer, see Fig. 5.2), which in turn leads to supression of superconductivity. In this context, we can consider Zn^{2+} ions (as well as Ga^{3+} ions, see below) as effective magnetic scatterers, because they substitute Cu^{2+} ions which have a local magnetic moment[1].

A much larger number of studies (see [5.10–12]) have been devoted to impurity substitution in YBCO systems, because these compounds allow a much larger variation of their composition. Immediately after the superconductivity with $T_c = 90$ K in YBCO was discovered, a large class of RBCO compounds with similar T_c was synthesized; they are obtained by substituing Y by lanthanides

---

[1] The formation of magnetic moments due to a Cu-site doping with Fe, Co, Ni, Zn, Ga, Al, in the LSCO system was observed in [5.116]

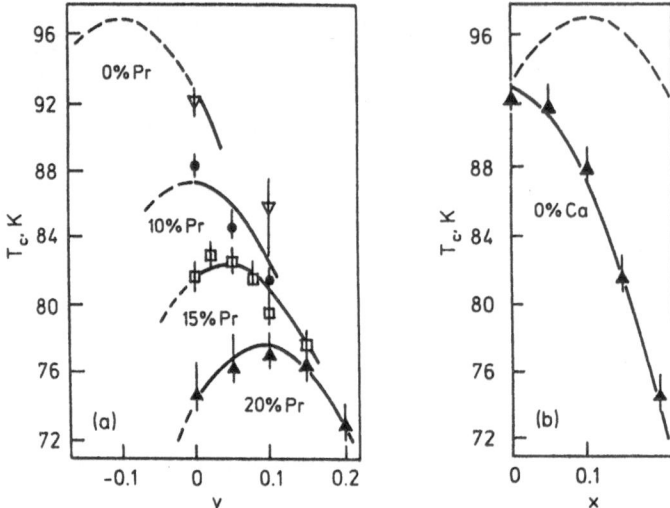

Fig. 5.5. The dependence of T_c in $Y_{1-x-y}Ca_yPr_xBa_2Cu_3O_{6.95}$ on the concentration of Ca at a fixed content of Pr (**a**), and on the concentration of Pr for $y = 0$ (**b**). Hatched is the T_c dependence on the concentration of holes without magnetic scattering [5.18]

Ln=La, Nd, ... (see Table 1.1) [5.12]. For them, no suppression of T_c in the compounds with rare-earth ions having a large magnetic moment has been observed, which indicates their weak coupling with the in-plane holes. The exceptions are Ce and Pr ions whose formal valence takes the value $z = +4$, which violates the isovalence of the substitution of Y ions and can change the concentration of carriers in CuO_2 planes. To test this hypothesis, the system $(Y_{1-x-y}Ca_y)Pr_xBa_2Cu_3O_{7-\delta}$ has been examined [5.18] and the $T_c(x, y)$ dependence investigated. Figure 5.5 demonstrates the $T_c(y)$ dependence for different x (a) and $T_c(x)$ at $y = 0$ (b). The authors suppose that substitution of Pr^{4+} for Y decreases the number of in-plane holes, and substitution of Ca^{2+} has the opposite effect, which allows one to investigate separately the dependence of T_c on the concentration of holes $n \propto (y - x)$ and on the magnetic scattering on the localized moments of Pr ($\mu \simeq 2.67\mu_B$). Indeed, it follows from Fig. 5.5a that $T_c(y)$ has a typical dependence $T_c(n)$ with a maximum at an optimal concentration of holes n for various concentrations x of Pr ions. The location of these maxima shifts to larger values of y under an increase of Pr content, which proves directly that the concentration of carriers $n \propto (y - x)$ decreases under an increase of x. At the same time, a decrease of the maximum value of $T_c(x, y)$ for the optimal n is observed, which indicates a decrease of T_c due to magnetic scattering leading to the destruction of Cooper pairs. Figure 5.5b shows the effect on $T_c(x, y = 0)$ of a change of the number of carriers n together with the magnetic scattering (solid curve), and the effect on T_c of changing $n \propto (-x)$ alone (hatched line). One can see that the maximum value of T_c in $YBa_2Cu_3O_{7-\delta}$ at $7 - \delta = 6.95$ is attained under a small decrease of the number of in-plane holes due to the tetravalent impurities (Pr^{4+} or, formally, a

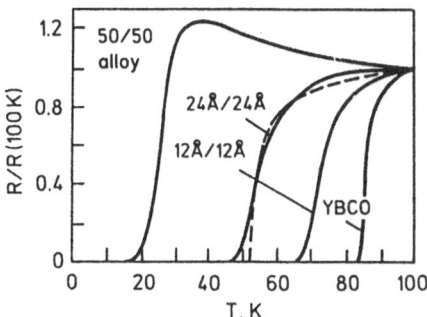

Fig. 5.6. Dependence of the reduced conductivity $R(T)/R(T = 100\,K)$ for superlattices of YBCO/PrBCO of thickness 12/12 (Å) and 24/24(Å), respectively, as compared to the pure YBCO film and the 50/50 (%) alloy [5.19]

negative concentration of the impurity Ca^{2+}, $y < 0$ in Fig. 5.5a). In the course of investigations of the effect of the impurity substitution on T_c, one must, obviously, strictly control the content of oxygen. As we have already discussed in Sect. 2.4, a change of oxygen content in YBCO has an essential effect on the number of carriers in CuO_2 planes and determines the value of T_c (Fig. 2.13).

Of a great interest for clarifying the role of Pr ions in the suppression of superconductivity in YBCO/PrBCO compounds are investigations of artificial lattices [5.19]. With the aid of the technique of sequential synthesis of YBCO and PrBCO layers, it proves possible to obtain films made up by alternating layers of $(YBCO)_n$ and $(PrBCO)_m$. In this case, it turns out that the transition temperature depends both on the thickness of the layer $(YBCO)_n$, and on the distance between the layers which is defined by the layer $(PrBCO)_m$. The latter plays the role of an insulating layer separating the superconducting layers $(YBCO)_n$. If the thickness of the Pr layer exceeds 96 Å (m = 8), one can neglect the connection between the superconducting Y layers. Figure 5.6 shows the dependence of the reduced conductivity of the film composed of layers of YBCO/PrBCO of the thickness 12/12 (Å) and 24/24 (Å), as compared to a pure YBCO film and the 50/50 (%) alloy. It is characteristic that even a YBCO layer consisting of two unit cells (24 Å) has $T_c > 50\,K$, while for a one-cell layer (12 Å) $T_c \simeq 10\,K$. The general dependence of T_c on the thickness d of a layer of YBCO (the number of unit cells m) is shown in Fig. 5.7, where the solid curve corresponds to the theoretical dependence as obtained in [5.20]. The gradual supression of T_c as the number of unit cells decreases is explained by the growth of fluctuations of the phase of the superconducting order parameter on crossover to a quasi-two-dimensional system, where a phase transition of the Berezinsky–Kosterlitz–Thouless type should be observed. At the same time, it should be noted that T_c for a superlattice with an equal number of Pr/Y layers proves much higher than for the solid solution $Pr_{0.5}Y_{0.5}BCO$ (e.g., according to Fig. 5.6, T_c is 50 and 70 K for the superlattices with n=m= 1, 2 respectively, as compared to $T_c = 20\,K$ for the 0.5/0.5 solid solution). These sublattices, when subject to external magnetic fields, demonstrate rather interesting, strongly anisotropic properties (see [5.19]). Their investigation again confirms the local nature of superconductivity in copper-oxide compounds which have a small correlation length of the order parameter (Sect. 4.3).

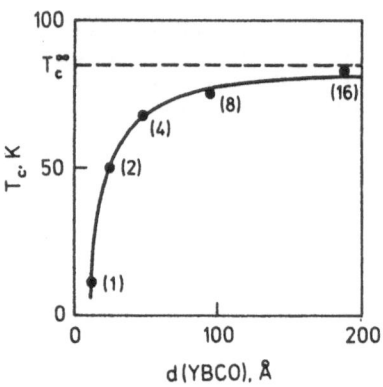

Fig. 5.7. The dependence of T_c on the thickness of the YBCO layer in the superlattice YBCO/PrBCO, T_c^∞ is the bulk transition temperature (after [5.20])

An investigation of the effect of substituting the Ba in YBCO by the rare-earth ions has been carried out for the $Ln(Ba_{2-x}Ln_x)Cu_3O_{7-y}$ compounds, where Ln=La, Nd, Sm, Eu, Gd. All the lanthanides produce an equal decrease in T_c with respect to the concentration of impurities x. This indicates a weak sensitivity of the superconducting transition to the appearance of magnetic moments on the Ba sites. In these experiments, the authors also observed a phase transition from the orthorhombic to the tetragonal phase (at $x \simeq 0.2$–0.3), which, however, did not have a significant effect on the superconducting properties (see [5.10]). In view of an often uncontrolled increase of oxygen content $7 - y > 7$ under the substitution of Ln for Ba and a complicated rearrangement of charge in the layers Ba – O4, Cu1 – O1, Cu2 – O2, 3, in these experiments, it does not prove possible to reach an unambiguous conclusion regarding the dependence of T_c on the concentration of holes in CuO_2 planes (see Fig. 2.12).

In the YBCO compounds, one observes a more complicated effect of the impurities which substitute for copper ions, than in LMCO compounds. First, in the YBCO compounds there are two nonequivalent copper positions: Cu1 and Cu2 (Fig. 2.12), whose substitution by impurities has a different effect on their electronic structure and superconductivity. Second, some impurities, e.g, Fe, Co effect the oxygen content and the short-range order in the Cu1 – O1 layer, which may change the number of carriers in the Cu2 – O2, 3 planes.

In one of the earlier experiments [5.21], the effect of substituting Cu by ions of 3d elements M^{2+} =Ti($3d^2$), Cr($3d^4$), Mn($3d^5$), Fe($3d^6$), Co($3d^7$), Ni($3d^8$) and Zn($3d^{10}$) was investigated. The authors observed that for $x = 0.1$ in $YBa_2(Cu_{1-x}M_x)_3O_{7-y}$ the strongest suppression of T_c is observed for the ions of Fe and Co, having the maximum magnetic moment, and also for the Zn ions. However, further investigations demonstrated that in the case of Fe and Co an extra annealing in an oxygen atmosphere restores the value of T_c for $x \leq 0.35$ almost completely, but does not relax the decrease of T_c by the impurities of Zn and Ni [5.22]. Such different effects of annealing on T_c in impurity compounds of YBCO is explained by the fact that at low concentrations the ions of Co and Fe preferentially occupy the positions of Cu1, and the ions of Zn and Ni the positions of Cu2. Therefore, annealing in oxygen restores the oxygen content and its

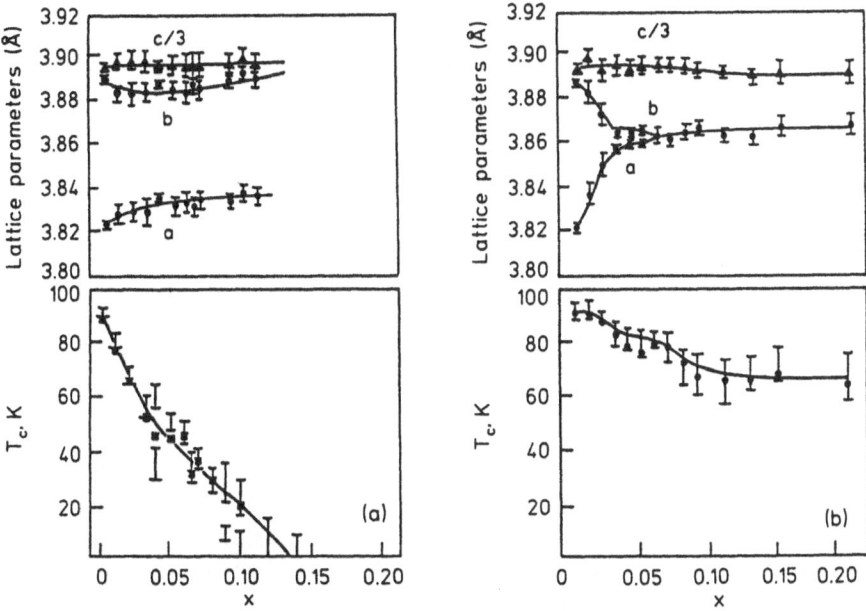

Fig. 5.8. The dependence of lattice parameters and T_c on the concentration of impurities in $YBa_2(Cu_{1-x}M_x)_3O_{7-y}$ (in at. %) for Zn (**a**), and Ga (**b**) [5.23]

coordination in the layer Cu1 – O1 for the Co and Fe impurities, but has no effect on the plane Cu2 – O2, 3 for the Zn and Ni impurities.

A more detailed analysis of the effect of diamagnetic impurities under a substitution for copper has been reported in [5.23], where the $Zn^{2+}(3d^{10})$ and $Ga^{3+}(3d^{10})$ impurities were used. Their ionic radii for a filled $3d$ shell, $R(Zn^{2+})= 0.75\,\text{Å}$, $R(Ga^{3+})= 0.62\,\text{Å}$ are close to $R(Cu^{2+})= 0.73\,\text{Å}$, which allows one to obtain single-phase samples in a wide range of impurity concentrations. Applying neutron diffraction demonstrated that Zn primarily occupies the positions of Cu2, and Ga the in-chain positions of Cu1. The dependencies of T_c and lattice constants are shown in Fig. 5.8 for Zn (a) and Ga (b) [5.23]. Zn impurities in the layers Cu2–O2, 3 have only a small effect on the parameters of the lattice, preserving the orthorhombic phase, but lead to a rapid suppression of T_c, as has already been noted for the LSCO compounds; see Fig. 5.4. Ga impurities, substituting in-chain copper, have a rather small effect on T_c, but already at a small concentration (6 %) lead to a transition to the tetragonal phase. The oxygen content remains close to optimal (6.8 and 7.0 for samples with Zn and Ga respectively), and therefore the transition to the tetragonal phase in the case of Ga is not related to the deficiency of oxygen. The resistance $\varrho(x)$ in the normal phase increases much faster for Ga impurities than for Zn, while they have the converse $T_c(x)$ dependence. These experiments clearly demonstrate that the main role in the appearance of superconductivity in copper-oxide compounds is played by the CuO_2 planes, where the specific properties of copper in $3d^9$ states are necessary to attain high-T_c.

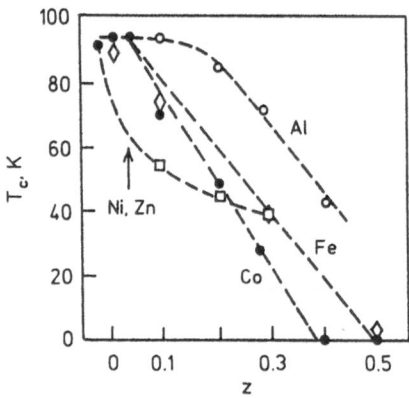

Fig. 5.9. T_c versus the concentration of impurities M = Ni^{2+}, Zn^{2+} and M = Al^{3+}, Fe^{3+}, Co^{3+} in $YBa_2Cu_{3-z}M_zO_{7-y}$ [5.24]

In Ref. [5.24], detailed studies of structural, magnetic and superconducting properties of the $YBa_2Cu_{3-z}M_zO_{7-y}$ compounds have been reported, both with paramagnetic, M=Ni, Fe, Co, and diamagnetic, M=Zn, Al impurities. It has been shown that the divalent impurities Zn^{2+}, Ni^{2+} do not change the content of oxygen in the sample, and, when occupying positions in CuO_2 planes, preserve the orthorhombic phase in the domain of single-phase states (for $z \leq 0.3$ or the concentration $x = z/3 < 0.1$). Al^{3+} impurities, as well as those of Co and Fe, being in the trivalent state, occupy in-chain Cu1 positions. They lead to an increase of oxygen content, filling the vacancies near the impurity site and leading to a transition to the tetragonal phase for $z < 0.1$. In Fig. 5.9, the dependence $T_c(z)$ for these impurity substitutions is shown. As has already been noted, substituting in-plane Cu2 by Zn impurities leads to a rapid decrease of T_c (Fig. 5.4). Substituting in-chain Cu1 by the trivalent impurities Al, Fe, Co at low concentrations ($x < 0.1$) has a much weaker effect on T_c. The same is true for Co and Fe impurities which have a large magnetic moment of $\mu \simeq 3.5\mu_B$. In this case, after the suppression of superconductivity the insulating phase immediately appears, escaping the normal, non-superconducting, phase. This is clearly seen in the temperature dependence of the resistance, $\varrho(T)$, which is shown in Fig. 5.10 for different concentrations of Fe and Co impurities. Apparently, both substitution of impurities for in-chain Cu1 and changing the oxygen content have the same effect on the charge transfer from the chains to the Cu2 plane, causing the transition from insulating to metallic phase which becomes superconducting. This is also confirmed by the universal dependence of T_c on the local structural parameters, such as the distance Cu1 – O4 (see Fig. 2.12), which, as was discussed in Sect. 2.4, defines the in-plane effective valence of copper (see Figs. 2.15, 2.16). The transition temperature T_c becomes zero for the distance Cu1 – O4 less than 1.82 Å, and for larger values they correlate linearly, no matter what the reason is – the Al impurities or oxygen content [5.11].

At the same time, it should be noted that in a series of experiments a more complicated T_c dependence on the concentration of impurities which substitute for copper in YBCO compounds has been observed. For example, in [5.25], with the aid

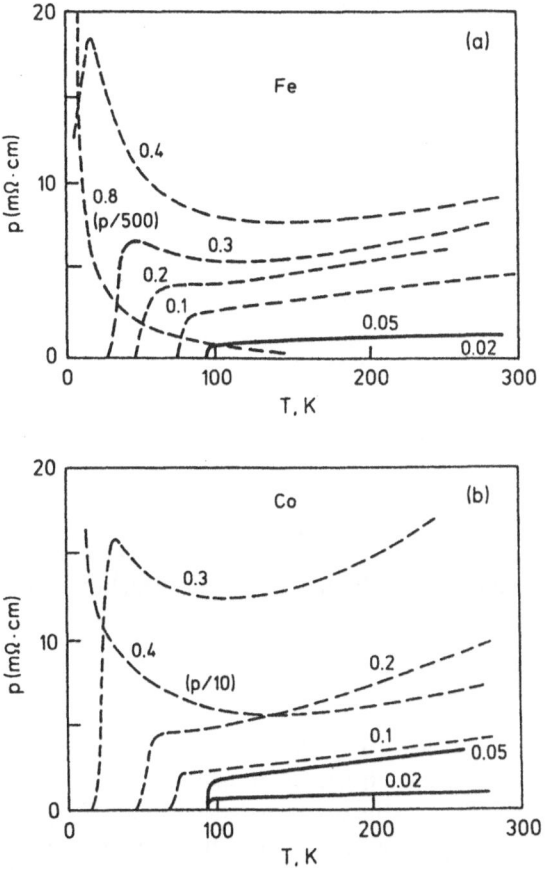

Fig. 5.10. Temperature dependence of the resistivity for different concentrations of Fe (a) and Co (b) impurities in $YBa_2Cu_{3-x}M_xO_{7-y}$ [5.24]

of neutron diffraction using the Ni isotopes in YBCO, it has been demonstrated that Ni impurities occupy the positions of Cu1 and Cu2 with equal probability, and a single-phase solution exists only in the range of small concentrations, $x < 0.07$. At the same time, the dependence $T_c(x)$ appears much weaker ($dT_c/dx \simeq 2(\text{K/at.}\%)$) than that reported in the other studies (see Fig. 5.9). Probably the discrepancy of the data for $T_c(x)$ is related to an insufficient control of the single-phase nature of the solution, so that the actual concentration of impurity substitution is smaller than the nominal one introduced in the course of the synthesis. In another study [5.26], with the aid of a special thermoprocessing, it also proved possible to obtain a very weak dependence $T_c(x)$ in the YBCO samples with Fe impurities: at a concentration $x = 0.12$, $T_c \simeq 90\,\text{K}$ (compare with Fig. 5.9). The authors observed that the orthorhombic phase is preserved up to $x = 0.15$, and Fe is distributed with equal probability over the Cu1 and Cu2 positions. These experiments, together with a series of others, demonstrate that in the process of substitution of Cu ions in YBCO compounds one should account for several factors affecting $T_c(x)$: distribution of impurities over the Cu1 or Cu2 sites, local changes of the structure, a possible charge transfer and a change of oxygen content.

The effect of impurity substitution on the superconducting properties of Bi-based compounds is similar to the aforementioned properties of LSCO and YBCO-compounds. For example, the substitution of Ca^{2+} by trivalent ions of Tm in $Bi_2Sr_2Ca_{1-x}Tm_xCu_2O_{8+y}$ leads to a gradual decrease of the formal valence of Cu (with the increase of oxygen content, $y > 0$), and to suppression of T_c [5.11]. The optimal number of the free carriers in the plane CuO_2 to attain high-T_c is achieved by instilling extra oxygen in the Bi-O planes, which causes a lattice modulation for these compounds. The studies of a substitution of $3d$-elements for copper in Bi-based compounds encounters experimental difficulties, and therefore at present these phenomena lack an adequate understanding.

Summarizing the studies of the effect of impurity substitution in copper-oxide compounds, we note the following results:

1. Isovalent out-of-plane substitutions ($Y^{3+} \rightarrow RE^{3+}$, $Sr^{2+} \rightarrow Ba^{2+}$, Ca^{2+} etc.) have a small effect on superconductivity, also in the case of paramagnetic impurities with a large magnetic moment. This indicates their weak coupling to the charge carriers in the CuO_2 planes (electronic isolation). The exception is Pr ions, which, besides a decrease of the hole density (or their localization) in YBCO-compounds, demonstrate a notable coupling of $4f$-electrons to the charge carriers in the CuO_2 planes.

2. Nonisovalent substitutions out of the CuO_2 planes, as well as changes of oxygen content or changes of local structure, change the formal valence of copper in the CuO_2 planes and have a significant effect on superconductivity. T_c has a typical dependence with a maximum (see Fig. 5.3) at the optimal number of the in-plane charge carriers.

3. The strongest suppression of superconductivity occurs upon the substitution of Zn^{2+} ions with the filled $3d$ shell for copper in CuO_2 planes. Very peculiar is a small suppression of T_c under a substitution of paramagnetic impurities of the Ni^{2+} type for copper[2].

5.3 Theoretical Electronic Band-Structure Studies

A very efficient method of investigating of the electronic structure of solids is the density functional method, which has been used by many groups to obtain a theoretical description of the electronic properties of copper-oxide compounds (see the review [5.27]). Underlying the method are self-consistent calculations of the charge distribution and the form of the potential inside the primitive cell on the basis of one-electron functions for the Bloch states in crystals. This method allows

---

[2] Investigations of the effects of Ni and Zn substitution for Cu in RBCO (R = Y, Nd, Gd, Dy and Er) confirmed this result [5.117].

"first principle" – without any fitting parameters – calculations of the ground state energy of the crystal, as well as an investigation of its dependence on deformations and displacements of ions. In particular, on the basis of the recent "frozen phonon" method one can determine several force constants in the dynamical matrix of the crystal. At the same time, in the framework of the density functional method one has to use the one-particle approximation to compute the correlation energy of electrons. The latter approximation may be justified only for $s - p$ metals with broad bands and strong screening. In transition metals with narrow d-bands an important role is played by Coulomb on-site correlations, and therefore the one-particle wavefunctions and the spectrum of one-electron excitations as computed in the frame of the density functional method are not directly related to the excitation spectrum of the many-particle electron system. We have already discussed this discrepancy between the theoretical band-structure calculations and the experimental spectrum of electronic excitations when comparing Figs. 5.1 and 5.2. Nevertheless, on the basis of the density functional method one can get a general picture of the several parameters of the electronic structure (ground state charge density distribution, hybridization parameters, and others) to an accuracy of several tenths of an eV.

Supplementing the results of band-structure calculations is the cluster method, which has been used successfully to analyze the electronic structure of copper-oxide compounds (see, e.g., [5.2, 28–30]). Investigating the electronic spectrum for a system of several atoms by the cluster method, one can consecutively account for the many-particle correlations. A certain problem for this method is accounting for surface effects and the effects of long-range Coulomb forces. In copper-oxide compounds of an ionic nature the Madelung potential plays an important role in determining the crystal field effects. Combining the results of both approaches – band theories and cluster calculations – it proves possible to find the basic interaction parameters in the electronic system, and to construct an effective many-particle Hamiltonian (see, e.g., [5.31–33]). The latter is fundamental to investigate the electronic properties (magnetic, transport etc.) in the low energy excitations scale. As usual, the effective Hamiltonian is used to construct different models of the superconducting transition. Let us consider the main results of these researches.

5.3.1 Electronic Band-Structure Calculations

In the very first band-structure calculations in $La_{2-x}M_xCuO_4$ compounds (see, e.g., [5.34]), it has been found that the most important contribution to the electronic density of states near the Fermi surface arises from the $pd\sigma$-band, constructed from $d(x^2 - y^2)$ states of Cu^{2+} and $p_\sigma(x, y)$ states of O^{2-} (see Fig. 5.1). The energy spectrum in La_2CuO_4 and the electronic density of states, both the total one and its partial components on separate ions, are shown in Fig. 5.11. Of the total of 17 bands formed by $Cu(3d)$–$O(2p)$ states (five $3d$ states on copper ions, and three $2p$-states on each of the four oxygen ions in the primitive cell of La_2CuO_4), only two have a large dispersion: bonding (B) and antibonding (A) $pd\sigma$ type bands, constructed from the orbitals $d(x^2 - y^2)$ and $p_\sigma(x, y)$. The Fermi surface crosses

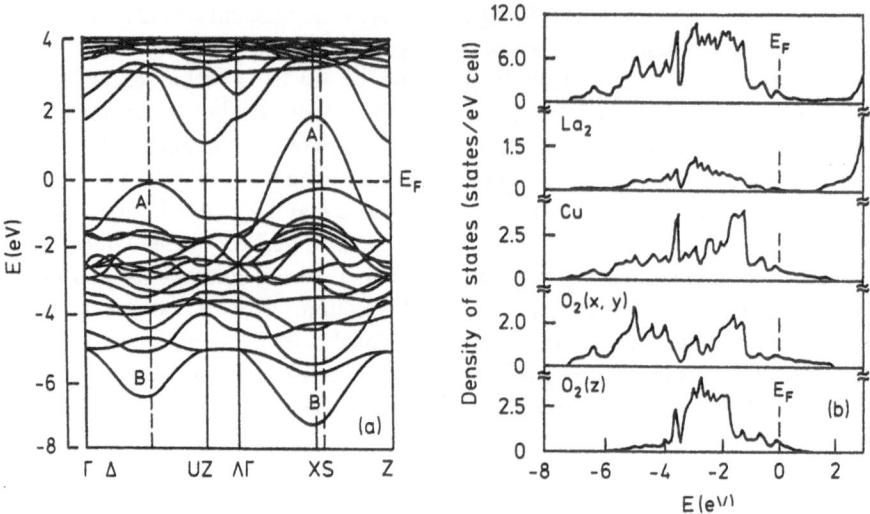

Fig. 5.11. Energy bands (**a**) and electronic density of states (**b**) in the tetr)nal phase of La$_2$CuO$_4$ [5.34]

the A-type band, and the other bands lie by far below the Fermi energy. Since $pd\sigma$-band electrons are localized in CuO$_2$ planes which are located far from each other, this band is an essentially two-dimensional one – the dispersion along the z-axis is small (see the line Λ from the center of the band Γ to the point $Z(0,0,1/2)$). In the tight-binding approximation, the two-dimensional A,B bands are described by the following formula:

$$E_{A,B}(\boldsymbol{k}) = \frac{1}{2}(E_p + E_d)$$
$$\pm \left[\left(E_p - E_d\right)^2 + (4t)^2(\sin^2 \frac{k_x a}{2} + \sin^2 \frac{k_y a}{2}) \right]^{1/2}, \qquad (5.2)$$

where, according to [5.34], $E_p \simeq E_d = -3.2\,\mathrm{eV}$ and $t =(\sqrt{3}/2)V_{pd\sigma}$, $V_{pd\sigma} = -1.8\,\mathrm{eV}$. The total width of the A–B band $W = 4t\sqrt{2} \simeq 9\,\mathrm{eV}$ proves large.

The La levels are weakly coupled to the states in Cu–O bands: the 5d level of La lies 1 eV above, and the 5p-level 15 eV below the Fermi level. Therefore, La may be viewed as an isolated ion with the charge +3, which, when replaced by a rare-earth ion with the same charge, has a minor effect on the electronic properties. In particular, the magnetic scattering of $pd\sigma$-band electrons on the magnetic moment of rare-earth ions is small and does not lead to a suppression of superconductivity.

A strong anisotropy of the $pd\sigma$-band leads to a quasi-two-dimensional nature of the Fermi surface. In the tight-binding approximation (5.2) for the half-filled A-band, the Fermi surface is defined by the equation $E_A(k_F)= 0$, or

$$\sin^2 \frac{k_x a}{2} + \sin^2 \frac{k_y a}{2} = 1, \qquad (5.3)$$

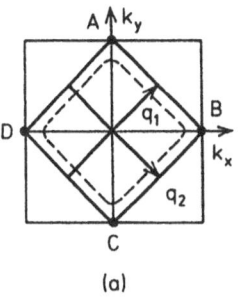

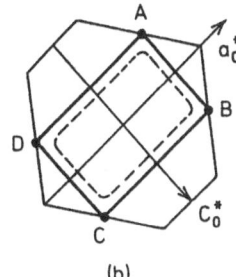

Fig. 5.12. Two-dimensional Fermi surface in the tetragonal (**a**) and orthorhombic (**b**) phases of La_2CuO_4

(a)   (b)

with the solution

$$|k_x| + |k_y| = \frac{\pi}{a}. \tag{5.4}$$

In Fig. 5.12a, the Fermi surface in the Brillouin zone ($\pm\pi/a, \pm\pi/a$) is shown by the lines AB, BC, CD, DA. At the points A, B,C, D the Fermi surface touches the boundary of the Brillouin zone, which leads to singular points in the electronic spectrum (saddle point type), and the van Hove singularities in the electronic density of states. A more detailed calculation for a real body-centered tetragonal lattice (see Fig. 2.4) demonstrates that touching the Fermi surface (Lifshitz transition) occurs at a finite concentration $x \simeq 0.15$ of divalent ions M=Sr, Ba in $La_{2-x}M_xCuO_4$ [5.35]. Hence, at the same concentration the Fermi surface passes through a van Hove singularity, and the density of states $N(0)$ has a maximum. They often relate the maximal value of T_c to this maximum too (see Sect. 7.4). However the total density of states on the Fermi surface proves small: according to [5.34], $2N(0)$= 1.2 states/eV (for the both directions of spin) for $x = 0$, and $2N(0)$= 1.9 states/eV for $x = 0.16$. In other studies, similar values of the density of states have been obtained [5.27].

The existence of flat regions AB, BC, CD and DA on the Fermi surface provides "nesting" – a high degree of congruence of the Fermi surface under a translations by the vectors $| k_x(1,2) |= 2k_F$, where k_F is the Fermi momentum and $k_x(1,2) = (\pi/a)(\pm 1, 1, 0)$ (2.2). Such a strong singularity usually leads to the appearance of an instability of the lattice with respect to a formation of the charge density wave (CDW) due to electron–phonon interaction (in the present case with the breathing mode). In this mode, the oxygen ions O1 shift in the basal plane along the Cu−O1 bonds in opposite directions in neighboring cells. As a result, a redistribution of charge on neighboring copper ions appears, forming a CDW with the wave vector $k_x(1)$ (or $k_x(2)$), and a gap in the electron spectrum opens on the Fermi surface. The density functional band-structure calculations really demonstrate the formation of a gap in the electronic spectrum under a freezing of the breathing mode [5.34]. However, the experimentally observed structural phase transitions in $La_{2-x}M_xCuO_4$ from the tetragonal (HTT) to orthorhombic (LTO) or low temperature tetragonal (LTT) phase are related to the condensation of tilting-type modes with the wave vectors $k_x(1)$ or $k_x(2)$ (Sect. 2.2.2). At the same time in the LTO-phase, which is related to the displacements of all the four O1 ions in the

CuO_6 octahedron, the gap in the electronic spectrum does not appear [5.36], and in the LTT-phase, with only two O1 ion displacements (Fig. 2.8), there appears a gap at the M-points: $(0, 1, 0)$ or $(1, 0, 0)$ [5.37]. In the opinion of the authors of [5.37], the formation of this gap and the almost twofold decrease of the density of states on the Fermi surface in the LTT-phase explains the anomalous behavior of several electronic properties (dielectric properties, suppression of superconductivity, an appearance of local magnetic moments).

At the transition from the HTT- to LTO-phase, the Fermi surface for the orthorhombic face-centered Brillouin zone (Fig. 2.4) becomes rather complicated [5.27]. For the two-dimensional model, the Fermi surface is shown in Fig. 5.12b. It is important to note that in the LTO-phase the distances from the points A, B, C and D to the center of the Brillouin zone remain equal, and therefore the van Hove singularities do not split.

Besides the band-structure calculations for the LTO- and LTT-phases, in [5.36, 37] the problem of the stability of the tetragonal phase and some phonons has been investigated on the basis of the total electronic energy calculation for the crystal and the frozen phonon method. It has been found that the breathing mode in LMCO is stable and has a high frequency, although, as we discussed above, according to band calculations, this mode may be unstable due to strong electron–phonon interaction. The dependence of the crystal energy on the displacements of ions under rotations of the CuO_6 octahedrons is described by a double-well potential (Fig. 2.9), which suggests the possibility of a phase transitions to the LTO- or LTT-phases.

Similar results have also been obtained for other copper-oxide compounds in the course of electronic band-structure calculations by the density functional method [5.27]. These calculations confirm a high degree of hybridization – strong covalent coupling of Cu $3d$ and O $2p$ states with a formation of a broad $pd\sigma$-band. The other electronic states have a weak hybridization, and the atoms out of the CuO_2 planes may be described in the ionic coupling approximation. In particular, in $YBa_2Cu_3O_{7-y}$ the Fermi surface crosses two antibonding bands (corresponding to two CuO_2 planes) and one quasi-one-dimensional band of the Cu1–O1 chains for $y = 0$ (see Fig. 2.12). The removal of oxygen O1 ions from the chains ($y = 1$) leads to a destruction of the latter band with an insignificant deformation of the $pd\sigma$-bands in the CuO_2 planes. As in the La_2CuO_4 case, according to these calculations $YBa_2Cu_3O_{7-y}$ is a metal (including $y = 1$) with sufficiently broad $pd\sigma$-bands and a rather small density of states: $2N(0) \simeq 5.5$(states/eV) per cell for the both spin directions. The density of states as related to one copper ion is the same as that obtained for La_2CuO_4.

The band-structure of $Bi_2Sr_2CaCu_2O_8$ looks very complicated ([5.27]). Besides the broad $pd\sigma$-band of Cu $3d$–O $2p$ states, which is typical of the CuO_2 planes, rather broad $pp\sigma$-band of Bi $6p(x, y)$ and O $2p(x, y)$ states contributes on the Fermi surface too. The total density of states per cell is about 3 states/eV, and, if related to one Cu atom, after subtracting the contributions of Bi – O states, it reduces to 1 states/eV. The band-structure of Tl-based compounds differs from that of Bi compounds only by the fact that on the Fermi surface, instead of the Bi – O band, a band constructed of Tl-6s-levels and O-2p-levels appears. The average density of states per copper atom is also not large, of order 1 state/eV, independent of the number of CuO_2 planes.

It is of some interest to compare the band-structure of cubic perovskites $Ba_{1-x}K_xBiO_3$, containing no Cu ions, with that of the aforementioned copper-oxide compounds. According to the calculations [5.38], the metallic properties of $BaBiO_3$ compounds are determined by a broad $sp\sigma$-band formed from the Bi(6s)-O(2p) orbitals. The maximum density of states, $2N(0) \simeq 0.8$ states/eV, is attained in the half-filled band in $BaBiO_3$. However, strong electron–phonon interaction with the "breathing" mode of the BiO_6 octahedron leads to a structural transition to the monoclinic phase with the formation of a CDW, and to the insulating state (Sect. 2.1). The minimal value $x = 0.25$ for which the cubic phase still exists defines the maximum $T_c = 30$ K. According to [5.38], $2N(0) = 0.6$ states/eV for $x \simeq 0.25$, and falls off rapidly with the growth of x. The role of Coulomb correlations in the broad $sp\sigma$-band should be not very important, and the appearance of superconductivity with $T_c \leq 30$ K in these compounds may be explained by accounting for the strong electron–phonon interaction of electrons in the $sp\sigma$–band with the "breathing" mode of the octahedrons BiO_6 (Sect. 7.4).

Thus, electronic band-structure calculations by the density functional method predict the existence of wide bands and rather small density of states on the Fermi surface. These calculations, where the Coulomb correlations are accounted for only in the one-particle mean-field approximation, fail to obtain an antiferromagnetic ground state with a sufficiently large insulator gap and significant local magnetic moments on copper sites in the CuO_2 planes [5.27]. According to [5.39], the main defect of the one-particle approximation in the density functional method is the neglect of a strong orbital polarization of local $3d$ states due to Coulomb correlations. To account for these effects, in [5.40] an additional local Coulomb interaction depending on the orbital state was introduced. On the basis of this method the authors of [5.40] succeeded in obtaining an antiferromagnetic ground state for La_2CuO_4 with the energy gap $E_G = 2.3$ eV and a magnetic moment on the copper sites of $\mu = 0.78\mu_B$. Although this method only approximately accounts for Coulomb correlations, the results obtained demonstrate that the density functional method may, with some modification, be successfully used to compute the ground state energy of electronic systems.

5.3.2 Effective Hamiltonians in Models of La_2CuO_4

The two most important features of the electronic structure of copper-oxide compounds – a strong hybridization of O $2p$ and Cu $3d$ states in the $pd\sigma$-band and strong Coulomb correlations of Cu $3d$ states – may be accounted for by constructing an effective Hamiltonian for the CuO_2 planes, which are the most important structural element in all the copper-oxide compounds. To define the parameters of this Hamiltonian, one can use the density functional method and some its modifications [5.31–33], as well as the method of cluster calculations [5.2, 28–30].

In the course of constructing an effective Hamiltonian, one has to restrict the number of electronic states considered, accounting for the rest of them by a corresponding renormalization of the model parameters. Therefore the parameters of the effective Hamiltonian depend on the number of the states selected, and they should be considered only in conjunction with the given set of electronic states. In particular, for the basis of one-electron states in the tight-binding approximation they use the atomic wave functions of the $3d$ and $2p$ states, which do not form a basis of orthogonal states in the crystal. Despite the several advantages of the tight-binding method with a non-orthogonal basis, it cannot be utilized to write down the effective Hamiltonian in the secondary quantization representation which requires an orthogonalized basis of one-electron states. The standard approach to constructing such functions in the form of Wannier functions (linear combinations of Bloch functions) is ineffective. Because of their essential delocalization, the Wannier functions inadequately describe the $3d-$states, which are strongly localized on the lattice sites.

In this context, more convenient seems to be the method of sequential orthogonalization of atomic wave functions, which has been developed by *Bogolubov* in the course of the construction of the polar theory of metals in 1946 [5.41]. In this method, the wave function $\Psi_{i\lambda}$ for a state λ on the $i-$th site of a lattice is written as a linear combination of the atomic function $\chi_{i\lambda}$ and the functions on the nearest sites j of the lattice:

$$\Psi_{i\lambda} = \chi_{i\lambda} + \sum_{j\beta} S_{ij}^{\lambda\beta} \chi_{j\beta} , \qquad (5.5)$$

where the coefficients $S_{ij}^{\lambda\beta}$ are chosen so as to fit the orthogonality condition $\langle \Psi_{i\lambda} \mid \Psi_{j\beta} \rangle = \delta_{ij}\delta_{\lambda\beta}$. This method proves effective for small overlaps of functions on neighboring sites, $\mid \langle \chi_{i\lambda} \mid \chi_{j\lambda} \rangle \mid \ll 1$. With the aid of this method, an effective Hamiltonian has been constructed [5.42] which accounts for $3d$ $(x^2 - y^2)$ states on Cu sites and $2p_\sigma(x, y)$ states on O sites in CuO$_2$ planes. In a more general approach, one can account for the other states: $3d$ $(3z^2 - r^2)$, $2p_\pi(x, y)$, $2p_\pi(z)$ as well. A "first principles" calculation of the parameters of the effective Hamiltonian is a hard problem, and therefore they are usually fixed with the aid of indirect calculations and a comparison with experimental data.

Theoretical models of copper-oxide compounds most frequently use the following three-band effective Hamiltonian, which was proposed by *Emery* and *Varma* et al. [5.43, 44]

$$H = \sum_{i\sigma} \varepsilon_i n_{i\sigma} + \sum_{ij\sigma} t_{ij} a_{i\sigma}^+ a_{j\sigma} + \frac{1}{2} \sum_{i\sigma} U_i n_{i\sigma} n_{i-\sigma}$$

$$+ \frac{1}{2} \sum_{i\neq j\sigma\sigma'} V_{ij} n_{i\sigma} n_{j\sigma'} . \qquad (5.6)$$

Here $n_{i\sigma} = a_{i\sigma}^+ a_{i\sigma}$ and $a_{i\sigma}^+(a_{i\sigma})$ are the creation (annihilation) operators for holes of spin σ on the sites i of the square lattice CuO$_2$, $\varepsilon_i = (\varepsilon_p, \varepsilon_d)$ are the energies of O $2p_\sigma(x, y)$-states and Cu $3d$ $(x^2 - y^2)$-states respectively, $t_{ij} = (t_{pd}, t_{pp})$ are the transfer integrals for $p - d$ and $p - p$ states on nearest Cu–O and O–O sites respectively; $U_i = (U_d, U_p)$ are the on-site Coulomb correlation energies for $3d$ and $2p$ states, and $V_{ij} = (V_{pd}, V_{pp})$ are the intersite Coulomb interactions. They choose the state $| 3d^{10}2p^6 >$ with filled $3d$ and $2p$ shells which contains no holes to be the vacuum state in (5.6). The Hamiltonian (5.6) can be rewritten in terms of electronic variables by introducing the operator of the number of electrons $N_{i\sigma} = (1 - n_{i\sigma})$. Theoretical studies of the model (5.6) will be discussed in Sect. 7.1.1.

To fix the parameters of the Hamiltonian (5.6), the results of calculations of the electronic spectrum in crystals of La_2CuO_4 under certain restrictions on the electronic density on some sites (constrained density-functional method) have been invoked [5.31, 32]. In view of a strong hybridization of O $2p$-states due to the direct coupling t_{pp}, in [5.31] the Anderson impurity model for $3d$ states was considered. In this model Cu $3d$ states, in view of their strong Coulomb correlation, are considered as impurity states in the broad O $2p$ band with a weak Coulomb interaction on the oxygen sites: $(U_p = V_{pd} = 0)$; the direct interaction between Cu $3d$ states is neglected. In the frame of the impurity model they have investigated the spectrum of one-particle excitations and found the ground state. It turned out to be magnetic with the symmetry $x^2 - y^2$ and an insulator gap $1 - 2\,eV$. Because the interaction of spins on copper sites is neglected in the impurity model, the ground state obtained does not have a long-range antiferromagnetic order. Studies of the spectrum of electron states in the system with an extra hole demonstrated that the two-hole states form a singlet of strongly correlated Cu $3d$ and O $2p$ holes in $(x^2 - y^2)$ states.

To determine the parameters of Coulomb correlation (U_i, V_{ij}) in (5.6), in [5.31] the dependence of the total electronic energy $E(n)$ versus the occupation number n for an isolated orbital not coupled to other orbitals has been considered. Computing this dependence on the basis of the density functional method and casting it to the form

$$E(n) = E_0 + \varepsilon_0 n + \frac{1}{2}Un(n - 1), \qquad (5.7)$$

allows one to determine the Coulomb parameter U. The parameters of the effective Hamiltonian (5.6), as obtained on the basis of the impurity model [5.31], are summarized in the first column Table 5.1.

In [5.32], the parameters of (5.6) have been computed accounting for a strong hybridization of Cu $3d$ and O $2p$ states in the density-functional method with additional constraints on the occupation number for the isolated orbital. Self-consistent calculations of the total energy (5.7), performed for a cluster of 2×2 CuO_2 unit cells, have allowed determination the Coulomb parameters in the Hamiltonian (5.6). A comparison of the band-structure as computed for La_2CuO_4 by the density functional method with the electronic spectrum of the model (5.6) in the mean field approximation has also been done. In this approximation one accounts only for single-particle states with energies ε_d^{MF} and ε_p^{MF}, for example

Table 5.1. Parameters of the effective Hamiltonian (5.6)

Parameters (eV)	[5.31]	[5.32]	[5.28]
$\Delta = \varepsilon_p - \varepsilon_d$	–	3.6	3.5
Δ^{MF}	1.2	1.3	–
t_{pd}	1.6	1.3	1.3
t_{pp}	0.65	0.65	0.65
U_d	8.5	10.5	8.8
U_p	4.1–7.3	4	6.0
V_{pd}	0.6	1.2	0

$$\varepsilon_{d\sigma}^{MF} = \varepsilon_d + U_d\langle n_{d-\sigma}\rangle + 4V_{pd}\sum_{\sigma'}\langle n_{p\sigma'}\rangle\,,\tag{5.8}$$

which are compared to one-particle states in the density-functional method. On the basis of an analysis of these calculations and computing $E(n)$ with n fixed, there is obtained a set of self-consistent parameters for (5.6) which are listed in the second column of the Table 5.1.

On the basis of cluster calculations [5.2, 28–30, 33], a rather detailed study of the excitations spectrum of the Cu $3d$–O $2p$ system has been carried out. Excitation spectrum in the clusters of CuO_4, Cu_2O_7 and Cu_2O_8 accounting for Cu $3d$ states with the symmetry $a_1 = d(3z^2 - r^2)$, $b_1 = d(x^2 - y^2)$, $b_2 = d(x, y)$, $e_g = \{d(x, z), d(y, z)\}$, and O $2p$ states of the same symmetry, have been found [5.28, 29]. In Fig. 5.13 several of the $3d$ orbitals on a copper ion and p_x, p_y orbitals in the plane of the cluster $(CuO_4)^{6-}$ are shown [5.28]. The single-hole O $2p$ states are represented by the appropriate combinations of the wave functions on the four oxygen sites in this cluster, e.g.

$$p(a_1) = \frac{1}{\sqrt{4}}\{p_1(x) + p_2(y) - p_3(x) - p_4(y)\}\,,$$
$$p(b_1) = \frac{1}{\sqrt{4}}\{p_1(x) - p_2(y) - p_3(x) + p_4(y)\}\,.\tag{5.9}$$

Regarding the p-d hybridization, the spectrum of single-hole excitations may be computed by a direct diagonalization of the p-d Hamiltonian, and then the matrix elements of the Coulomb interaction for two-hole states may be found. In cluster calculations [5.28], the parameters of the model Hamiltonian defined on the basis of a comparison of computed and experimental photoemission spectra; the values obtained are given in the third column of Table 5.1. They agree quite well with constrained band-structure calculations described above, with V_{pp} being set at 0.

The most interesting result of the cluster calculations is the revelation of two-holes states inside a CuO_4 unit cell. A large exchange energy for $3d(x^2 - y^2)$ and O $2p_\sigma(x, y)$ holes, given by

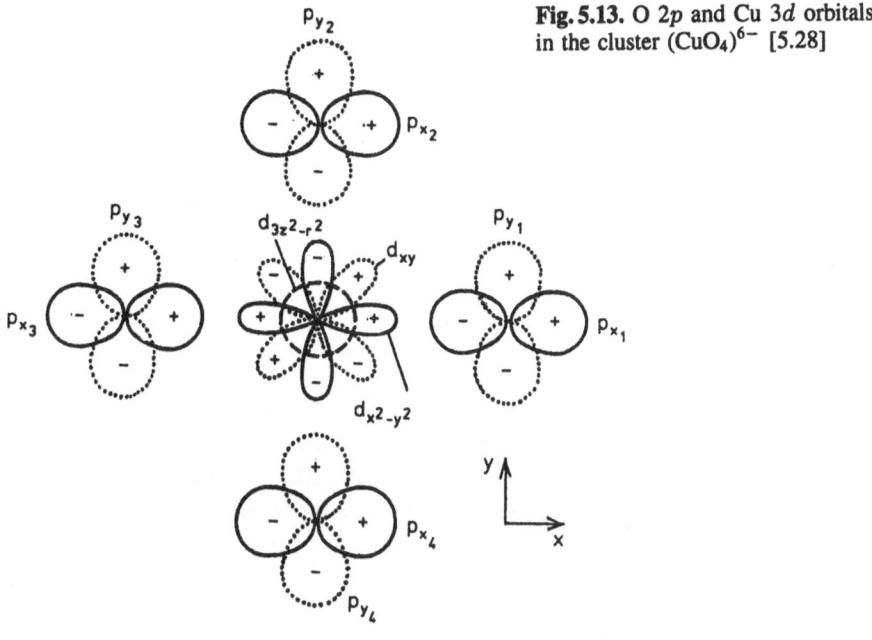

Fig. 5.13. O $2p$ and Cu $3d$ orbitals in the cluster $(CuO_4)^{6-}$ [5.28]

$$J_{pd} = E_S - E_T \simeq -2t_{pd}^2 \left(\frac{1}{\Delta + U_p} + \frac{1}{U_d - \Delta} \right) \simeq 1\,\text{eV}. \tag{5.10}$$

(for the state $p(b_1)$ in (5.9): $J(b_1) = 4J_{pd} \simeq 3-4\,\text{eV}$) leads to an essential reduction of the energy of the singlet $E_S(^1A_{1g})$ with regard to the triplet $E_T(^3A_{1g})$ state. As a result, the singlet state of Cu $3d$ and O $2p$ holes, $|\, \Phi(2)S\rangle$, with a large binding energy may be viewed as a spinless charged quasi-particle on a background of one -hole states at the copper sites in the CuO_2 planes [5.45]. Estimates of the hybridization energy between the singlet $|\, \Phi(2)S\rangle$ and the one-hole $|\, \Phi(1)\sigma\rangle$ states on the nearest and next-nearest copper sites, which have been obtained for Cu_2O_7 and Cu_2O_8 clusters [5.29], show that this energy is not enough to destroy the singlet state. Therefore, at low doped hole concentrations, despite the sufficiently large width of the singlet band, the spectrum of low energy excitations in the CuO_2 plane may be described by the one-band effective Hamiltonian:

$$H_{t-J} = -\sum_{i,j} t_{ij} c_{i\sigma}^+ c_{j\sigma} + J_{nn} \sum_{ij} (S_i S_j - \frac{1}{4} n_i n_j). \tag{5.11}$$

The first term describes a motion of the singlet over the single-occupied lattice sites with the transfer integral for the nearest neighbor sites $t \simeq 0.4\,\text{eV}$, and for the next-nearest sites (along the diagonal in the plane) $t' \simeq -0.2\,\text{eV}$. The creation operators $c_{i\sigma}^+$ (annihilation $c_{i\sigma}$) describe the creation of a composite Fermi-particle on a site i, that is an annihilation of the state $|\, \Phi(1)\sigma\rangle$ and a creation of $|\, \Phi(2)S\rangle$, and therefore they operate in the space of singly-occupied states:

$$c_{i\sigma}^+ = a_{i\sigma}^+ (1 - n_{i-\sigma}), \quad c_{i\sigma} = a_{i\sigma}(1 - n_{i-\sigma}), \quad n_{i\sigma} = a_{i\sigma}^+ a_{i\sigma}. \tag{5.12}$$

The second term describes antiferromagnetic exchange for spins $S = 1/2$ on the copper sites with an indirect exchange interaction:

$$J_{nn} = \frac{4t_{pd}^4}{\Delta^2} \left(\frac{1}{U_d} + \frac{2}{2\Delta + U_p} \right). \tag{5.13}$$

Its value, according to the data of Table 5.1, is estimated as $J \simeq 0.17\,\mathrm{eV}$ (with $U_{pd} \simeq 1\,\mathrm{eV}$ [5.29]). Similar results for the parameters of the $t - t' - J$ model (5.11) have been obtained [5.33], where clusters of larger size in a CuO_2 plane have been examined (up to Cu_5O_{16}): $t = 0.41\,\mathrm{eV}$., $t' = -0.07\,\mathrm{eV}$, $J_{nn} = 0.13\,\mathrm{eV}$. The exchange interaction for the next nearest neighbors is small: $J_{nnn} \simeq 0.02\,\mathrm{eV}$, and may be neglected. Close values of the parameters of the Hamiltonian (5.11) are obtained by electronic doping, i.e. introducing an extra electron in the CuO_4 unit cell [5.29, 33]: $t_e = -0.4\,\mathrm{eV}$, $t'_e = 0.1\,\mathrm{eV}$. Note that studies of the band-structure of quasi-particle states (ignoring spin correlations) demonstrate that the excitation spectra have an indirect gap $E_{gap} = 1.5 - 2\,\mathrm{eV}$, while a direct gap related to charge transfer Δ is $3 - 4\,\mathrm{eV}$ [5.29]. Taking account of the interaction of quasi-particles with spin fluctuations, given by the second term in (5.11), can essentially change the nature of the quasi-particle spectrum (Sect. 7.1.2). We will consider the reduction of the effective three-band Hamiltonian (5.6) to the single-band t-J Zhang-Rice model (5.11) in greater detail in Sect. 7.1.1.

The effect of apex oxygen in CuO_5 pyramids and in CuO_6 octahedrons on the electronic spectrum has been investigated [5.30] for clusters with two copper ions, Cu_2O_9 and Cu_2O_{11}. With a sufficiently large difference, $\Delta\varepsilon_A = \varepsilon(p_z) - \varepsilon(p_\sigma)$, between the energy levels of the apex, $\varepsilon(p_z)$, and planar $\varepsilon(p_\sigma)$ oxygen, the singlet state for the doped in-plane hole $| \Phi(2)S \rangle$ is stable. However, when these levels come closer the hybridization of one-hole states on the apex, $2p_z$, and in-plane, $2p_\sigma$, oxygen begins and the singlet state is destroyed. The stability of the singlet state $| \Phi(2)S \rangle$ is defined by the value of the splitting energy $2t$ for the ground (A_{1g}) and excited (B_{1u}) states of the system of the singlet $| \Phi(2)S \rangle$ and the Cu $3d$ hole $| \Phi(1)\sigma \rangle$ on the nearest site. Cluster calculations of [5.30] demonstrate that the splitting energy $2t$ correlates linearly with the difference $\Delta\varepsilon_A$, which just indicates the increase of stability of the singlet state with respect to an increase of $\Delta\varepsilon_A$. At the same time comparing the maximum temperatures of the superconducting transition T_c with $2t$ or $\Delta\varepsilon_A$ for different copper-oxide superconductors shows a certain correlation: superconductivity appears at $t \geq 0.22\,\mathrm{eV}$ ($\Delta\varepsilon_A \geq -1\,\mathrm{eV}$), and T_c attains the maximum values for $t = 0.3 - 0.35\,\mathrm{eV}$ (for $\Delta\varepsilon_A \simeq 5\,\mathrm{eV}$). A much smaller effect on the stability of the singlet (the value of $2t$) is caused by a change of the energy difference $\Delta = \varepsilon_p - \varepsilon_d$ for the in-plane O $2p$ and Cu $3d$ states, and $\Delta\varepsilon_d = \varepsilon_d(3z^2 - r^2) - \varepsilon_d(x^2 - y^2)$ for multiplet $3d$ states. An increase of Δ leads to decrease of the exchange energy (5.10) and the coupling energy of the singlet $2t$. Bringing the multiplet levels closer (a decrease of $\Delta\varepsilon_d$) also destabilizes the singlet $A_{1g}(b_1 b_1)$. Thus, cluster calculations [5.30] unambiguously suggest that T_c correlates with the stability of the singlet state: destabilization of the latter at the coupling energies at $2t \leq 0.44\,\mathrm{eV}$ leads to a disappearance of superconductivity. It may be that at sufficiently high concentrations of doped holes, under a filling of the band of the quasi-particle singlet states, their destabilisation also takes place. This phenomenon could explain the disappearance of superconductivity in copper-oxide superconductors at high concentration of carriers, as shown in Fig. 5.3.

5.4 Experimental Studies of the Electronic Structure

Direct information on the electronic structure of solids may be obtained using electronic spectroscopy. To investigate valence (occupied) states X-ray photoelectron spectroscopy may be used [5.46]. In this method, photoionisation spectra, both the X-ray photoemission (XPES) and ultraviolet photoemission (UPES) are studied. The latter directly removes electrons from the valence band near the Fermi surface. Studies of X-ray emission spectra (XES), which appear under the filling of a γ-quantum created hole in the core of an atom, caused by an electron from the valence band, also permit investigation of the density of states in the valence band. Inverse photoemission spectra (IPES, or BIS–Bremsstrahlung-isochromat spectra) which appear with capture of electrons at levels in the conduction bands also provide information regarding the nature of the band spectrum above the Fermi energy. Auger spectroscopy (AES–Auger electron spectra) makes it possible to study more complicated two-hole states in the valence band.

With the aid of photoemission spectroscopy (PES) it is only possible to investigate, however, surface layers of a sample of thickness about $20 - 30$ Å. This fact complicates the analysis of the data obtained because the surface properties of a material may differ from those of the bulk, and besides the measurement results depend on the quality of the surface and the changes of its properties under the effect of irradiation. Therefore experiments on measuring energy-loss of fast electrons (EELS – electron-energy-loss spectroscopy) would seen more reliable. On transition of an electronic beam through 1000 Å films it is possible to investigate the bulk properties of the sample.

Apart from the PES which studies one-electron excitations spectra, optical methods probe the nature of band spectra near the Fermi surface. Measuring optical reflectivity and absorption spectra provides information on collective excitations in the system: charge density fluctuations and magnetic excitations. Next we turn to some results of these experiments.

5.4.1 High-Energy Scale Spectroscopy

Photoemission spectra are described by differential photoionization cross sections [5.46]:

$$\frac{d\sigma(i \to f)}{d\Omega} \propto \frac{1}{h\nu} \varrho(E_F) |\langle \Psi_f | e^{i\boldsymbol{q}\boldsymbol{r}} (e\nabla) | \Psi_i \rangle|^2 , \tag{5.14}$$

where $h\nu$, e and $\boldsymbol{q}$ are the energy, polarization and wave number of the incident photon (γ−quantum) respectively, Ψ_i and Ψ_f are the wave functions of the initial and final states of the system respectively and $\varrho(E_F)$ is the density of final states. The energy conservation law holds for photoexcitation:

$$h\nu = E + E_f(\boldsymbol{k}) - E_i, \tag{5.15}$$

where E is the kinetic energy of the photoelectron, and the difference between the final $E_f(\boldsymbol{k})$ and initial E_i energies for the emitted electron in the state $\boldsymbol{k}$ defines the binding energy $E_B = E_f(\boldsymbol{k}) - E_i$ of the electron on a core level in the given ion. An additional approximation is usually involved which ignores the wave number dependence in (5.14): $\exp(i\boldsymbol{qr}) \simeq 1$, because the photon wavelength is usually much larger the atom size r_0: $\lambda = 2\pi/q \gg r_0$. Only the dipole transitions $i \rightarrow f$ changing the angular momentum by $\Delta l = \pm 1$ (e.g., $s \rightarrow p, p \rightarrow d$ transitions) is permitted. If the photon energy $h\nu \gg E_f$, one can use one-particle approximation for the final state and write down the angle averaged intensity of the photoionisation process as follows:

$$I(h\nu, E) \sim \sum_{l,\alpha} \sigma_l^\alpha(E, h\nu)\varrho_l^\alpha(E), \tag{5.16}$$

where $\varrho_l^\alpha(E)$ is the partial local density of states of a complex of bound atoms (or one atom) of the type α, σ_l^α is the photoionization cross-section of this complex and l is the angular momentum. The latter relation allows one to measure the local density of states, weighted by the corresponding cross-section of the transition under study.

In the very first experiments which investigated photoemission spectra in ceramic samples of LSCO and YBCO, a weak dependence of copper spectra on doping was found [5.47]. This directly confirmed the picture of the electronic spectrum under doping shown in Fig. 5.2, according to which, under doping, holes appear in the O $2p$ band. The same experiments have detected strong correlations in the $3d$ band, in view of which the two-hole states of $Cu^{3+}(3d^8)$ appeared shifted by about 10 eV to the higher energy range. In order to explain the photoemission spectra, cluster calculations have been performed [5.47] and have given one of the first estimates of the parameters of the effective Hamiltonian of the p-d model (5.6).

In Fig. 5.14 for comparison we show the $Cu2p_{3/2}$ core-level absorption spectra in CuO, Cu_2O and $La_{2-x}Sr_xCuO_4$. For monovalent copper Cu^{1+} in Cu_2O compounds there is only one peak at the binding energy of about 932 eV which corresponds to the final state $Cu2p^53d^{10}$ – hole on the core level $2p$. For divalent copper Cu^{2+} the main peak at 933 eV corresponds to the final state of Cu $2p^53d^{10}$ O $2p^5$, in which the core hole state Cu $2p^5$ is screened due to charge transfer from the oxygen atom O $2p^5$. The satellite at 942 eV corresponds to the two-hole state Cu $2p^53d^9$. The trapezoidal form of the satellite is related to multiplet splitting of $3d$ states. The 9 eV shift of the satellite represents the Coulomb repulsion of two holes. Estimates of the Coulomb repulsion energy, as evaluated for the $3d$ shell, yield $U_d \simeq 7$ eV. The ratio of the areas under the satellite line and the main peak controls the charge transfer from oxygen to copper. The estimates demonstrate that

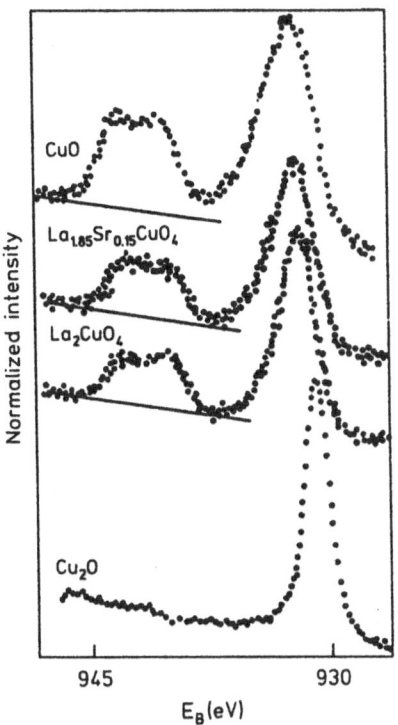

Fig. 5.14. Absorption spectra for the Cu $2p_{3/2}$ versus the binding energy E_B [5.48]

the charge on copper in the ground state $3d^9$ is approximately 9.4. Weak dependence of the XPES of the Cu $2p_{3/2}$ spectrum on the content of strontium indicates directly a localized nature of $3d$ copper holes and an absence of $Cu^{3+}(3d^8)$ states in the metallic phase of $La_{2-x}Sr_xCuO_4, x > 0$.

The most interesting results concerning hole states in cooper oxides have been obtained by the EELS method in Karlsruhe [5.3, 48–52] [3]. Measuring absorption spectra of fast electrons with incident energy $E = 170\,\text{keV}$ with high resolution, $\Delta E = 0.4\,\text{eV}$, in films of thickness $1000\,\text{Å}$ has allowed a high-precision investigation of the local density of hole states. Under a small momentum transfer, the main contribution arises from dipole transitions with $\Delta l = \pm 1$, and therefore the studies of the absorption edge under excitations of O $1s$ level ($E_B = 528\,\text{eV}$) or Cu $2p_{3/2}$ level ($E_B = 931\,\text{eV}$) allows investigation of hole states in O $2p$ and Cu $3d$ bands respectively.

Electron energy-loss spectra under excitation of the O $1s$ level in polycrystalline films of LSCO and YBCO have been investigated [5.49]. In insulator samples there was no absorption near the Fermi level, but under doping the absorption intensity essentially increased (proportionally with the concentration of holes in the O$2p$ band). In Fig. 5.15, the results of more elaborate measurements [5.50]

---

[3] Recent results of EELS and X-ray absorption spectroscopy of cuprate superconductors are reviewed by Fink et al. [5.118].

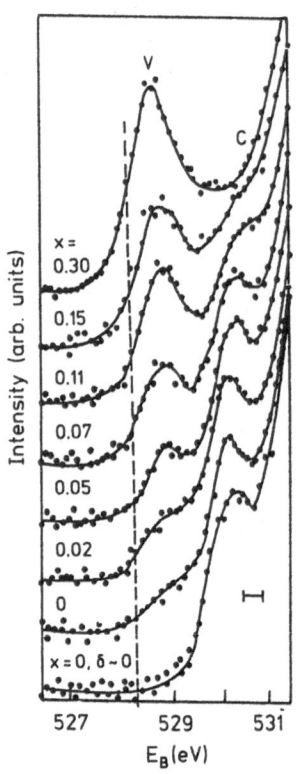

Fig. 5.15. The electron energy-loss spectra (EELS) at the O $1s$ absorption edge of La$_{2-x}$Sr$_x$CuO$_{4+\delta}$ [5.50]

of the the electron energy-loss spectra in La$_{2-x}$Sr$_x$CuO$_{4+\delta}$ are shown. The stoichiometric compound ($x = 0$, $\delta \simeq 0$), gives the lower curve in Fig. 5.15. The absorption peak at $E_c = 530.2\,$eV is ascribed to a small mixture of O2p states in the upper Hubbard band (see Fig. 5.2). According to (5.1), the wave function of Cu^{2+} has a mixture (β) of hole O2p states in this band, determining the intensity of the order β^2 of this absorption peak in conduction band (C). Appearance of a hole in the O 2p band (due to excess oxygen $x = 0$, $\delta > 0$, or doping by Sr $x > 0$) leads to a shift of the absortion spectrum to the lower energies domain with the edge at $E_F = 528.5\,$eV, corresponding to the Fermi energy. The Fermi level, defined with respect to the binding energy in photoemission spectra, lies 1.5 eV below the conduction band E_C, correspond to $\Delta_{pd} = 1.8 - 2\,$eV, the charge transfer gap. With increase of Sr concentration, the intensity of the peak near the valence band (V) at $E_V \simeq 529\,$eV gradually increases, and that of the peak at E_C gradually decreases. At the same time the insulator-metal transition at $x = 0.06$ occurs smoothly, which may be related to forming a metallic state due to delocalization of holes at $x > 0.06$. Thus, hole O 2p band gradually appears under doping near the lower edge of the insulator p-d gap. At larger concentrations of Sr, $x \geq 0.3$, the peak at E_C almost completely disappears, and the system transfers to a state with normal metallic properties where superconductivity no longer occurs (see Fig. 5.3). Probably, at large concentrations of holes the correlation splitting of

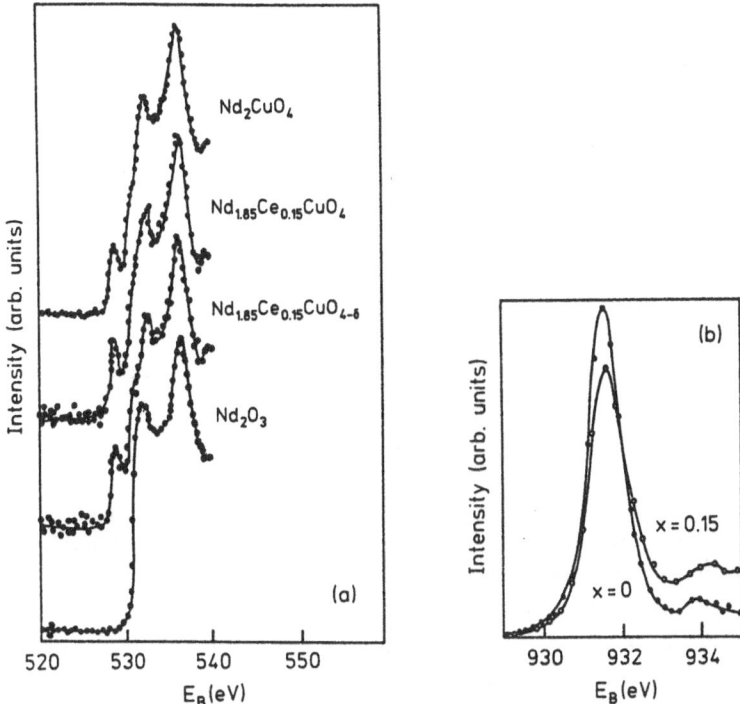

Fig. 5.16. EELS of $Nd_{2-x}Ce_xCuO_{4-\delta}$ at the O $1s$ absorption edge (**a**), and in $Nd_{2-x}Th_xCuO_{4-\delta}$ at the $Cu2p_{3/2}$ edge (**b**) [5.51]

CuO $pd\sigma$-band disappears, and the system becomes a normal metal with a broad $pd\sigma$-band which is shown on the right hand side of Fig. 5.2. Such a picture of the evolution of the valence and conduction bands in LSCO under doping is confirmed by studies of soft X-ray absorption spectra on the oxygen K-level [5.53].

In Fig. 5.16a,b, the results of analogous studies of absorption spectra for n-type doped systems $Nd_{2-x}Ce_xCuO_{4-\delta}$ are shown. Energy-loss spectra for O $1s$ levels are doping independent. The absorption peak at $E \simeq 528.5$ eV, preceding the main absorption peak, is caused by the contribution of O $2p$ hole states to the upper conduction band due to p-d hybridization, as in LSCO compounds. The intensity of this pre-peak is independent of doping, which suggests that the contribution of O $2p$ states in the upper Hubbard band is constant. At the same time, the intensity of the $Cu2p_{3/2}$ loss peak (Fig. 5.16b) under doping by Th^{4+} ions decreases by about 15 % at $x = 0.15$, indicating a fall of $3d$ holes concentration under electron doping.

Investigations of the symmetry of hole states appearing under doping have been performed using the EELS method for YBCO and $Bi_2Sr_2CaCu_2O_8$ [5.49]. Measuring the absorption edge at O $1s$ and Cu $2p_{3/2}$ levels was performed on single-crystals or oriented films, revealing the dependence of absorption on the orientation of the momentum transfer q. If q is directed along the crystallographical

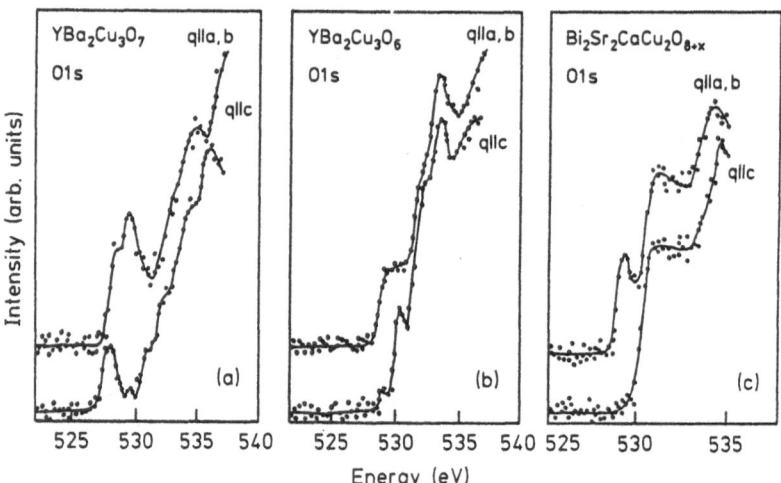

Fig. 5.17. Electron energy-loss spectrum (EELS) at the O $1s \rightarrow$ O $2p$ edge for momentum transfer q parallel to and perpendicular to the c-axis in single crystal films of $YBa_2Cu_3O_7$ (**a**), $YBa_2Cu_3O_6$ (**b**) and $Bi_2Sr_2CaCu_2O_8$ (**c**) [5.52]

axis c, the transitions of core electrons to unoccupied hole states with orientation of orbitals along the z-axis, of O $2p_\pi(z)$ or Cu $d(3z^2 - r^2)$ type can be examined; and with q lying in the (a, b) plane, to the states with orbitals oriented in this plane, of the type O $2p_\sigma(x, y)$ or Cu $3d(x^2 - y^2)$. Measuring the ratio of absorption intensities $I_{x,y}/I_z$ for the hole states leads to an estimate of the contributions of the corresponding orbitals. In Fig. 5.17 and 5.18 the results of the experiment on exciting the core levels O $1s$ and Cu $2p_{3/2}$ are shown for $YBa_2Cu_3O_7$ (a), $YBa_2Cu_3O_6$ (b) and $Bi_2Sr_2CaCu_2O_8$ (c) [5.52].

The simplest case is Bi/2212. For the Fermi level the energy $E_B = 528.8 \text{ eV}$ corresponds to the absorption edge O$1s$ at $q \parallel a, b$. The absence of absorption at $q \parallel c$ above the absorption edge E_B (see Fig. 5.17c) indicates the absence of hole states of O $2p_\pi(z)$ type. The absorption at $q \parallel a, b$ is related to hole states in CuO_2 planes and BiO layers with the symmetry of $2p(x, y)$. The nature of bonding, σ or π, cannot be found from this experiment.

Investigations of absorption under excitations of the Cu $2p_{3/2}$ level for $q \parallel a, b$ and $q \parallel c$ (see Fig. 5.18c) demonstrate that besides in-plane hole states, $3d$ $(x^2 - y^2)$, there is a small admixture of the order 10–15 %, of out-of-plane states, most probably of the symmetry $3d$ $(3z^2 - r^2)$.

An interpretation of absorption spectra for $YBa_2Cu_3O_{7-y}$ is more complicated as both the CuO_2 planes and the CuO_3 chains contribute to the density of states near the Fermi level. Band calculations show [5.27] that a contribution to the absorption spectrum at O $1s$ may be due to the states near the Fermi surface, arising both from the orbitals $2p_\sigma(x, y)$ in the CuO_2 planes and the orbitals $2p_\sigma(z)$ on O4 oxygen ions coupled to the in-chain Cu1 $3d(z^2 - y^2)$ states. However shifts of O $2p$ levels in the crystal field for different sites O1, O2, O3 and O4 must be taken

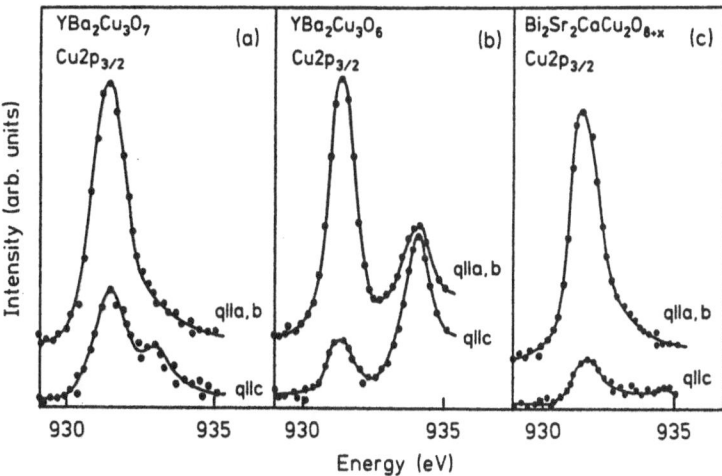

Fig. 5.18. EELS at the Cu $2p_{3/2} \rightarrow$ Cu $3d$ edge [5.52]

into account (see Fig. 2.12). A comparison with tests on the absorption intensities $I_{x,y}/I_z$ for the energies $E < 931\,\text{eV}$ has shown that the main contribution to the O $1s$ absorption comes from the hole O $2p$ states in the plane of the σ-bond (Cu2 – O2,3 or Cu1 – O4). The existence of holes on O $2p$ orbitals out of the σ-bond plane, of type $p_\pi(z)$ in CuO$_2$ planes, or O1 $2p_\pi(x)$ in the chains, is at variance with the experimental data [5.52].

Absorbtion spectra for the Cu $2p_{3/2}$ level in YBa$_2$Cu$_3$O$_7$ (Fig. 5.18a) demonstrate that an approximately equal density of holes is located on σ-orbitals $3d(x^2 - y^2)$ for in-plane Cu2, and $3d(z^2 - y^2)$ on in-chain Cu1. A small asymmetry of the spectrum (and a shoulder for $q \parallel c$) relates to the transition to the state Cu $2p^5 3d^{10}$ – O $2p^5$, when screening of the core hole Cu $2p^5$ due to charge transfer from oxygen O $2p^5$ occurs. This asymmetry and the shoulder for $q \parallel c$ disappear in the spectrum of YBa$_2$Cu$_3$O$_6$, which signals the absence of these states. At the same time, the abrupt fall of intensity for $q \parallel c$ in YBa$_2$Cu$_3$O$_6$ (Fig. 5.18a) and the appearance of a peak at $E_B = 934\,\text{eV}$ which is characteristic of Cu^{1+} (see Fig. 5.14) indicate that removal of in-chain oxygen causes the filling of hole states on the orbital Cu1 $3d(z^2 - y^2)$, and Cu1 passes to a monovalent state with a small number of unoccupied $3d$ states, mostly of the $3d$ ($3z^2 - r^2$) type.

Thus, experimental studies of XPES, EELS and other high-energy photoelectron spectra make it possible to clarify many important features of the electronic structure of high-T_c superconductors [5.118]. The most important results are strong Coulomb correlations in the copper $3d$ band, an appearance of new bands inside the p-d gap: for p-type doping, near the top of the O $2p$ band, or, for n-type doping, near the bottom of the Cu $3d$ band. These results confirm the band-structure shown in Fig. 5.2, characteristic of semiconductors with a charge transfer gap, $\Delta_{pd} = \varepsilon_p - \varepsilon_d < U_d$. The new band, while providing metallic properties at small concentrations of impurities, corresponds to the "filled gap" model under doping which is shown in Fig. 5.2. An important role in forming it is played by many-particle correlations in the $pd\sigma$-band. However, they relax to a large extent at sufficiently high concentrations of carriers ($x \geq 0.3$), the correlation $pd\sigma$-band smears and the system passes to a state with the normal metallic properties. In this "normal metal" state superconductivity no longer appears. The experimental studies presented had been significant in finding the parameters of the $pd\sigma$ model effective Hamiltonian.

5.4.2 Studies of the Fermi Surface

The methods of photo-electron spectroscopy in the ultraviolet energy range, $h\nu = 20 - 30\,\text{eV}$, have been effective in the studies of the electronic structure of copper-oxide superconductors near the Fermi energy E_F for $|E_F - E| \leq 0.5\,\text{eV}$. The angular resolved photoemission spectra (ARPES), and inverse photoemission (ARIPES) with high resolution of $30-60\,\text{meV}$ have made it possible to investigate the Fermi surface in momentum space [5.54–56].

According to (5.16), the intensity of photoemission is determined by the partial density of states $\varrho_l^\alpha(E)$, which is directly related to the one-particle density of states $A(\boldsymbol{k}, \omega)$ for quasi-particles having the momentum $\boldsymbol{k}$ and the energy $\omega = E - E_F$. For sufficiently large quasi-particle lifetimes, the function $A(\boldsymbol{k}, \omega)$ with the given $\boldsymbol{k}$ has a maximum at the energy $\omega = \varepsilon(\boldsymbol{k})$, describing a one-particle state (a hole in ARPES for $\varepsilon(\boldsymbol{k}) < E_F$, and an electron in ARIPES for $\varepsilon(\boldsymbol{k}) > E_F$). The intensity of this maximum is determined by the weight $Z_{\boldsymbol{k}}$ of the one-particle states in the one-electron Green function (see Sect. 7.1.1). A one-particle maximum depending on $\boldsymbol{k}$ is seen on a broad background for a range of several eV. Probably, this background is built into the photoionization process itself and related to the incoherent excitations which arise with the abrupt removal of an electron ("shake-off spectrum" [5.57].)

The most detailed studies of the excitation spectra near the Fermi surface have been performed for single crystals of $Bi_2Sr_2CaCu_2O_8$ which have a stable surface. Due to small energy of ultraviolet radiation the specimen surface should not degrade under the effect of radiation, and will correctly represent the specimen's bulk properties. Besides, the quasi-twodimensional nature of the electronic spectrum in Bi compounds – an absence of dispersion along k_z – essentially simplifies the measurements of the photoemission spectra as a function of the components of the momentum k_x, k_y in the CuO_2 plane. The ARPES [5.54] and ARIPES [5.55] data have allowed them to determine the one-particle energies $\varepsilon(\boldsymbol{k})$ for the scattering vector $\boldsymbol{k}$ in certain directions of the Brillouin zone. The results of these experiments are shown in Fig. 5.19, where the dark squares correspond to ARPES measurements, and the lights to ARIPES [5.3]. In the same figure the electron dispersion curves are shown for the high-symmetry directions of the Brillouin zone $\Gamma-X$ and $\Gamma - M - Z$ (Γ is the center of the Brillouin zone), as computed by the density functional method. Obviously, the experimental dependence $\varepsilon(\boldsymbol{k})$ is similar to the theoretical curves with respect to its shape, but has a much smaller dispersion. This decrease may be related to a large effective mass of quasi-particles,

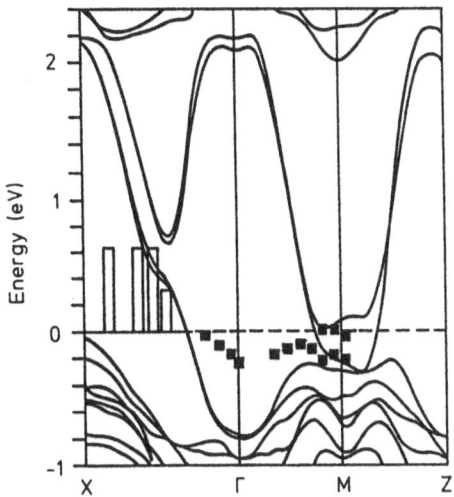

Fig. 5.19. Band-structure of $Bi_2Sr_2CaCu_2O_8$, and experimental one particle states in the filled (*dark squares*) and empty (*light rectangles*) bands [5.3]

the order of value $m^* \simeq 4m$. The estimate $m^* \simeq 2m$ for the direction $\Gamma{-}X$ has been obtained [5.56] and, under certain assumption concerning the nature of the background in the photoemission spectra, the decay $\Gamma_k(\omega)$ – the inverse lifetime of quasi-particles – has been estimated. It turned out that it is best described by the linear function of the energy $\omega = |\varepsilon(k) - E_F|$, not by the second power-type dependence $\Gamma_k \propto \omega^2$ which is characteristic of the normal Fermi liquid.

ARPES and ARIPES measurements have thus found a hole-type Fermi surface which is sufficiently large in the k-space and its form coincides with those calculated by the density functional method. At the same time, there are some indications of an anomalous behavior of the quasi-particle excitations spectrum ($\Gamma_k(\omega) \propto \omega$) near the Fermi surface, which suggests alternative models of Fermi liquid–marginal Fermi liquid or the Luttinger liquid (in which $Z_k \to 0$ on the Fermi surface). These models are discussed in greater details in Sect. 7.1.1. In order to find out which one of these electronic liquid models best describes the normal state of copper-oxide compounds, photoemission spectra should be measured with a higher resolution near the Fermi surface [4].

Investigations of the temperature dependence of photoemission spectra in Bi compounds have revealed the formation of a superconducting gap at $T < T_c$. Figure 5.20 shows a high-resolution photoemission spectrum of a single crystal of $Bi_2Sr_2CaCu_2O_8$ at temperatures above and below $T_c = 83\,K$ [5.58], as compared to the spectrum of gold. With a fall of temperature below T_c, the density of states above the Fermi energy $E_F = 0$ decreases, and that below the Fermi energy increases (shown dashed), which may be related to an opening of a gap on the Fermi surface. The shift of the absorption edge to the higher binding energy range by $\Delta = 30 - 40\,meV$ enables the value of the gap Δ to be found. At the same time the ratio $2\Delta/kT_c \simeq 8$ exceeds more than twice the universal BCS theoretical

---

[4] Studies of the Fermi surface by ARPES in $Nd_{1-x}Ce_xCuO_4$ compound are presented in [5.113].

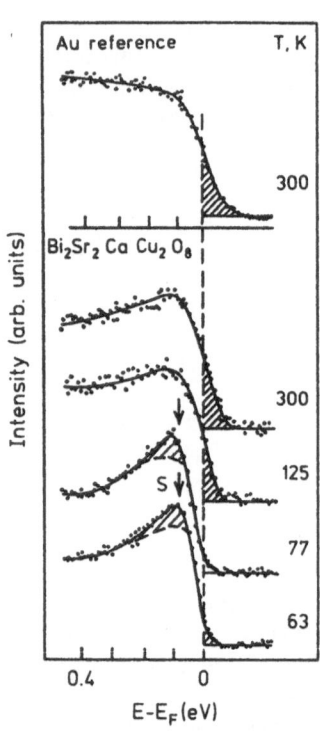

Fig. 5.20. A photoemission spectrum in Bi$_2$Sr$_2$CaCu$_2$O$_8$ above and below T_c = 83 K, as compared to the spectrum of a gold film [5.58]

value of 3.5. Similar results have been obtained [5.59] in the course of high resolution 30 meV ARPES measurements. The value of the gap Δ = 24 meV was also larger than it should be according the BCS theory, $2\Delta/kT_c \simeq 7$. Measuring the gap for several symmetry directions in Bi$_2$Sr$_2$CaCu$_2$O$_8$ reveals the unambigious presence of a non-s-wave component in the superconducting condensate which is compatible with the $d(x^2 - y^2)$ or $d(xz) + d(yz)$ symmetry of the gap [5.60].

As well as the photoemission spectra, angular correlation of positron annihilation radiation (ACAR) can be used to study the form of the Fermi surface in metals. When entering a solid, positrons are quickly thermolized and annihilated with electrons of the metal. There usually appear two $\gamma-$quanta with the energies of about 0.5 MeV, flying in almost opposite directions. Measuring the angular distribution $N(p_x, p_y)$ with respect to the momenta p_x, p_y of the out-flying $\gamma-$quanta, the electron-positron momentum distribution $\varrho(\boldsymbol{p})$ can be determined:

$$N(p_x p_y) \propto \int \varrho(\mathbf{p}) dp_z , \qquad (5.17)$$

where in the one-particle approximation

$$\varrho(\boldsymbol{p}) = \sum_n \int \left| e^{-i\boldsymbol{p}\boldsymbol{r}} \Psi^+(\boldsymbol{r}) \Psi_n(\boldsymbol{r}) \right|^2 d^3r . \qquad (5.18)$$

Here $\Psi_n(r)$ and $\Psi^+(r)$ are the electron and positron wave functions, the summation is carried out over all the filled electron states n ([5.61] and the references therein). To reproduce the electronic density of states in the momentum space, a single crystal should be measured in several crystallographical directions. This method has proved very useful in analyzing Fermi surfaces of several transition metals, but has encountered difficulties in interpreting of the results for copper-oxide superconductors [5.61, 27]. For example, in the course of ACAR experiments in YBCO compounds, only a small difference in electron momentum density for insulator and metallic states of YBCO has been observed. A modulation of the positron wave function in the crystal field of the ionic lattice of YBCO contributes significantly to the momentum distribution function (5.18). Since the positron annihilation probability is proportional to the electronic density of states on the Fermi surface, the study of small changes in the momentum density (5.18) caused by a small concentration of doped holes in metallic phase of YBCO requires high statistical precision measurements. Additional difficulties for YBCO are due to preferential concentration of positrons near the Cu1 – O1 chains, which hinders measuring electronic density of states in the CuO_2 planes.

ACAR experiments for untwinned crystals of metallic phase YBCO revealed a clear image of the part of the Fermi surface corresponding to the chains which was hard to obtain in crystals with a twin structure [5.62]. An ARPES, as well as ACAR, study of the Fermi surface in $YBa_2Cu_3O_{6.9}$ has been performed [5.63]. Comparing the results of these two studies confirms the existence of a large hole Fermi surface in the CuO_2 plane, as predicted by band-structure calculations.

Additional confirmation of the existence of a Fermi surface in YBCO compounds has been obtained through observations of the magnetic susceptibility oscillations in strong magnetic fields – the de Haas–van Alphen effect. Studies of the magnetic susceptibility χ of powders of $YBa_2Cu_3O_{6.9}$ oriented along the c-axis in the fields up to $B = 100$ T, at $T = 4.2$ K have been done [5.64]. The analysis of χ dependence on $1/B$ has shown three characteristic oscillation frequencies, corresponding to three small cross-sections of the YBCO Fermi surface as computed by the density functional method [5.27]. However, the value of the cyclotron mass m_H^* (for the cyclotron frequency $\omega_c = eB/m_H^*c$) for these orbits turned out to be $3 - 4$ times larger than the band masses obtained in theoretical calculations. To remove this discrepancy one must assume a strong renormalization of mass, $m^* = (1+\lambda)m$ with the coupling constant $\lambda = 2 - 3$ due to a strong interaction of electrons with phonons or other excitations in the system. Approximately the same band mass renormalization is observed in the aforementioned ARPES experiments, and in measuring the electronic heat capacity (see Sect. 4.1). It has not however proved possible to observe large Fermi surface cross-sections, corresponding to antibonding states in the $pd\sigma$-band for the CuO_2 planes and characteristic of all copper-oxide superconductors (see, e.g., curves A in Fig. 5.11 for the band spectrum of LSCO). It is believed that observing them requires stronger magnetic fields [5.64].

Thus, independent experiments – measuring the angular resolution of photoemission spectra (ARPES and ARIPES), angular correlations at positron annihilation (ACAR), and the de Haas–van Alphen effect – confirm the existence of a Fermi surface in copper-oxide superconductors coinciding shape with that computed by the density functional method. The properties of quasi-particle excitations near the Fermi surface may, however, essentially differ from those predicted by theoretical calculations based on the one-particle approximation in the density functional method: an essential $2-4$ fold increase of the band mass and lifetimes of quasi-particles uncharacteristic of the Fermi liquid.

5.4.3 Optical Electron Spectroscopy

Unlike photoelectron spectroscopy of single-particle states, optical studies of electronic spectra provide information on the spectrum of collective electron-hole pair excitations with the energy of several eV near the Fermi surface. In particular, by means of infrared spectroscopy methods, several fundamental properties of oxide superconductors have been defined, such as effective mass and density of the free carriers, spectrum and lifetimes of quasi-particle excitations caused by their interaction with the lattice. Studies of reflection spectra in the low frequency range, $\hbar\omega \leq 2\Delta$, confirmed the appearance of a gap 2Δ in the spectrum of quasi-particles in the superconducting state, and has led to estimation of the value of the gap and its temperature dependence $\Delta(T)$.

The first experiments carried out on ceramic samples led to contradictory results, due to strong anisotropy of conductivity in copper-oxide compounds and inadequate surface quality. Further experiments with single crystals have removed some discrepancies in the data, although there still remain many unsolved problems [5.65]. The most interesting results were obtained for untwinned crystals of YBCO [5.66]. We will discuss some of these results.

Infrared spectroscopy methods can be used to determine complex permittivity function in the long-wave limit ($q = 0$) [5.65]:

$$\varepsilon(\omega) = \varepsilon_1(\omega) + \frac{4\pi i}{\omega}\sigma_1(\omega), \tag{5.19}$$

where $\varepsilon_1(\omega)$ is its real part. The imaginary part $\varepsilon_2(\omega)$ is written in terms of frequency dependent real part of the conductivity $\sigma_1(\omega)$ which determines absorption of radiation at the frequency ω. These functions are coupled by the Kramers-Kronig relation:

$$\varepsilon_1(\omega) = 1 + 8P \int\limits_0^\infty \frac{\sigma_1(z)}{z^2 - \omega^2} dz, \tag{5.20}$$

where P is the principal value of the integral.

The total absorption coefficient obeys the so-called f-sum rule,

$$\int\limits_0^\infty \sigma_1(\omega)d\omega = \frac{\pi}{2}\frac{ne^2}{m} = \frac{1}{8}\omega_p^2, \tag{5.21}$$

where n is the total electron density, e and m are, respectively, the mass and the charge of a free electron and ω_p is the plasma frequency. To study separate components of the absorption spectrum, for example due to only intraband transitions of free carriers, they frequently use the partial sum rule. It determines the effective number of carriers inside a primitive cell of volume v_0:

$$N_{\text{eff}}(\omega) = \frac{2mv_0}{\pi e^2} \int_0^\omega \sigma_1(z)dz \,, \tag{5.22}$$

which participate in optical transitions with frequencies less than ω. In particular, accounting in (5.21) only for the free carriers, contributing to the low-frequency (intraband) absorption, $\omega < \omega_g$, the effective optical plasma frequency can be found:

$$\omega_p^{*2} = \frac{4\pi ne^2}{m^*} = \frac{4\pi e^2}{m} \frac{N_{\text{eff}}(\omega_g)}{v_0} \,, \tag{5.23}$$

where m^* is an effective (optical) mass of free carriers.

When processing experimental data a certain model for conductivity σ_1 is used, the simplest being the Drude model:

$$\sigma_1(\omega) = \frac{\omega_p^2}{4\pi} \frac{\tau}{1 + \omega^2\tau^2} \,, \tag{5.24}$$

where $1/\tau = \Gamma/\hbar$ is the inverse lifetime of quasi-particles – charge carriers. To account for the lifetime dependence on the frequency, and the renormalization of mass of free carriers due to their interaction with the lattice, the generalized Drude model is used:

$$\sigma_1(\omega) = \frac{\omega_p^2}{4\pi} \frac{m}{m^*(\omega)} \frac{\tau^*(\omega)}{1 + \omega^2\tau^*(\omega)^2} \,. \tag{5.25}$$

The parameters of the generalized Drude model are related to characteristic parameters of the spectrum of quasi-particle excitations – charge carriers [5.67]:

$$\lim_{\omega \to 0} \frac{m^*(\omega)}{m} = 1 - \text{Re} \left. \frac{\partial \Sigma(\omega)}{\partial \omega} \right|_{\omega=0} = 1 + \lambda \,, \tag{5.26}$$

$$\frac{1}{\tau^*(\omega)} = \frac{m^*(\omega)}{m} \frac{1}{\tau(\omega)} \,, \quad \frac{1}{\tau(\omega)} = -2\,\text{Im}\,\Sigma(\omega) \,, \tag{5.27}$$

where $\Sigma(\omega)$ is the self-energy operator of the one-particle Green function (Sect. 7.1.1). If the main contribution to the self-energy operator arises from the electron–phonon interaction, then the parameters of (5.26), (5.27) (at $T = 0$) are given by:

$$\lambda = 2 \int_0^\infty \frac{dz}{z} \alpha^2(z)F(z) \,, \tag{5.28}$$

$$\frac{1}{\tau(\omega)} = 2\pi \int_0^\omega dz\, \alpha^2(z)F(z) \,, \tag{5.29}$$

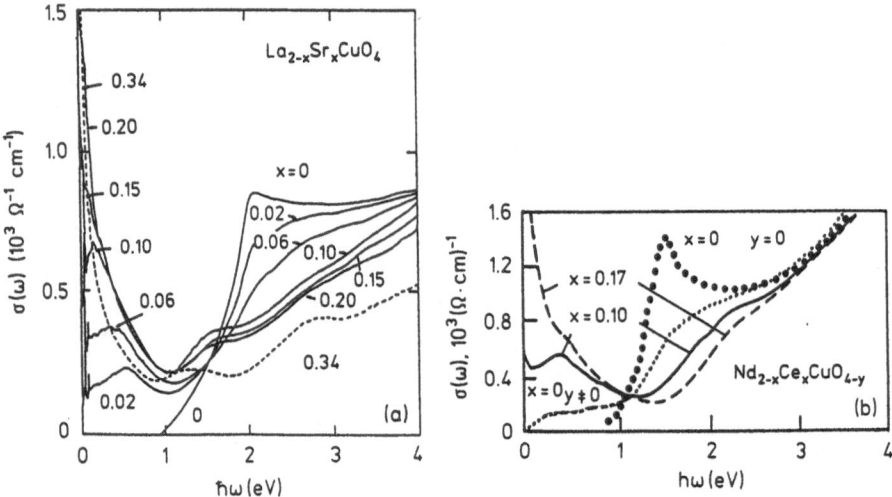

Fig. 5.21. Optical conductivity $\sigma_1(\omega)$ for La$_{2-x}$Sr$_x$CuO$_4$ (a), and Nd$_{2-x}$Ce$_x$CuO$_{4-y}$ (b) [5.68]

where $\alpha^2(\omega)F(\omega)$ is the Eliashberg function where, for the transport properties (5.26), (5.27), the matrix element of electron–phonon interaction $\alpha^2(\omega)$ includes the additional factor $(1-\cos\theta)$, where θ is the scattering angle of quasi-particles under the relaxation of their momentum. However, formulas (5.26) and (5.27) only hold at low temperatures, $kT \ll \hbar\omega$, and for frequencies much less than the characteristic frequency of the lattice vibrations. Outside this region a rather complicated dependence of the parameters of the generalized Drude model on frequency and temperature is seen, due to the retarding nature of electron–phonon interaction [5.67]. In the case of copper-oxide compounds, the interaction of quasi-particles – charge carriers with dynamic spin fluctuations can also contribute essentially to their scattering (Chap. 3).

Studies of frequency dependence of infrared reflection on single crystals and films of copper-oxide compounds have thrown light on the transition from charge-transfer insulator to metallic state under a change of concentration of charge carriers (holes or electrons) in high-T_c superconductors. Figure 5.21 shows the measured optical conductivity $\sigma_1(\omega)$ for single crystals of La$_{2-x}$Sr$_x$CuO$_4$ (a), and Nd$_{2-x}$Ce$_x$CuO$_{4-y}$ (b), dependent on doping, for an external field parallel to the conducting planes CuO$_2$ [5.68]. In the insulator phase ($x = 0$) absorption appears only at photon energies $\hbar\omega > 1\,\mathrm{eV}$, which indicates the existence of an insulator optical gap. In accordance with the general structure of the electronic spectrum as shown in Fig. 5.2, this gap is caused by charge transfer from the filled O $2p$ type band to the upper Hubbard Cu $3d$ subband. Under doping (p-type in LSCO and n-type in Nd–Ce–Cu–O compounds) the intensity of this absorption decreases, but there appears an absorption at lower energies, $\hbar\omega < 1\,\mathrm{eV}$, which increases more rapidly than the concentration of carriers. An absorption characteristic of

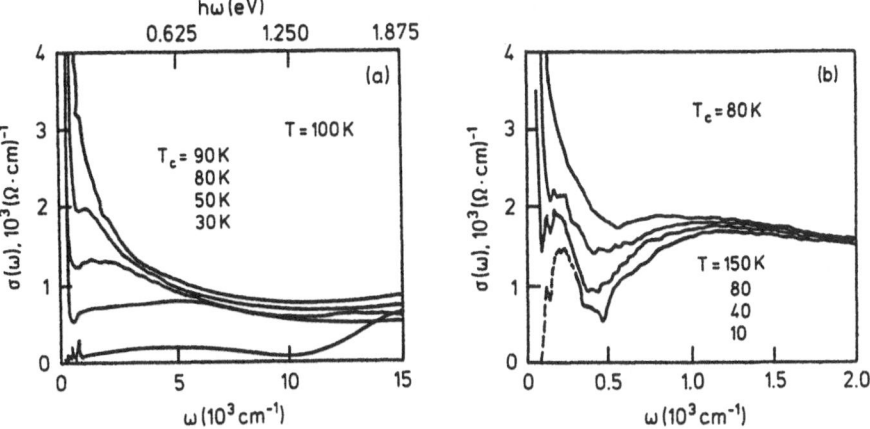

Fig. 5.22. Optical conductivity $\sigma_{ab}(\omega)$ for single cristals of YBCO: **(a)** at $T = 100$ K for samples with $T_c = 90, 80, 50$ and 30 K, and an insulator sample; **(b)** for a sample with $T_c = 80$ K at $T = 150, 80, 40$ and 10 K [5.69]

metals, in the form of Drude peak (5.24) also appears near $\omega = 0$. The results obtained suggest the existence of new states inside the gap related to O $2p \rightarrow$ Cu $3d$ charge transfer. They agree with the measured electron energy-loss spectra (see Fig. 5.15) for LSCO. A strong dependence of optical conductivity on doping, as shown in Fig. 5.21, is observed only for absorption in the CuO_2 plane – for the polarization of light $E \parallel c$ the absorption spectrum preserves its insulator nature up to significant concentrations of carriers [5.68].

The same dependence of the optical conductivity on the concentration of carriers is observed in $YBa_2Cu_3O_{7-y}$ compounds [5.69] and in $Bi_2Sr_2(Ca - Nd - Y)Cu_2O_8$ [5.70]. A detailed study of the temperature and frequency dependence of the optical conductivity in the frequency range $30 - 20\ 000\,\mathrm{cm}^{-1}$ ($4\,\mathrm{meV} - 2.5\,\mathrm{eV}$) has been performed on several single crystals of $YBa_2Cu_3O_{6+x}$ [5.69]. Figure 5.22 shows the measured in-plane conductivity $\sigma_{ab}(\omega)$: (a) at $T = 100$ K for five samples with different oxygen content (superconducting with $T_c = 90, 80, 50$ and 30 K, and an insulator sample with $x = 0.2$) and (b) for samples with $T_c = 80$ K at $T = 150, 80, 40$ and 10 K. The conductivity of the insulator sample is characterized by an optical gap, $\hbar\omega_g \simeq 1.75\,\mathrm{eV}$, determined by charge transfer O $2p \rightarrow$ Cu $3d$. With an increase in the number of carriers (and T_c) an increase of absorption takes place in the mid-infrared region at $\hbar\omega \leq 1\,\mathrm{eV}$, accompanied by an increase in the intensity of the Drude absorption peak at $\omega \simeq 0$.

The analysis of frequency and temperature dependence of $\sigma_{ab}(\omega)$ at different concentrations of carriers leads to the conclusion that two components exist in the infrared absorption. A rather narrow Drude component (5.24) with a typical width $\Gamma \simeq kT$ affects only a small part of the effective number of free carriers (5.22) participating in absorption. For example, for a sample with $T_c = 90$ K, the estimation (5.23) of N_{eff} from the value of the optical plasma frequency $\hbar\omega_p^* =$

1.5 eV which is fixed by the value of $\sigma_1(0)$ in (5.24), yields $N_{eff} \simeq 0.14$ per Cu2 site, which is essentially less than the anticipated number of free carriers $\delta \simeq 0.33$ ($n = \delta/v_0$). The second component which also contributes to N_{eff} is related to absorption in the mid-infrared region, $kT < \hbar\omega \leq 1$ eV, whose intensity, like that of the Drude peak, rapidly increases with the number of free carriers δ (see Fig. 5.22a). Additional confirmation of the two-component model was obtained when trying to describe the frequency and temperature dependence of $\sigma_{ab}(\omega)$ with the aid of the generalized Drude model (5.25). Assuming that the parameters of the model (5.26) and (5.27) may be estimated by an appropriate choice of the function $\alpha^2(\omega)F(\omega)$ in (5.28) and (5.29), the frequency dependence of $\sigma_{ab}(\omega)$ in the limit of high and low frequencies may be examined. To describe the low frequency Drude component a sufficiently weak coupling, $\lambda \simeq 0.4$ in (5.28) at $\hbar\omega_p^* \simeq 1.5$ eV must exist. However, in the frequency range $\hbar\omega \geq 50$ meV the absorption essentially less than the experimental value of $\sigma_{ab}(\omega)$ and the fitting in the high frequency range (at $\lambda \propto 2$ and $\hbar\omega_p^* \simeq 2.15$ eV) gives an essentially larger value of $\sigma_{ab}(\omega)$ in the low frequency domain. From this analysis it was concluded that the two components of the optical absorption $\sigma_1(\omega)$ have a different nature. The low-frequency, Drude part of the absorption may be related to a relaxation of coherent, translational motion of quasi-particles with the effective mass $m^*/m = \delta/N_{eff} \simeq 2 - 3$. Absorption in the mid-infrared region is caused by noncoherent phenomena of charge carrier excitations, which become especially noticeable with a decrease of temperature.

A different interpretation of the nature of the second absorption component in the frequency range $\hbar\omega \simeq 250$ meV has been proposed [5.71, 72]. Measurements on untwinned crystals of YBCO have independently revealed the conductivities $\sigma_a(\omega)$ and $\sigma_b(\omega)$ for polarization E parallel to the a and b axes, respectively. Figure 5.23 shows these results for crystals with different values of T_c (with different oxygen content) [5.72]. The conductivity along the a-axis does not have the characteristic increase in the frequency range $\hbar\omega \geq 50$ meV, which could be related therefore to the absorption due to in-chain charge carriers along the b-axis. The latter is defined by the difference $\sigma_{1b} - \sigma_{1a}$ shown in the insert in Fig. 5.23. There is a stronger dependence of the conductivity on doping (or on the value of T_c) for the difference $\sigma_{1b} - \sigma_{1a}$ compared to the in-plane conductivity σ_{1a}. This indicates a little change of the number of in-plane carriers under doping (from 0.25 per cell of CuO_2 at $T_c = 93$ K, down to 0.15 at $T_c = 56$ K [5.72]). Since the in-plane conductivity $\sigma_{1a}(\omega)$ does not demonstrate a noticeable structure in the mid-infrared region, $\hbar\omega \simeq 0.25$ eV, its frequency dependence has been described on the basis of the one-component Drude model in the generalized form (5.25). Figure 5.24 shows the effective mass $m^*(\omega)$ and the relaxation rate $1/\tau^*(\omega)$ for $\sigma_{1a}(\omega)$ shown in Fig. 5.23b. The observed, almost linear, ω-dependence of $1/\tau^*(\omega)$ is not characteristic of the usual Fermi liquid in normal metals where $1/\tau \propto \omega^2$.

Optical conductivity has been computed on the basis of the theory of strong electron–phonon coupling accounting for the retardation effects of this interaction [5.67]. Good agreement has been obtained for measured reflection coefficients [5.71], and the effective rate of relaxation $1/\tau^*(\omega)$ (5.27) in the generalized

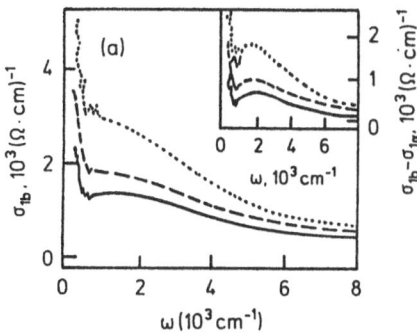
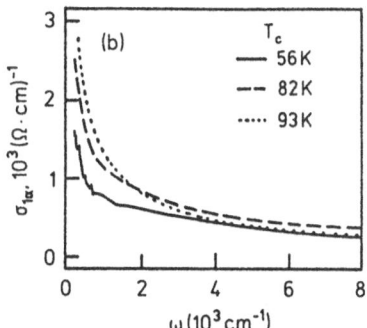

Fig. 5.23. Optical conductivity of untwinned crystals YBCO with T_c = 56, 82 and 93 K for the polarization $E \parallel b$, $\sigma_{1b}(\omega)$ (a) and $E \parallel a$, $\sigma_{1a}(\omega)$ (b), as well as their difference $\sigma_{1b}(\omega) - \sigma_{1a}(\omega)$ (*insert*) [5.72]

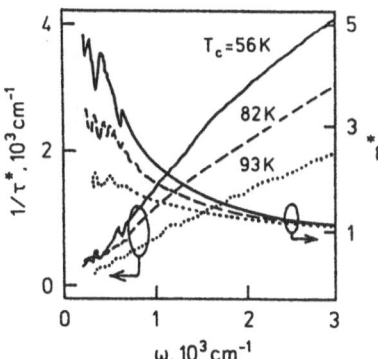

Fig. 5.24. Frequency dependence of the effective mass $m^*(\omega)$ and the relaxation rate $1/\tau^*(\omega)$ for the conductivity $\sigma_{1a}(\omega)$ shown in Fig. 5.23 (b) [5.72]

Drude model has demonstrated an almost linear frequency dependence in a broad frequency range. Thus doubts are raised about the existence of the additional channel of relaxation in the mid-infrared region for charge carriers in CuO_2 planes in YBCO crystals. At the same time, in the LSCO [5.68] and Bi-Sr [5.70] compounds a rapid growth of absorption $\sigma_1(\omega)$ at a small density of carriers in the mid-infrared region is observed. A model of the mid-gap band which appears under doping (see Fig. 5.2), may explain this additional "interband" absorption appearing near the almost filled O $2p$ band in the mid-infrared region.

Studies of infrared absorption in $BaBiO_3$ crystals under doping by K and Pb demonstrate a rather rapid transition from insulator to metallic state, together with the usual Drude peak. The anomalous behavior of $\sigma(\omega)$ in the mid-infrared region in copper-oxide superconductor is probably related to the existence there of antiferromagnetic fluctuations of copper spins (Chap. 3), absent in compounds on the basis of $BaBiO_3$. Additional studies should reveal the mechanism of infrared absorption in copper-oxide superconductors.

In infrared reflection experiments on the superconducting transition in copper-oxide compounds, with formation of the superconducting gap $\Delta(T)$ in the spectrum of quasi-particles, the infrared conductivity $\sigma_1(\omega)$ should decrease in the frequency range $\hbar\omega < 2\Delta(T)$ – the binding energy of a Cooper pair. At $T = 0$ the whole intensity $\sigma_1(\omega)$ at $\hbar\omega < 2\Delta(0)$ should collapse into a narrow peak at zero frequency:

$$\sigma_s(\omega) = \frac{\pi n_s e^2}{m}\delta(\omega)\,, \tag{5.30}$$

where n_s is the density of superconducting electrons. Note that the density, i.e. the ratio n_s/m, also determines the penetration depth of magnetic field in the London limit λ (4.10). Since for copper-oxide superconductors the relation $2\Delta \gg \hbar/\tau$ ("clean", or London, limit) holds, the remaining "normal" absorption component σ_1 ($\hbar\omega > 2\Delta$), according to the simple Drude formula (5.24), should constitute only a small portion of the superconducting contribution (5.30). Therefore a discovery of a classical threshold dependence of $\sigma_1(\omega)$ at $\hbar\omega = 2\Delta$ during transition to the superconducting state in copper-oxide superconductors poses some difficulties. The existence of additional absorption in the mid-infrared region which overlaps the energy range 2Δ makes an observation of the superconducting transition difficult.

As an example, Fig. 5.22b shows the experimental data for the superconductor YBCO with $T_c = 80$ K. With decrease of temperature in the spectrum $\sigma_1(\omega)$ there is an obvious fall in the energy range $\hbar\omega \simeq 54$ meV, and a threshold in absorption at 20 meV. However, the singularities in the range of 54 meV do not shift with respect to variations of T_c and they exist above T_c. This does not agree with the behavior of the superconducting gap in the BCS theory. A threshold in absorption at a lower energy $\simeq 20$ meV, although dependent on T_c, is small compared to the BCS theory for samples with high-T_c. Due to lack of precision in the calculation of $\sigma_1(\omega)$ by the reflection coefficient in the low frequency range, $\hbar\omega < 50$ meV, and also anomalous behavior of the abovementioned singularities, an unambiguous conclusion regarding the superconducting gap is not possible.

More definite conclusions regarding the formation of a superconducting gap in untwinned crystals of YBCO have been reached [5.71, 72, 66]. Figure 5.25 shows the dependence $\sigma_{1a}(\omega)$ at different temperatures for three superconductors YBCO with $T_c = 93$ K (a), 82 K (b), and 56 K (c). The dotted line shows the dependence at temperatures close to T_c ($T = 90$ (a), 80 (b) and 60 K (c)). With $T_c = 93$ K a rapid fall of absorption occurs under transition to superconducting states for frequencies $\omega < 500$ cm^{-1}. At the same time, 90 % of absorption at the normal state for these frequencies disappears in the superconducting phase. This indicates a superconducting gap in the CuO$_2$ plane in YBCO of value $2\Delta/kT \simeq 8$. For samples with a smaller oxygen content (lower T_c) the temperature dependence $\sigma_{1a}(\omega)$ is less sharp, and the fall at 500 cm^{-1} is also observed at $T < T_c$. Comparing the temperature dependence of σ_{1a} ($\omega = 500$ cm^{-1}) for samples with $T_c \simeq 60$ K with the spin-lattice relaxation rate both for ^{17}O and ^{63}Cu (see Sect. 3.2.2) reveals the proportionality $\sigma_{1a}(\omega) \propto 1/T_1T$ [5.72]. This suggests an interrelation of the formation of a gap in the spectrum of spin antiferromagnetic fluctuations, and the

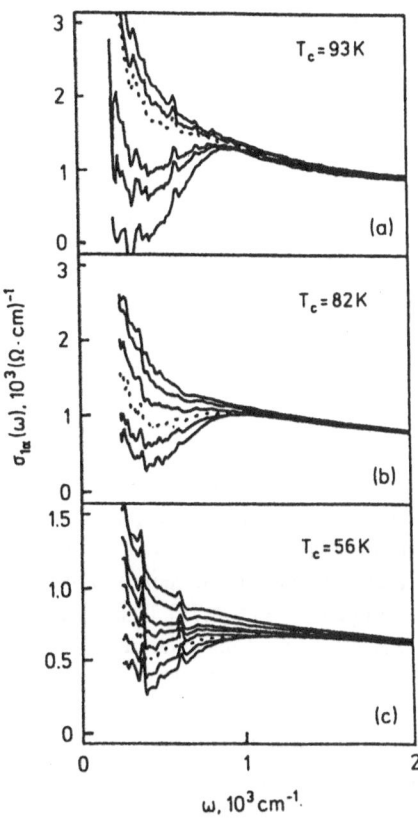

Fig. 5.25. The in-plane conductivity $\sigma_{1a}(\omega)$ of YBCO for samples with $T_c = 93$ K (a), 82 K (b), and 56 K (c) under temperature decrease; the dashed line corresponds to $T \simeq T_c$

transition to the superconducting state under pairing of carriers in CuO_2 planes (see Fig. 3.6)[5].

Some interesting data regarding the behavior of $\sigma_{1a}(\omega)$ under transition to superconducting state have been obtained in the low frequency range, $\hbar\omega \ll 2\Delta, T$ [5.73]. The ratio of absorption in the superconducting $\sigma_{1s}(\omega)$ and normal $\sigma_{1n}(\omega)$ phases, respectively at $T \leq T_c$ has a peak which is reminiscent of the Hebel-Slichter peak for the rate of spin-lattice relaxation in the conventional superconductors due to the growth of the density of states under a gap formation (see Sect. 3.2.2). More detailed comparative measurements of $\sigma_1(\omega)$ for single crystals of $Bi_2Sr_2CaCu_2O_6$ with $T_c = 82$ K and for a non-superconducting sample of $Bi_2Sr_2CuO_6$ have shown that this peak in the ratio $\sigma_{1s}(\omega)/\sigma_{1n}(\omega)$ is related to the decrease of the relaxation rate $1/\tau$ of the Drude part of conductivity under a transition to the superconducting state [5.73]. A rapid decrease of the relaxation rate of charge fluctuations $1/\tau$ in $\sigma_1(\omega \to 0)$ at $T < T_c$, and of the spin-lattice relaxation rate for spin fluctuations, again indicate a strong interaction of these subsystems in copper-oxide superconductors.

---

[5] Measurements [5.114] of the infrared reflectance of Ni-doped $YBa_2Cu_3O_{7-y}$ films show very gapless spectra that leads the authors to suggest a non-BCS type superconductivity.

As well as the infrared absorption in copper-oxide superconductors, the Raman scattering of light has also been studied [5.74]. Its intensity is proportional to the value of charge fluctuations in the system, and it contains both the lattice component related to phonons, and the electronic component related to free charge carriers. Raman scattering on phonons will be considered in Chap. 6, and the electronic scattering below. Since its intensity is defined by Im $[1/\varepsilon(\omega)]$, according to (5.19) the Raman scattering, as well as infrared absorption, is a probe of the frequency dependent conductivity $\sigma_1(\omega)$. However, unlike infrared absorption, the Raman scattering should disappear at $\omega \to 0$ due to complete screening of long-wave charge fluctuations. In the relaxation time approximation (5.24) the intensity of Raman scattering is described by the relation [5.74]:

$$I_s(\omega) \propto \frac{\omega\tau}{1+\omega^2\tau^2}[1+n(\omega)], \qquad (5.31)$$

where $n(\omega)$ is the Bose-Einstein distribution and ω is the frequency shift of the Stokes component. Experimental studies of Raman scattering in copper-oxide superconductors have found an essential electronic contribution to the scattering in a broad frequency range [5.74]. The most interesting data have been obtained for untwinned crystals of YBCO with different oxygen content [5.75]. The (Z,Z) intensity of Raman scattering (the incident and scattered photons polarized along the z-axis of the crystal) were described by a formula of the type (5.31) with an almost frequency-independent relaxation time (for $\omega > 500\,\mathrm{cm}^{-1}$). At the same time, the Raman continuum disappeared in insulator samples, which confirms its electronic nature. The Raman scattering for (X,X) or (Y,Y) polarization is of a quite different nature. Its intensity is frequency independent outside the phonon frequency range, and little changes is seen under doping with change of oxygen content. This indicates a different nature of Raman scattering of light for (Z,Z) polarization and in the planes: (X,X), (Y,Y) [5.75]. In insulator samples for (X,X) polarization an intensive two-magnon peak is seen, which disappears under transition to metallic state with an increase of oxygen content. The lack of frequency dependence of the intensity (5.31) may be explained if a linear dependence of relaxation rate on frequency, $1/\tau \propto \omega$, is assumed which is observed in infrared absorption (see Fig. 5.24). However, strong Raman scattering for (X,X) polarization both in the insulating and the conducting phases does not connect this scattering to the free carriers contribution. Such a strong difference of scattering for in-plane, (X,X) and (Y,Y), and (Z,Z) polarizations may be related to an essential contribution to Raman scattering of antiferromagnetic spin fluctuations in CuO_2 planes. They are directly observed in the form of a two-magnon peak in insulator samples of YBCO, and they transform to diffuse scattering in a broad frequency range in the metallic phase, supplementing the scattering on the electronic charge density fluctuations. A similar picture of a strong coupling of electronic and spin degrees of freedom of charge carriers in CuO_2 planes is confirmed both in magnetic scattering (see Sects. 3.1.3, 3.2.2, 3.3.2) and infrared absorption experiments.

Studies of electronic Raman scattering under a transition to superconducting state indicate a gap in the charge density fluctuations spectrum. For example, under transition to the superconducting state in single crystals of $YBa_2Cu_4O_8$, two gaps have been found: a smaller one for the (Y, Y) polarization, and a larger one for the (X, X) polarization, perpendicular to the chains [5.74]. The average value of the gaps proved close to the universal value $2\Delta/kT_c \simeq 3.5$ of the BCS theory. Further studies are required to clarify the nature of the electronic Raman scattering and to explain the anisotropy of the gap in the spectrum of charge density (and, probably, also spin) fluctuations in the CuO_2 planes in copper-oxide superconductors [6].

5.5 Transport Properties

Much attention has been given to the studies of transport properties of copper-oxide compounds, i.e. conductivity, Hall effect, thermopower and heat conductivity, because these studies reveal the nature of electronic excitations near the Fermi surface. Experimental data regarding the transport properties of high-temperature superconductors are given in reviews [5.76, 77]. A theoretical analysis of the transport properties of LSCO and YBCO compounds on the basis of electronic band-structure calculations is given in [5.27, 78] and the results obtained in this field, including the latest studies, are discussed below.

The transport coefficients should be defined. Under the effect of external electric E_β and magnetic B_γ fields and in the presence of a temperature gradient $\nabla_\beta T$, the current in the sample in the linear approximation with respect to the external fields is given by

$$j_\alpha = \sigma_{\alpha\beta} E_\beta + \sigma_{\alpha\beta\gamma} E_\beta B_\gamma + V_{\alpha\beta} \nabla_\beta T. \tag{5.32}$$

In the isotropic relaxation time τ approximation, the electroconductivity tensor may be written in the form [5.78]

$$\sigma_{\alpha\beta} = \frac{\tau}{4\pi} (\Omega_p^2)_{\alpha\beta}, \tag{5.33}$$

where the tensor of plasma frequencies reads

$$(\Omega_p^2)_{\alpha\beta} = \frac{4\pi e^2}{v_0} \sum_k v_{k\alpha} v_{k\beta} \delta(\varepsilon_k) \equiv 4\pi e^2 2N(0) < v_\alpha v_\beta >. \tag{5.34}$$

Here v_0 is the volume of the primitive cell, $2N(0)$ is the density of electronic states (per atom) on the Fermi surface, and $v_{k\alpha} \equiv v_{n\alpha}(k) = \partial \varepsilon_n(k)/\partial(\hbar k_\alpha)$ is the group velocity for the quasi-particles with spectrum $\varepsilon_n(k)$; n is the band index. Similar but more complicated expressions define the tensors $\sigma_{\alpha\beta\gamma}$ and $V_{\alpha\beta}$ [5.78]. In this approximation, the anisotropy of transport coefficients is determined by the anisotropy of the Fermi surface.

---

[6] An anisotropic gap in electronic Raman scattering was observed in Bi- and Tl-based compounds [5.119]. A theoretical interpretation of electronic Raman scattering pointing to $d(x^2 - y^2)$ pairing is given by Devereaux et al. [5.120].

For the orthorhombic (tetragonal) structure with the coordinate axes x, y, z chosen along the crystallographic axes a, b, c, the tensors σ and V are diagonal with three (two) independent elements $\sigma_{\alpha\alpha}$ and $V_{\alpha\alpha}$ ($\alpha = x, y, z$). The third-rank tensor $\sigma_{\alpha\beta\gamma}$ vanishes unless all three indices are distinct, and it satisfies Onsager relation: $\sigma_{xyz} = -\sigma_{yxz}$, etc. This leads to three (two) independent Hall coefficients for the orthorhombic (tetragonal) symmetry:

$$R_{xyz}^H = \frac{\sigma_{xyz}}{\sigma_{xx}\sigma_{yy}},$$

(5.35)

and analogous expressions for R_{yzx}, R_{xzy}. The thermoelectric power tensor, and Seebeck coefficient $S_{\alpha\beta}$, are defined under the condition $j_\alpha = 0$, when an electric field arises due to a temperature gradient:

$$E_\alpha = S_{\alpha\beta}\nabla_\beta T = -(\sigma^{-1})_{\alpha\gamma}V_{\gamma\beta}\nabla_\beta T.$$

(5.36)

In orthorhombic lattices, the tensor S_{aa} is diagonal, as well as $\sigma_{\alpha\alpha}$ and $V_{\beta\beta}$.

Resistivity $\varrho_{\alpha\beta}$, rather than conductivity is usually measured

$$E_\alpha = \varrho_{\alpha\beta}j_\beta,$$

(5.37)

known as voltage drop E_α at the given current j_β. In particular, for a magnetic field B_z and an external electric field E_x

$$\varrho_{xx} = \sigma_{xx}^{-1}, \quad \varrho_{yx} = \sigma_{xyz}B_z/\sigma_{xx}\sigma_{yy},$$

(5.38)

(in the approximation linear in B). Thus, the Hall coefficient (5.35) is determined by the relation

$$R_{xyz} \equiv R_H = \frac{E_y}{B_z j_x} = \frac{1}{B_z}\varrho_{yx}.$$

(5.39)

Besides the Hall coefficient, the Hall angle θ_H is defined by

$$\tan\theta_H = \frac{E_y}{E_x} = \frac{\varrho_{yx}}{\varrho_{xx}} = \frac{\sigma_{xyz}B_z}{\sigma_{yy}}.$$

(5.40)

Assuming that the resistance ϱ_{xx} along the external field E_{xx} is determined by the transport relaxation time τ_{tr} for quasi-particles with the effective mass m^*, while the transverse component ϱ_{yx} is due to the motion of quasip-articles caused by the Lorentz force $[v_k \times B]_y$ with the effective mass m_H and the relaxation time τ_H, the following relations apply:

$$\varrho_{xx} = \frac{m^*}{ne^2}\frac{1}{\tau_{tr}}, \quad \tan\theta_H = \frac{\varrho_{yx}}{\varrho_{xx}} = \omega_c\tau_H,$$

(5.41)

where $\omega_c = eB_z/cm_H$ is the cyclotron frequency. If effective masses and relaxation times of the longitudinal and transverse motions are equal then the Hall coefficient reads

$$R_H = \frac{1}{ne}.$$

(5.42)

In this one-band approximation $n_H = 1/R_H e$ determines the density of free carriers, holes for $R_H > 0$ and electrons for $R_H < 0$.

5.5.1 Resistivity

A typical feature of copper-oxide compounds is the strong anisotropy of their transport properties. Therefore representative results can be obtained only on high quality single crystal samples. Figure 5.26 shows the measured in-plane resistivity ϱ_{ab} and resistivity perpendicular to the plane ϱ_c for several single crystals of LSCO (a) and YBCO (b) [5.79]. The anisotropy of the resistivity at a room temperature for samples with maximum T_c ($x = 0.15$ in LSCO and $y \simeq 0.07$ in YBCO) attains the value $\varrho_c/\varrho_{ab} \simeq 200(\text{LSCO}) - 30(\text{YBCO})$. With reduction of temperature and a decrease of hole concentration in CuO_2 planes the anisotropy grows due to a more rapid growth of ϱ_c. Another peculiarity of the resistivity is its linear temperature dependence over a wide range, first observed in experiments with ceramic samples [5.80]

$$\varrho_{ab}(T) = \alpha T + \beta. \tag{5.43}$$

A linear dependence for $\varrho_c(T)$ is observed only for high quality samples at optimum doping. With a decrease of hole concentration $\varrho_c(T)$ takes the semiconducting nature, $d\varrho_c/dT < 0$. In overdoped samples of LSCO ($x \simeq 0.3$) with a large concentration of holes a characteristic metallic conductivity $\varrho \propto T^2$ for ϱ_{ab} and ϱ_c is seen. Such a strong anisotropy and the change of type of dependence $\varrho_c(T)$ for a decrease of carrier concentration argues in favor of a different nature of ϱ_{ab} and ϱ_c [5.79].

Some interesting results regarding the high quality untwinned samples of YBCO have been obtained [5.81]. The temperature dependence of the resistivity ϱ_α along the three axes ($\alpha = a,b,c$) of an orthorhombic crystal is given in Fig. 5.27. Linear temperature dependence (5.43) exists along all three axes, the in-plane anisotropy at room temperature being $\varrho_a/\varrho_b \simeq 2$ and $\varrho_c/\varrho_a \simeq 35$, $\varrho_c/\varrho_b \simeq 75$. The anisotropy of in-plane resistivity indicates a large contribution to conductivity of the Cu – O chains along the b-axis in YBCO. The estimate of this anisotropy as $\sigma_b - \sigma_a/\sigma_b \simeq 0.6$ coincides with the measured anisotropy of the conductivity in the infrared frequency range [5.71,72] (see Fig. 5.23). It is interesting to note that for high quality samples the "residual" resistivity β in (5.43) obtained by extrapolating $\varrho(T)$ to the domain $T < T_c$ turns out to be negative: $\beta_a \simeq -14$ ($\beta_b \simeq -3$) $\mu\Omega\cdot\text{cm}$. This assumes the necessity to change the linear asymptotic (5.43) to a stronger one of the type $\varrho \propto T^2$ at $T \to 0$ for the normal phase, as is observed for non-superconducting samples of LSCO at $x = 0.3$ (see Fig. 5.26).

Comparing these experimental data with the band-structure calculations [5.78, 27] in which it was supposed that the main contribution to transport current relaxation comes from the scattering of electrons by phonons, the transport relaxation time at high temperatures is estimated by

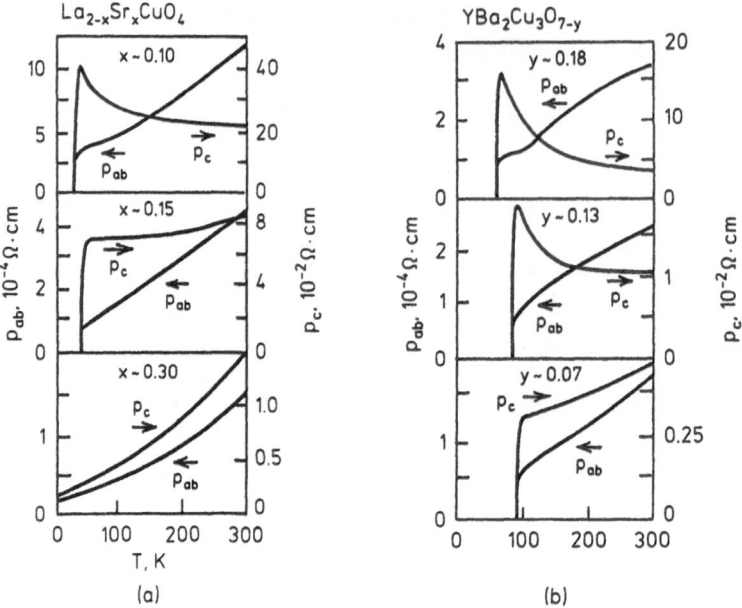

Fig. 5.26. Temperature dependence of the in-plane ϱ_{ab} and perpendicular to the plane ϱ_c resistivity for single crystals of LSCO (**a**) and YBCO (**b**) [5.79]

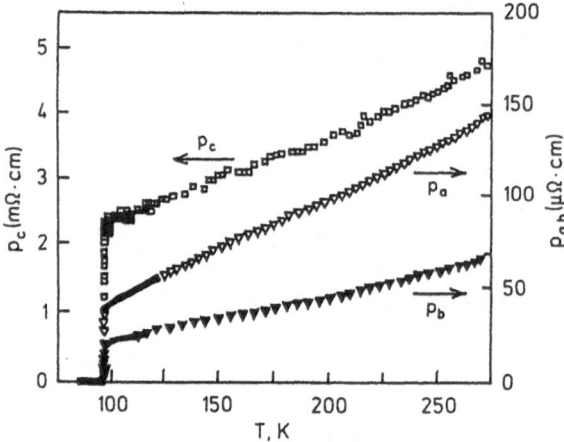

Fig. 5.27. Temperature dependence of the resistivity ϱ_α along the three axes of orthorhombic crystals of YBCO [5.81]

$$\frac{\hbar}{\tau_{tr}} = 2\pi\lambda_{tr}kT\left(1 - \frac{\hbar^2\langle\omega^2\rangle}{12(kT)^2} + \cdots\right). \qquad (5.44)$$

The electron–phonon coupling constant λ_{tr} is determined by the function $\alpha_{tr}^2 F(\omega)$ with the matrix element of electron–phonon interaction α_{tr}, which determines the relaxation of quasi-particle momentum (see (5.29)). Using the experimental phonon density of states $F(\omega)$ for LSCO and YBCO (Chap. 6), a linear dependence of the resistivity (5.43) was obtained in a broad temperature range $T > T_c$. However, the absolute values of the resistivity, or the slope $d\varrho/dT \propto \lambda_{tr}\Omega_p^2$, proves to be several times less than the experimental values. So, for LSCO at Sr concentration $x = 0.15$ [5.78] it was found that $\lambda = 0.65$, $\hbar\Omega_{pxx} = \hbar\Omega_{pyy} = 2.9$ eV and $\varrho_{xx} = \varrho_{yy} = 80\,\mu\Omega\cdot$cm, $\varrho_{zz}/\varrho_{xx} \simeq 28$ at $T = 295$ K. For YBa$_2$Cu$_3$O$_7$ it has been found that $\lambda = 0.32$, $\hbar\Omega_{pxx}(\hbar\Omega_{pyy}) = 2.9(4.4)$ eV and $\varrho_{xx}(\varrho_{yy}) \simeq 37(16)\mu\Omega\cdot$cm, $\varrho_{zz}/\varrho_{yy} \simeq 16$ at $T = 295$ K. Comparing these data with those shown in Figs. 5.26 and 5.27, we find that the theoretical results [5.78] for the in-plane resistivity are approximately five times less than the experimental ones, at a 5–8-fold lower anisotropy ϱ_c/ϱ_{ab}. As has been noted [5.78], this discrepancy between theory and experiment could be improved if using: 1) larger values of λ_{tr}, 2) smaller Ω_p and 3) additional scattering mechanism. However, an essential increase of λ_{tr} contradicts large free path lengths $l = v_F\tau_{tr} \gg a$ up to $T \simeq 1000$ K in LSCO [5.80]. The relaxation time in YBCO $\hbar/\tau_{tr} \simeq 2kT$ as observed in infrared reflection experiments also leads to the estimate $\lambda_{tr} \sim 0.3$ (see Sect. 5.4.3). Accounting for the experimental values [5.82] of plasma frequencies $\hbar\Omega_p = 0.8$ eV (LSCO) and 1.4 eV (YBCO), a two to three times reduction for theoretical plasma frequencies is acceptable. However, to obtain complete agreement with experiment, λ_{tr} should be increased up to $1.5 - 2$ [5.83].

It seems more probable that some other scattering mechanism of quasi-particles exists. In particular, the existence of strong antiferromagnetic spin fluctuations provides a sufficiently intensive scattering of the carriers. Calculations of the resistivity aroused by spin fluctuations [5.84] yield results close to those observed in experiments. At the same time, in the temperature range $T > T_s = \hbar\omega_s/k$ there is a linear dependence $\varrho_s(T)\propto T$, and at $T \ll T_s$ a quadratic function $\varrho_s(T)\propto T^2$ appears. The characteristic energy of spin fluctuations $\hbar\omega_s$ for low concentrations of carriers is small, and therefore a linear temperature dependence is observed in a broad temperature range. In overdoped samples with high concentrations of carriers, $\hbar\omega_s$ takes the values of the electronic energy scale, and a large domain with a quadratic dependence of $\varrho(T)$ appears. An important feature of the theoretical calculations [5.84] is that they consistently account for strong Coulomb correlation on copper sites in the frame of the effective Hamiltonian (5.6). The non-Fermionic nature of the hole creation and annihilation operators for the subband of the singly occupied states on copper sites leads to the appearance of an inelastic spin scattering of copper holes. This spin scattering, which is described by dynamic spin susceptibility (3.34), gives at least half of the total value of the resistivity[7].

The aforementioned anisotropy of the resistivity and its linear temperature dependence, typical of copper-oxide compounds, also appears in Bi- and Tl-based compounds. An especially high anisotropy of order $\varrho_c/\varrho_{ab} \simeq 10^5$ is observed in Bi$_2$Sr$_2$CaCu$_2$O$_8$ [5.85]. A high degree of linearity of temperature dependence of resistivity is typical of Bi compounds. For Tl-based compounds the

---

[7] Evidence for dominant contribution of spin scattering for charge carriers in YBCO compound has been found in [5.115].

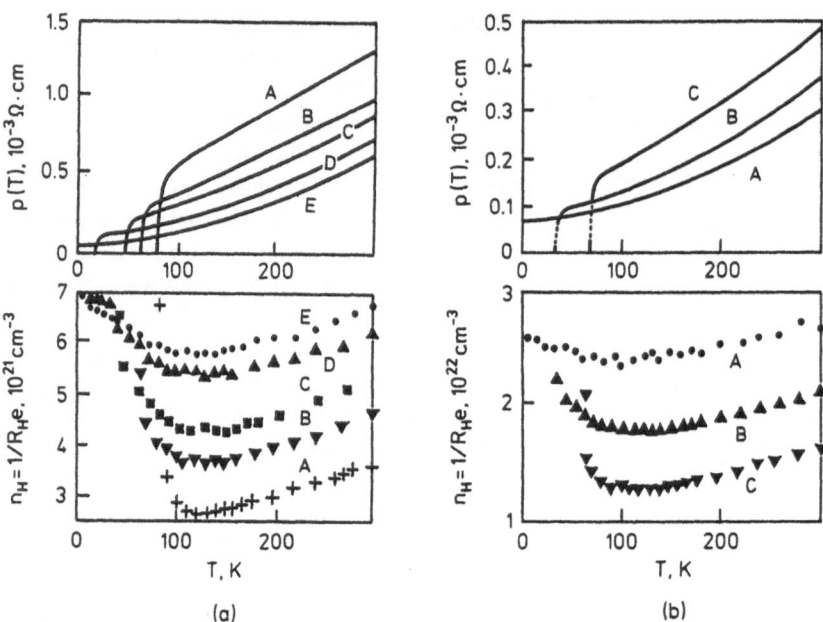

Fig. 5.28. Temperature dependence of the resistivity and the inverse Hall coefficient $n_H = 1/R_H e$ for Tl–2201 (**a**) and Tl–1212 (**b**) [5.87]

value of anisotropy reaches $\varrho_c/\varrho_{ab} \simeq 6 \cdot 10^2$ at room temperature even for non-superconducting samples of Tl$_2$Ba$_2$CuO$_{6+\delta}$ [5.86].

Since Tl-based compounds allow for a gradual increase of the concentration of holes until superconductivity disappears (see Fig. 5.3b), it is of some interest to study this transition from superconducting to normal metals. Figure 5.28 gives the resistivity $\varrho(T)$ and the inverse Hall coefficient $n_H = 1/R_H e$ for samples of Tl$_2$Ba$_2$CuO$_{6+\delta}$ (Tl–2201) (a) and TlBa$_2$CaCu$_2$O$_{7-\delta}$ (Tl–1212) (b) [5.87] at different concentrations of holes. A smooth change of temperature dependence of the resistivity, from a linear to quadratic one, is observed under a gradual increase of n_H and a decrease of T_c until a transition to the superconducting state disappears, $T_c = 0$. It has been noted [5.87] that such a smooth variation of the properties of Tl compounds with a growth of the concentration of carriers (holes) leads to the conclusion that the superconducting and normal properties of these compounds should be described by the same microscopic model.

5.5.2 Hall Effect

The measurements of the Hall coefficient (5.39) in copper oxide compounds have revealed several unusual properties of these compounds in the normal phase [5.77]. Studies of the Hall coefficient as a function of doping according to the simple one-band model (5.42) allow the concentration of free carriers $n_H = 1/R_H e$ and their type: $R_H > 0$ for holes and $R_H < 0$ for electrons. Fig. 5.29 shows the dependence

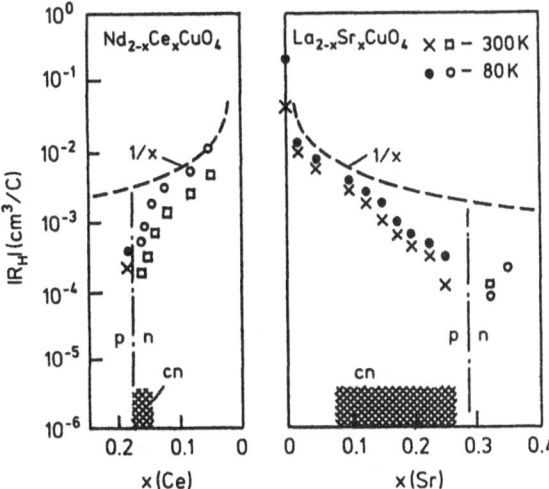

Fig. 5.29. The Hall coefficient $|R_H|$ as a function of the Ce concentration in $Nd_{2-x}Ce_xCuO_4$, and that of Sr in $La_{2-x}Sr_xCuO_4$ [5.88]

of R_H on the concentration of Sr in LSCO and Ce in NdCeCuO compounds [5.88]. At small Sr concentrations $R_H > 0$, but at $x = 0.3$ it changes the sign, $R_H < 0$, which indicates the change of the type of carrier. A similar change of the R_H sign from $R_H < 0$ to $R_H > 0$ occurs in NdCeCuO as the concentration of Ce grows. It is important to note that the sign inversion of R_H takes place in both compounds on the boundary of the disappearance of superconductivity at a redundant concentration of carriers – holes in LSCO and electrons in NdCeCuO. The absolute value $|R_H| \propto 1/n_H$ obeys the $1/x$ dependence only at small impurity concentrations, when $n_H \propto x$. In YBCO compounds a rapid fall of $n_H = 1/R_H e$ during the transition to insulating phase at the concentration of oxygen $x < 6.4$. A smoother dependence of n_H on the concentration of carriers is observed in Bi and Tl compounds, which implies a sufficiently simple relation: $n_H \propto p$, where p is the concentration of carriers in CuO_2 planes [5.77].

When comparing the band calculations of Hall coefficient for LSCO and YBCO compounds [5.27, 78] with experimental data [5.77] both the absolute value and the sign of $R_{xyz} > 0$, $R_{yzx} < 0$, $R_{zxy} < 0$ do not contradict the data obtained for single crystal samples. The band-structure calculations also predict the change of sign of R_{xyz} in LSCO as the concentration of Sr increases. It has been noticed [5.78] that these results confirm the picture of a Fermi liquid with a definite Fermi surface in the ground state of oxide superconductors. As already mentioned in Sect. 5.3, the band calculations do not describe the insulating phase of copper-oxide compounds, caused by strong Coulomb correlations on the copper sites. In particular, the dependence $R_H \propto (1/x)$ at $x \to 0$ cannot be obtained in band calculations.

The Fermi liquid picture encounters the greatest difficulties when used to try to explain the anomalous temperature dependence of the Hall coefficient. In the copper-oxide superconductors the dependence of the type $R_H \propto 1/(T+T_0)$ in the metallic conductivity domain [5.77] is seen. Attempts to explain this temperature dependence by a multiband structure of copper-oxide compounds (when the simple model (5.42) fails) encounter difficulties. A similar linear temperature dependence $n_H \propto T$ for compounds with quite different topology of Fermi surface, and a small sensitivity of this dependence to the concentration of carriers, does not explain the temperature dependence of R_H as a specific compensation of the contributions from the different types of carriers in a multiband model.

Important results have been obtained [5.89] which clarify the temperature dependence of the Hall coefficient R_{xyz} for polycrystals of YBa$_2$Cu$_{3-x}$Zn$_x$O$_{7-\delta}$ with different concentrations of Zn. As noted in Sect. 5.2, the Zn impurities replace Cu ions in CuO$_2$ planes and lead to a strong suppression of T_c (see Fig. 5.4). Figure 5.30 gives the temperature dependence of the in-plane resistivity and the inverse Hall coefficient $1/R_H$ (5.39) at different concentrations of Zn impurities x [5.89]. As x increases the residual resistivity grows too, but the temperature dependence $\varrho_{xx} \propto T$ remains almost the same. At the same time, $n_H = 1/R_H e$ at low temperature does not change much while the temperature dependence $n_H \propto T$ is suppressed as the concentration of impurities grows. Since the Hall coefficient (5.39) is determined by the transverse component of the resistivity ϱ_{xy} which is related to the transverse relaxation time τ_H, and the longitudial resistivity ϱ_{xx} is determined by the transport relaxation time τ_{tr}, such a different behavior of these components may be explained by different scattering mechanisms for longitudinal and transverse (Hall) currents. The measured Hall angle which, according to (5.41), defines the transverse relaxation time τ_H, obeys the simple law [5.89]:

$$\cot \theta_H = \frac{1}{\omega_c \tau_H} = aT^2 + bx \,. \tag{5.45}$$

By this formula, the contributions from elastic scattering on impurities ($\propto x$) and nonelastic scattering ($\propto T^2$) sum to give the rate of transverse relaxation. Thus these experiments seem to confirm the existence of two relaxation times

$$\frac{\hbar}{\tau_{tr}} \simeq 2kT \,, \quad \frac{\hbar}{\tau_H} \simeq kT \left(\frac{kT}{W} \right) \,, \tag{5.46}$$

where the effective band for quasi-particles is $W \simeq 800$ K [5.89]. This difference in relaxation times explains the temperature dependence of the Hall coefficient, which, according to (5.39), (5.43) and (5.45) is equal to

$$\frac{1}{R_H} = \frac{B_z}{\varrho_{xx}} \cot \theta_H = B_z \frac{bx + aT^2}{\alpha + \beta T} \,. \tag{5.47}$$

In the high temperature and weak impurities scattering region $n_H = 1/R_H e \propto T$, and for T$\to$ 0 and a high impurity concentration the temperature dependence of n_H is suppressed.

According to the *Anderson* theory [5.90], such a strong difference of relaxation times (5.46) is explained by the fact that only the spinons participate in the Hall (transverse) current. Their interaction determines the quadratic temperature dependence of the transverse relaxation rate characteristic of Fermi particles with

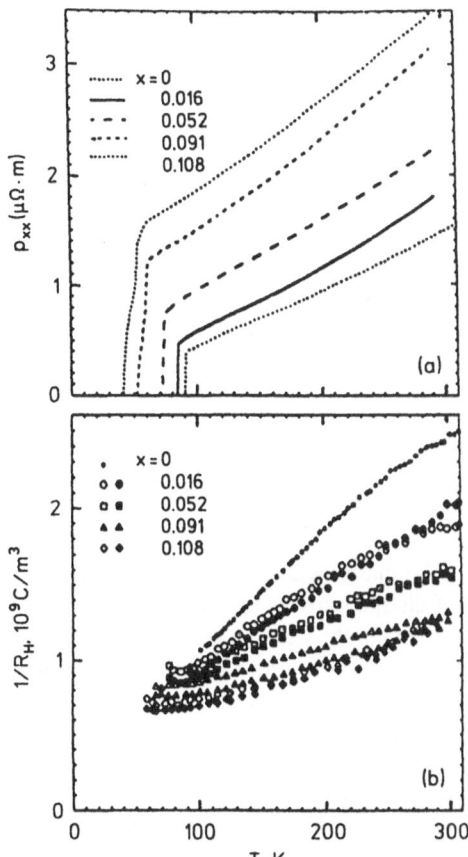

Fig. 5.30. Temperature dependence of the resistivity (**a**) and the inverse Hall coefficient $1/R_H$ (**b**) in YBCO crystals at different concentrations of Zn impurities x [5.89]

Fermi energy W. The longitudinal current is determined by the relaxation of spinless holes (holons) which are not affected by a magnetic field. It is also possible that the simple isotropic relaxation time approximation used to compute the longitudinal and transverse resistivities in (5.41) fails for a system of quasi-particles. Corrections to the Boltzmann equation may play an important role and qualitatively change the temperature dependence of the transverse resistivity [5.91]. An explanation of the temperature dependence $\cot\theta_H \propto T^2$ on the basis of a multiband model for CuO_2 planes which accounts for strong Coulomb correlations has also been proposed [5.92].

Some rather peculiar properties have been discovered in the measurements of magnetoresistivity in non-superconducting single crystals Bi$-$2201 [5.89]. In the whole temperature range T> 20 K where $\varrho_a \propto T$ one observes an anomalous behavior of magnetoresistivity which proves to be isotropic and positive. In this case the scattering of holes on spin fluctuations plays an important role [5.89].

5.5.3 Thermopower and Heat Conductivity

Phenomena unusual for standard metals were also found in copper-oxide supercon-ductors during studies of thermopower and heat conductivity–transport properties related to heat transfer. Regarding results of thermopower studies [5.93], accord-ing to (5.36), the thermopower or Seebeck coefficient S is defined by the ratio of the voltage drop ΔV to a temperature difference ΔT on the sample boundaries, $S = \Delta V / \Delta T$. The voltage drop arises due to the change of diffusion rate along and opposite to the temperature gradient (diffusion contribution S_d), and also due to electron drag by a current of non-equilibrium phonons (phonon-drag contribution S_g). Introducing an electron energy ε dependent conductivity $\sigma(\varepsilon)$, the diffusion contribution is represented in the form

$$S_d = \frac{k}{e} \frac{1}{\sigma} \int\limits_{-\infty}^{\infty} d\varepsilon \left(-\frac{\partial f(\varepsilon)}{\partial \varepsilon} \right) \sigma(\varepsilon) \frac{\varepsilon}{kT} \equiv \frac{k}{e} \frac{\varepsilon_{av}}{kT}, \tag{5.48}$$

where $f(\varepsilon)$ is the Fermi distribution, at the energies ε taken with respect to the Fermi level μ_F. If the average energy of charge carriers ε_{av} is equal to the heat energy kT, then $S_d \simeq k/e = 87\mu V/K$. In metals, the average electron energy near the Fermi surface is determined by the parameter $\varepsilon_{av} \simeq (kT)^2/\mu_F$, and therefore $S_d \propto kT$ and much less than the standard value k/e. In systems having compli-cated electronic spectra the thermopower may have a nonlinear temperature de-pendence and an arbitrary sign. The phonon-drag contribution S_g has an essential nonlinear temperature dependence. Its value is determined by the electron–phonon interaction. In the superconducting phase, $S = 0$ since the voltage drop is zero, $\Delta V = 0$, but certain thermoelectric effects remain [5.93].

Thermopower in copper-oxide superconductors for low carrier concentrations has sufficiently large values, $S \simeq k/e$, but quickly decreases as the carrier con-centration grows. Figure 5.31 shows the measured thermopower in single crystals LSCO in the conducting plane (S_{ab}) and perpendicular to it (S_c) [5.93]. The co-efficient S measured on polycrystals of LBCO or LSCO behaves similarly to S_{ab}. S_{ab} has a strongly nonlinear temperature dependence with a typical maximum near $T \simeq 100$ K. The dependence $S_{ab}(T)$ for single crystals of YBCO has a more irregular nature given in Fig. 5.32.

In samples with small in-plane hole concentrations n_h, $S_{ab} > 0$ and is large. As oxygen content increases and n_h rises, S_{ab} decreases and becomes negative. Both in LSCO and YBCO compounds, the thermopower perpendicular to the plane behaves more regularly, $S_c \propto T$. Characteristic of the $S_{ab}(T)$ dependence in single crystals of Bi- and Tl-based compounds is a maximum in the range $T \simeq 100-150$ K, followed by an almost linear decreasing temperature dependence: $S_{ab} \propto -T$ [5.93, 94].

The existence of a strongly nonlinear temperature dependence and an irreg-ular nature of S_{ab} in YBCO crystals may be caused by several factors. One of them, which is observed in normal metals, is strong electron–phonon interaction

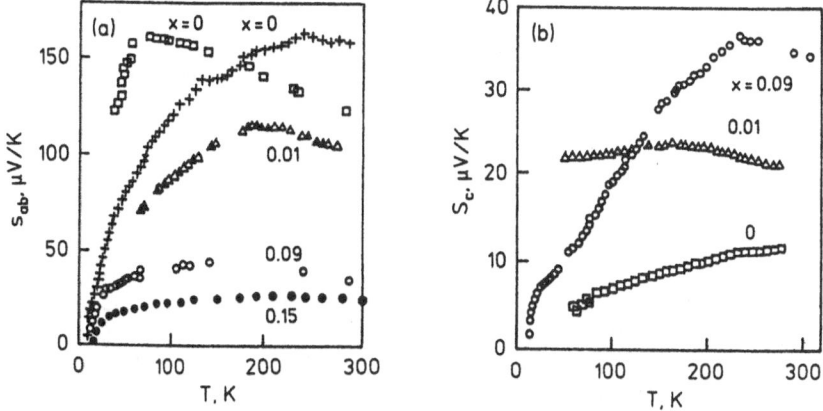

Fig. 5.31. Thermopower in the conducting plane (S_{ab}) and perpendicular to it (S_c) for single crystals of La$_{2-x}$Sr$_x$CuO$_4$ [5.93]

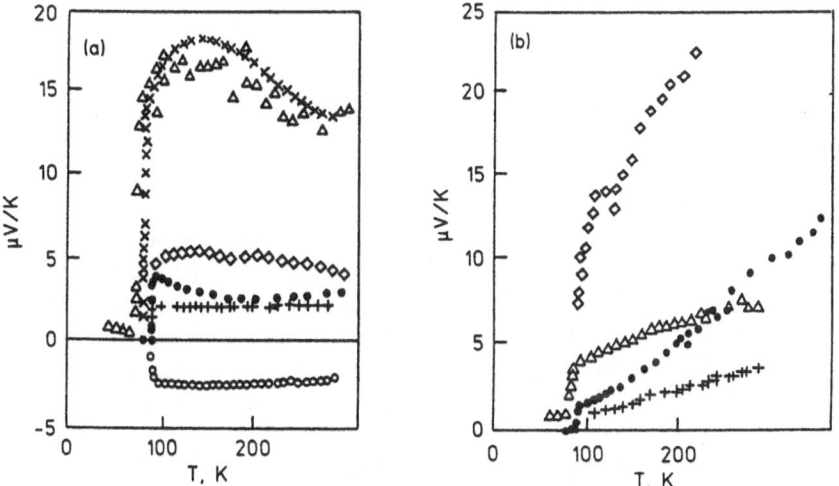

Fig. 5.32. Thermopower S_{ab} and S_c for single crystals of YBa$_2$Cu$_3$O$_{7-y}$ [5.93]

[5.93, 94]. Accounting for the effective mass renormalization of the carriers, the diffusion contribution (5.48) may be written as:

$$S_d = [1 + a\lambda(T)]AT,   \tag{5.49}$$

where $\lambda(T)$ is a temperature dependent effective electron–phonon coupling constant. As already discussed in Sect. 5.4.3, $\lambda(T)$ may be computed if the Eliashberg function $\alpha^2(\omega)F(\omega)$ is known (see (5.26) and (5.28)). $\lambda(0)$ determines mass renormalization at $T \rightarrow 0$, and at high temperatures $\lambda \ll 1$. This changes the slope of the temperature dependence of the diffusion contribution S_d (5.49). Estimates [5.94] demonstrate that in order to explain the temperature dependence $S_{ab}(T)$

in single crystals of YBCO on the basis of (5.49) a strong coupling should be assumed, in some cases $\lambda \simeq 5$.

Another reason for a strong nonlinearity of $S(T)$ may be phonon-drag effect, the contribution S_g. This mechanism was proposed [5.95] when the dependence $S_{ab}(T)$ was studied for several samples of YBa$_2$Cu$_3$O$_{7-\delta}$ with a high oxygen content. It was found that $S_{ab} < 0$ at $\delta = 0.02$ and 0.01, and $S_{ab} > 0$ at $\delta = 0.12$ and 0.15. The sign is determined by the sum of a positive contribution of in-plane holes [5.95], and a negative contribution of in-chain electrons. Thus, a decrease of in-chain oxygen content at an increase of δ leads to suppression of the electronic contribution, and the sign of S_{ab} changes from negative to positive. Computing $S_{ab}(T)$ in the frame of band theory [5.78] for YBCO and LSCO compounds yields $S_{ab} < 0$ and $S_c > 0$ with an almost linear temperature dependence. However, the absolute value of $S_\alpha(T)$ depends strongly on the choice of model for $\sigma(\varepsilon)$ in (5.48).

The most interesting result of studies [5.95] is the discovery of strong phonon drag effects. In high temperature ranges, $T > \theta_D/4 \simeq 150\,\mathrm{K}$, the experimental dependence $S_{ab}(T)$ is described by the formula

$$S_{ab}(T) = S_d + S_g = AT + B/T. \qquad (5.50)$$

The second term S_g which describes the phonon drag effects is much larger than the diffusion term S_d. With a decrease of temperature, $T < \theta^* \simeq 160\,\mathrm{K}$ a sharp decrease of $|S_{ab}(T)|$ is observed. This can be related to freezing out of the "umklapp" processes of electron–phonon scattering, $S_g^U \propto \exp(-\theta^*/T)$. Switching off these scattering processes for holes in CuO$_2$ planes produce a large positive contribution to $S_g(T)$, which has been observed for all four samples, independent of the sign of $S_{ab}(T)$. (For normal scattering processes, the direction of motion of the charge carriers dragged by phonons coincides with the direction of phonon diffusion, and for the "umklapp" scattering processes the direction is opposite to that of the phonon diffusion.) Thus according to [5.95] the anomalous behavior of thermopower $S_{ab}(T)$ for YBCO systems – the change of sign of S_{ab} and a strong nonlinear dependence with a maximum of $|S_{ab}(T)|$ at $T \simeq 100 - 150\,\mathrm{K}$ – is explained by the standard multiband theory of metals, accounting for strong phonon drag effects (for holes in the CuO$_2$ planes and electrons in the Cu1 – O1 chains) by nonequilibrium phonons. At the same time one should assume a sufficiently strong interaction of charge carriers with optical phonons in the frequency range $\hbar\omega \simeq 10 - 30\,\mathrm{meV}$.

The peak of $S_{ab}(T)$ at $T_c < T < 150 - 200\,\mathrm{K}$ in many copper-oxide superconductors obviously indicates a strong electron-phonon coupling which leads to the phonon drag effect and, probably, to a renormalization of mass and relaxation time of the free carriers. The standard band theory, which does not allow for the effects of strong electron–phonon interaction (or an interaction to some other bosonic excitations – spin fluctuations), cannot explain the strong nonlinear dependence $S_{ab}(T)$.

Definite confirmation of the important role of phonons in thermodiffusion in copper-oxide superconductors was obtained during measurements of heat conductivity. In all samples of a sufficient quality an essential growth of heat conductivity below the superconducting phase transition is observed. In normal metals, where the principal contribution arises from electrons, the heat conductivity falls at $T < T_c$ due to freezing out of the free carriers. In copper-oxide superconductors, where phonons make the major contribution to the heat conductivity, the decrease of the relaxation rate of phonons with their scattering on the free carriers leads to an increase of the heat conductivity. For example, elaborate measurements of the heat conductivity $\kappa_{ab}(T)$ in several single crystals of YBCO [5.96] demonstrate that the electronic contribution in normal state is only 30 %, while the dominant mode of phonon relaxation is electron–phonon interaction. By the estimates given, the coupling constant for acoustic phonons is $\lambda_{tr} \simeq 0.25$, which coincides with the estimates of λ_{tr} for resistivity in Sect. 5.5.1. Impurities, as well as other defects in the sample, lead to suppression of the peak in $\kappa_{ab}(T)$ at $T < T_c$, which confirms the model of electron–phonon relaxation proposed above. Thus, the small concentration of carriers in copper-oxide superconductors leads to a dominant role of phonons in the transport data related to heat transfer, which explains the anomalous behavior of thermopower and heat conductivity.

In general, study of transport properties in cuprates has suggested the following picture of free carrier transport depending on their concentration:

1. In the low concentration region, $\delta < 0.05$, they are a system of spinless holes which strongly interact with the antiferromagnetic background of copper spins; the holes localize upon a decrease of temperature.

2. In the mid-concentration region, $0.1 < \delta < 0.25$, the holes make up a rather special system of quasi-particles – "strange metal" – which exhibits a series of unusual transport properties: resistivity varies linearly with temperature, a strong temperature dependence of Hall coefficient (in clean samples) and an anomalous behavior of magnetoresistivity are observed. Sufficiently strong spin antiferromagnetic fluctuations preserved in this region play an important role in the scattering of quasi-particles. The most probable model for the system of quasi-particles in this case is a system of strongly correlated singlet hole states on copper and oxygen sites, which has been described in Sect. 5.3.2 in the frame of t-J Zhang-Rice model (5.11).

3. At high concentrations of carriers, $\delta > 0.3$, in the region of "normal metal", the transport properties of copper-oxide compounds are characteristic of conventional metals. Dynamical spin fluctuations have an itinerant nature, and the system of Cu $3d^9$ and O $2p^6$ electrons can be considered as a Fermi liquid with a well defined Fermi surface.

5.6 Superconducting Gap

Investigations of the superconducting gap in the quasi-particle excitation spectrum is a direct probe of the mechanism of high temperature superconductivity. The symmetry of the superconducting order parameter and its temperature dependence are directly determined by the nature of superconducting pairing. Reviews of the experimental studies of the superconducting gap are available [5.97–100]. Some of these experiments have already been covered in the previous chapters and will be briefly mentioned again in this section, mostly regarding tunneling experiments.

Already in the first experiments with cuprates an appearance of superconducting pairs at $T < T_c$ was found. These experiments include measuring flux quanta in superconducting rings of ceramic YBCO under random fluctuations of magnetic flux [5.101]. The value of the flux quanta $\phi_0 = hc/2e$ indicates unambiguously that the superconducting current is transferred by electron pairs having the charge $2e$. This result was confirmed later in observations of a flux lattice, measuring the Shapiro steps in the non-stationary Josephson effect at the bias voltage $V = \hbar\omega/2e$, etc. Here the superconducting order parameter is represented by the correlation function of Cooper pairs:

$$\Delta_{\alpha\beta}(\mathbf{k}) \propto \langle c_{\alpha\mathbf{k}} c_{\beta-\mathbf{k}} \rangle, \tag{5.51}$$

where $\mathbf{k}$ is the wavevector of an electron (hole), and $\alpha(\beta)$ is spin (or pseudo-spin if accounting for the spin-orbital coupling). The existence of a Cooper pair condensate with a zero total momentum (5.51) has been confirmed in the Andreev's reflection experiments [5.97, 98].

Usually the conventional pairing, which is also called s-wave pairing, can be distinguished from the unconventional one. For the unconventional pairing the symmetry of the parameter (5.51) is lower than the symmetry of the point group of a crystal. The 2×2 matrix in spin indices (5.51) may be written as

$$\Delta_{\alpha\beta}(\mathbf{k}) = i[\Delta(\mathbf{k}) + \hat{\sigma}\mathbf{d}(\mathbf{k})]\sigma_y, \tag{5.52}$$

where $\hat{\sigma} = (\sigma_x, \sigma_y, \sigma_z)$ and σ_α are the Pauli matrices. The scalar function $\Delta(\mathbf{k}) = \Delta(-\mathbf{k})$ describes singlet pairing (spin of the Cooper pair $S = 0$), and the vector function $\mathbf{d}(\mathbf{k}) = -\mathbf{d}(-\mathbf{k})$ corresponds to triplet pairing (with $S = 1$). We will further consider only the singlet pairing since measuring the Knight shift in high temperature superconductors (see Sect. 3.3.1) indicates only this form of pairing. The conventional pairing in this case corresponds to the fully symmetric representation $^1A_{1g}$ for which the gap function $\Delta(\mathbf{k})$ has the same symmetry as the Fermi surface. For the unconventional pairing the symmetry of $\Delta(\mathbf{k})$ is lower than that of the Fermi surface.

For simplicity, let us consider a square lattice in a CuO$_2$ plane with a period a. Then the conventional pairing is described by the function

$$\Delta_s(\mathbf{k}) = \psi_{0s} + \psi_{1s}(\cos ak_x + \cos ak_y), \tag{5.53}$$

where ψ_{0s} is a (complex) order parameter for the conventional s-pairing (with an isotropic gap), and ψ_{1s} is an order parameter for the so-called "extended" s-wave pairing. The unconventional pairing is described by an order parameter with a lower $^1B_{1g}(\psi_{1d})$ or $^1B_{2g}(\psi_{2d})$ symmetry

$$\Delta_d(\boldsymbol{k}) = \psi_{1d}(\cos ak_x - \cos ak_y) + \psi_{2d}\sin ak_x \sin ak_y . \tag{5.54}$$

This type of pairing is known as the d-wave pairing. It is interesting to note that for a square lattice having a half-filled Brillouin zone the Fermi surface is determined by the relation $|k_x| + |k_y| = \pi/a$, or $\cos ak_x + \cos ak_y = 0$, to which ψ_{1s} in (5.53) does not contribute. For d-wave pairings the gap (5.54) on the Fermi surface changes its sign and becomes zero in four points (of the type $(\pi/2, \pi/2)$ for ψ_{1d} and of the type $(0, \pi)$ for ψ_{2d}). In a three dimensional lattice with a lower symmetry the order parameter $\Delta(\boldsymbol{k})$ may have a rather complicated dependence in $\boldsymbol{k}$-space. For example, for the orthorhombic symmetry D_{2h} in YBCO the "extended" s-wave pairing of $^1A_{1g}$ symmetry is described as the function

$$\Delta_s(\boldsymbol{k}) = \psi_a \cos ak_x + \psi_b \cos bk_y + \psi_c \cos ck_z , \tag{5.55}$$

where a, b and c are the lattice constants. Therefore in low symmetry crystals it is hard to distinguish experimentally anisotropic s-wave pairing of the type (5.55) from pairings of a lower symmetry [5.102].

The appearance of an order parameter in the superconducting phase leads to a modification of the quasi-particle excitation spectrum. According to the general BCS theory, the spectrum of quasi-particles is described by the function

$$E(\boldsymbol{k}) = [\varepsilon(\boldsymbol{k})^2 + |\Delta(\boldsymbol{k})|^2]^{1/2} , \tag{5.56}$$

where $\varepsilon(\boldsymbol{k})$ is the excitation energy in the normal phase measured with respect to the Fermi energy. In the more general Eliashberg theory the changes of the quasi-particle spectrum are described by a complex frequency dependent function $\Delta(\omega)$. The appearance of a gap in the quasi-particle spectrum leads to changes of density of states near the Fermi surface which are described by the following relation

$$N_s(\omega)/N_n(\omega) = \text{Re}\,[\omega/\sqrt{\omega^2 - \Delta^2(\omega)}] , \tag{5.57}$$

where $N_{s,n}(\omega)$ is the density of states in the superconducting (s) and normal (n) phases, for the energy ω measured with respect to the Fermi energy. Usually it is this parameter which determines the changes of the physical properties of the system under the transition to the superconducting state. Some experiments aimed at determining the gap function will be considered further.

5.6.1 Tunneling Experiments

Tunneling experiments provide direct information regarding changes of electronic density of states near the Fermi surface at the superconducting phase transition. In the case of one-particle tunneling from a superconductor (S) through an insulator layer to normal metal (N) a break-up of Cooper pairs occurs, and the differential conductivity is directly determined by the densities of one-particle states in the superconductor $N_s(\omega)$ and in the metal $N_n(0)$

$$dI/dV \propto |t|^2 N_s(\omega = eV) N_n(0), \tag{5.58}$$

where t is the tunneling matrix element and V is the bias voltage at S–N contact. According to (5.57), $N_s(\omega) = 0$ at $|\omega| < \Delta$ and $N_s(|\omega| \geq \Delta)$ has a sharp maximum. However this simple picture of tunneling which has been confirmed for conventional superconductors strongly distorted when applied to copper-oxide superconductors. A series of anomalies which have been observed in earlier experiments with ceramic samples have been related to some extrinsic effects, such as the poor quality of a sample surface, formation of Coulomb gap etc. [5.98].

Further experiments revealed a series of "intrinsic" anomalies which are inherent in copper-oxide superconductors. One of them, a finite conductivity at zero bias voltage and a nonsymmetric nature of the curve dV/dI for $V > 0$ and $V < 0$, are related to a large gap value Δ which is comparable with the potential barrier for tunneling through an insulator layer. In this case, the differential conductivity is determined by the density of states $N_s(\omega)$ weighted by the distribution function of tunneling barriers $D(E)$ [5.98]. Besides, accounting for a finite quasi-particle lifetime leads to a complex gap $\Delta(\omega) = \Delta_1(\omega) + i\Delta_2(\omega)$, which determines a smearing of the density of states in the superconducting phase and a nonzero value of $N_s(\omega = 0)$. The latter effects have been considered in [5.103]. Another reason for smearing of the density of states in copper-oxide superconductors may be the large anisotropy of the gap, which arises naturally both in the case of the "extended" s-wave pairing and in the case of the unconventional d-wave pairing (5.54). In some experiments "two gaps" were observed which may also be related to a strong anisotropy – minimum and maximum values of the gap $\Delta(\boldsymbol{k})$. All these features hinder an unambiguous interpretation of tunneling experiments in copper-oxide superconductors [5.98]. Some concrete measurements will now be discussed.

Thorough investigations of tunneling spectra in three dimensional superconductors on the basis of $BaBiO_3$ [5.100] have demonstrated that in them the energy gap and the excitation spectra correspond to weak coupling BCS-type superconductors at $2\Delta/kT_c = 3.5$. However, attempts to estimate the coupling constant λ on the basis of measuring the function $\alpha^2(\omega)F(\omega)$ by the fine structure of d^2I/dV^2 [5.100] have not given the desired result. It was found that details of the spectrum d^2I/dV^2 depend on the sample and behave irregularly, which does not allow them to be unambiguously related to the Eliashberg function, and the value of the coupling constant λ (5.28) is insufficient to obtain high-T_c. This raises doubts about the electron–phonon mechanism of superconductivity in this compound [5.100]. At the same time, a series of theoretical calculations of T_c in $K_xBa_{1-x}BiO_3$ lead to a satisfactory description of these superconductors on the basis of the electron–phonon model and, qualitatively, correctly reproduce the results of tunneling experiments [5.104].

Great success has been achieved with the aid of scanning tunneling spectroscopy in which the tunneling current flows from a metallic needle to a superconductor surface through a small gap which acts as an insulator layer. By moving the needle parallel to the surface, the conductivity of different layers inside one primitive cell can be examined, while by varying the distance from the needle to the surface, the tunneling current can be regulated. High-quality tunneling spectra for clean surfaces of Bi/2212 have been obtained [5.105, 106]. The temperature dependence of the gap has been investigated. Its value at zero temerature was expressed by $2\Delta/kT_c = 7 - 9$. On the basis of this method it was also found that BiO layers are of a semiconducting nature with the energy gap 0.1 eV whose value depends on the oxygen content [5.107]. On transition to the superconducting state BiO layers preserve their semiconducting nature which indicates the two dimensional nature of superconductivity in these compounds.

It is more convenient to perform measurements of the temperature dependence of the gap on a stationary flat tunnel junction. Well defined tunneling spectra have been obtained [5.108] using high-quality $Au/Bi_2Sr_2CuO_6/$ $Bi_2Sr_2CaCu_2O_8$ single crystal ($T_c = 98$ K) junction. The temperature dependence of the gap $\Delta(T)$ is similar to that of the weak coupling in the BCS theory, although the ratio $2\Delta/kT_c = 5.8$ significantly exceeds the conventional value of 3.5. Point contact spectroscopy for a tunneling junction of the semiconductor GaAs-superconductor $Bi/22(n-1)n$ was used [5.109] to investigate the fine structure of d^2V/dI^2. It was suggested that phonons play a dominant role in the superconducting pairing in these compounds with a strong coupling at $2\Delta/kT_c = 6 - 8$. The phonon structure of the $I - V$ characteristics was also observed in other compounds: in LSCO and $EuBa_2Cu_2O_7$ at large values of $2\Delta/kT_c \simeq 10$ [5.110], in $Nd-Ce-Cu-O$ at $2\Delta/kT_c = 3.8 - 4.0$ [5.111]. Computing the Eliashberg function by the data on the spectrum d^2V/dI^2 suggests that phonons contribute significantly to the superconducting pairing. High quality tunneling spectra for break junction of a single crystal Bi-2212 were also obtained [5.112].

5.6.2 Other Experiments

An essential decrease in the density of one-particle states at the superconducting phase transition which has been discover in tunneling experiments with copper-oxide superconductors has been confirmed in other experiments as well. First, the measurements of thermodynamic parameters – the heat capacity at $T < T_c$, the jump of the heat capacity $\Delta C(T_c)$ indicate the existence of the conventional pairing. As noted in Sect. 4.2, the residual heat capacity linear in temperature (see (4.26)) at $T < T_c$ is most likely of an "extrinsic" nature which is irrelevant to the unconventional pairing and to a "novel" nature of superconductivity in copper-oxide superconductors. Measuring the heat capacity jump and the critical fluctuations above and below T_c (see Fig. 4.4) confirm the existence of a two-component order parameter withing the framework of the anisotropic Ginzburg-Landau theory. Studies of temperature dependence of the magnetic field penetration depth $\lambda(T)$ (see Chap. 4.3) are still controversial (see footnote [4] in Chap. 4).

The more direct methods for investigating the gap and its symmetry are based on measuring the quasi-particle excitation spectrum (5.56) in the superconducting phase. As noted in Sect. 5.4.2, investigations of the temperature dependence of the photoemission spectra in Bi compounds allow one to observe the formation of the superconducting gap at $T < T_c$ (see Fig. 5.20). The in-plane value of the gap proves greater than the conventional value of 3.5 in the weak coupling BCS theory. Investigations of infrared conductivity (Sect. 5.4.3) unambiguously indicate an essential decrease in the density of states under superconducting transition (see Fig. 5.25), although the data relating to the value of the gap and its symmetry are still contradictory. Raman electronic scattering reveals an existence of two gaps with respect to the polarization of the incident light; one of them is greater than the conventional value, and the other is less (see Sect. 5.4.3).

A change in the density of one-particle states under transition to the superconducting state is also observed in investigations of the interaction of electrons (holes) with spin and phonon degrees of freedom in a crystal. The measurements of the Knight shift and the rate of spin-lattice relaxation also confirm the singlet pairing and the formation of a gap on the Fermi surface (see Sect. 3.3). The lack of the Hebel–Slichter peak at $T < T_c$ for ^{17}O nuclei in YBCO, and a fall in relaxation rate more rapid than in conventional weak coupling superconductors (see Fig. 3.10a), are frequently interpreted as arguments in favor of an unconventional pairing with a gap having nodes on the Fermi surface. However, accounting for the finite lifetime of quasi-particles, leading to a complex gap, smears the density of states (5.57) and suppresses the Hebel-Slichter peak [5.103]. An anisotropy of the gap may also change the density of states (5.57). In the next chapter we will consider the effect of the superconducting transition to a renormalization of phonon frequencies and their lifetimes. Now we just observe that experiments on single crystals of $RBa_2Cu_3O_{7-y}$ demonstrate that in their (ab) plane there is a superconducting gap $2\Delta/kT_c = 5$ [5.74].

Summarizing our consideration of the superconducting gap in copper-oxide compounds, we note the following items:

1. Below the superconducting transition temperature T_c a condensate of Cooper pairs in the singlet state arises.

2. A number of the experiments indicate the conventional (s-type) pairing with an anisotropic gap $\Delta(\mathbf{k})$, although some experiments are better described within the framework of the unconventional (d-type) pairing.

3. A variation in the values of the gap in experiments of different types may be explained by suggesting a strong anisotropy of the gap (or "two" gaps), with the typical value $2\Delta_{ab}/kT_c \simeq 8$ for the in-plane wavevectors (k_x, k_y), and $2\Delta_c/kT_c \simeq 3.5$ for k_z [5.98].

4. A series of tunneling experiments reveal that the fine structure in the d^2V/dI^2 spectra correlates with the density of phonon states. Estimates of the Eliashberg function $\alpha^2F(\omega)$ in this case indicate an essential phonon contribution to the superconducting pairing.

6. Lattice Dynamics and Electron–Phonon Interaction

In conventional superconductors the studies of phonon spectra were of significance in developing the microscopic theory based on the electron–phonon mechanism of pairing [6.1]. Much attention has therefore been paid to the investigations of the phonon spectra of high temperature superconductors aimed at finding a certain phonon contribution to the formation of the Cooper pair condensate.

The first goal was to estimate the effect of electron–phonon interaction of free carriers on the lattice dynamics of oxide compounds. It is well known that this interaction leads to a renormalization of ionic frequencies Ω_{qj} (q is a wave vector, j is a polarization index) according to the following relation

$$\omega_{qj}^2 = \Omega_{qj}^2 + 2\Omega_{qj}\Sigma_{qj}(\omega = \omega_{qj}). \tag{6.1}$$

Here ω_{qj} is a renormalized phonon frequency, $\Sigma_{qj}(\omega)$ is the polarization operator. To the second order in electron–phonon interaction in normal phase it may be written as

$$\Sigma_{qj}(\omega) = \sum_{pmn} |g_j(p,m;p+q,n)|^2 \frac{f_m(p) - f_n(p+q)}{\varepsilon_m(p) - \varepsilon_n(p+q) + \omega}, \tag{6.2}$$

where $g_j(p, m; p+q,n)$ is the matrix element of the electron–phonon interaction, $f_m(p)$ is the Fermi distribution function for electrons of energy $\varepsilon_m(p)$, p is a wave number and m is a band index. The real part of the polarization operator (6.2) determines a softening of the ionic phonon frequencies due to electron–phonon interaction: $\Delta\omega_{qj} = \omega_{qj} - \Omega_{qj} = \mathrm{Re}\,\Sigma_{qj}(\omega_{qj}) < 0$, and its imaginary part determines phonon decay as $\gamma_{qj} = -\,\mathrm{Im}\,\Sigma_{qj}(\omega_{qj} + i\delta)$. The latter function is directly related to the Eliashberg function [6.1]

$$\alpha^2(\omega)F(\omega) = \frac{1}{\pi\hbar N(0)} \frac{1}{N} \sum_q \frac{\gamma_{qj}}{\omega_{qj}} \delta(\omega - \omega_{qj}), \tag{6.3}$$

where $N(0)$ is the density of electronic states (per atom and spin) on the Fermi surface. This relation allows computation of the coupling constant λ in the BCS theory

$$\lambda = 2 \int_0^\infty \frac{\alpha^2 F(\omega)}{\omega} d\omega = \frac{2}{\pi\hbar N(0)} \frac{1}{N} \sum_{qj} \frac{\gamma_{qj}}{\omega_{qj}^2}, \tag{6.4}$$

given the phonon decay γ_{qj} due to electron–phonon interaction.

Thus, with the aid of the studies of phonon spectra, their dependence on temperature and the concentration of free carriers under doping, the magnitude of electron–phonon interaction in copper-oxide superconductors can be estimated. In Sect. 6.1 the results of experiments on inelastic neutron scattering which investigate the phonon spectra throughout the Brillouin zone will be discussed. Optical experiments, especially changes of phonon spectra at the superconducting transition, are discussed in Sect. 6.2. Some theoretical models of lattice dynamics, and in particular an anharmonic model of the soft mode in LMCO crystals, are considered in Sect. 6.3.

6.1 Phonon Spectra – Neutron Scattering

A most complete study of phonon spectra of copper-oxide superconducters was performed with the aid of the inelastic neutron scattering method [6.2, 3]. The first experiments on polycrystalline samples measured the generalized phonon density of states $G(\omega)$, equal to the sum of partial contributions weighted with the neutron scattering cross-sections of separate atoms. Measuring $G(\omega)$ allows one to determine the general picture of the phonon density of states $F(\omega)$ in the Eliashberg function (6.3) to construct different models for this function. Later, after synthesizing sufficiently large single crystals, the phonon dispersion curves were measured throughout the Brillouin zone for the main crystallographical directions for the crystals of Nd_2CuO_4, La_2CuO_4 and $YBa_2Cu_3O_x$ [6.2, 3][1].

Measurements on polycrystalline samples of $La_{2-x}Sr_xCuO_4$ and $Bi_2Sr_2(Ca_{1-x}Y_x)Cu_2O_8$ have not found an essential change of $G(\omega)$ at transition from the dielectric ($x = 0$ in LSCO and $x = 1$ in Bi/2212) to metallic state. In metallic samples, no notable changes in $F(\omega)$ were seen on passing to the superconducting state [6.2]. A stronger deformation of phonon spectra was observed in $YBa_2Cu_3O_x$ compounds at a change of oxygen content $6 < x < 7$, as well as upon doping with Pr, Zn impurities which suppress superconductivity. Figure 6.1 shows the generalized phonon density of states $G(\omega)$ for three samples of YBCO with the changes of the composition indicated above [6.2]. The phonon spectra undergoes greatest deformation at an increase of oxygen content, which is accompanied by the transition to metallic state (Fig. 6.1c). Then a filling of O1 positions along the chains occurs (see Fig. 2.12), which changes the number of bonds for Cu1 – O1 and O1 – O4 atoms. This changes the spectrum in the low frequency range, 15 and 20 meV, which are related to vibrations of Cu1, O1, O4 atoms [6.4]. However, an essential increase in the density of states in the frequency range $40 - 60$ meV and 80 meV, caused by in-plane vibrations of the atoms O2, O3 and Cu2 cannot be explained by these structural changes alone. Investigations of $G(\omega)$ with continuous adding of oxygen from $x = 6$ to $x = 7$ have led to the conclusion that in this frequency range, for the vibration modes related to stretching and bending of the bonds Cu2 – O2, O3 in the planes, the effects of

---

[1] More recent results on electron–phonon coupling in the cuprate superconductors studied by inelastic neutron scattering are presented in [6.37].

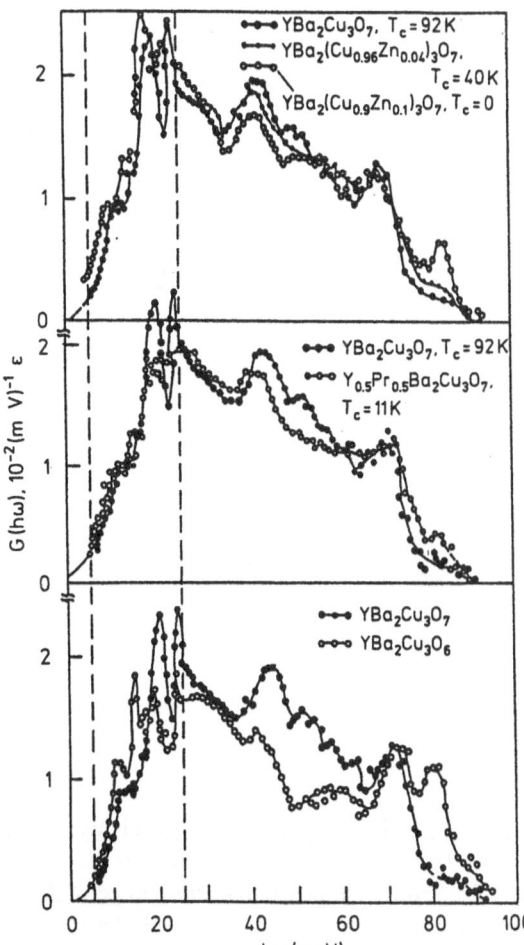

Fig. 6.1. Generalized phonon density of states in $YBa_2Cu_3O_7$ when substituting Cu by Zn (**a**), Y by Pr (**b**), and for a change of oxygen content: $O_6 \rightarrow O_7$ [6.2]

strong electron–phonon interaction are seen [6.4]: with an increase in the in-plane carrier density an essential softening of the corresponding force constants occurs. Subsequent measurements of phonon dispersion curves (see Fig. 6.2) confirmed these conclusions [6.2]. A similar deformation of phonon spectra is observed at substituting Cu by Zn (Fig. 6.1a), or Y by Pr (Fig. 6.1b). Since in this case the structural changes are insignificant, the deformation of phonon spectra may be related to changes of electron–phonon interaction, although in the case of Pr the density of free carriers changes too (see Sect. 5.2, Fig. 5.5).

The phonon dispersion curves of single crystals of $La_{2-x}Sr_xCuO_4$ have been studied in great detail [6.2, 3, 5]. The first experiments investigated the soft mode at the structural phase transition from tetragonal to orthorhombic phase (see Sect. 2.2). Figure 6.3 shows phonon dispersion curves for this mode (symmetry Σ_4) together with its temperature dependence (insert) [6.5]. Measuring the scattering intensity confirmed the soft mode model as a rigid rotation of a CuO_6

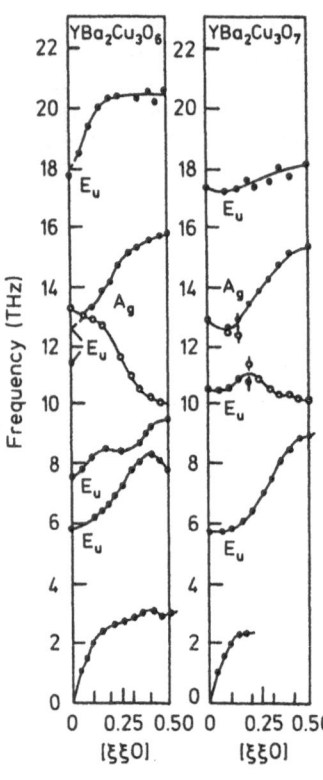

Fig. 6.2. Phonon dispersion curves for longitudinal (•) and transverse (o) modes in $YBa_2Cu_3O_x$, ($x = 6$ (**a**) and $x = 7$ (**b**)), symmetry Σ_1 [6.3]

octahedron round the Cu ion. In general, both the temperature dependence of the soft mode frequency above and below T_0 and the critical scattering near the transition temperature T_0 looked typical of the structural phase transitions in perovskites with a soft rotation mode. A strong interaction of the soft mode with the acoustic vibrations was found, and the related essential anharmonic effects [6.2]. As we noted in Sect. 2.2, the latter are due to tilting of the CuO_6 octahedra in a strongly anharmonic two-well potential (see Fig. 2.9). The data of acoustic measurements, in particular a significant softening of the elastic coefficient C_{66} (by $30 - 40\%$ [6.6]) at $T < T_0$, have confirmed a strong coupling of the acoustic modes with the anharmonic soft mode. A theoretical model to describe the soft mode and its interaction with deformations will be considered below in Sect. 6.3.

Further comparative studies of phonon dispersion curves in the frequency range up to 20 THz $\simeq$ 82.6 meV in dielectric and metallic samples of LSCO and YBCO have revealed the role of electron–phonon interaction in these crystals. In dielectric state (Fig. 6.2a and 6.4c) the phonon dispersion looks typical of ionic crystals. In particular, an essential splitting of the longitudinal and transverse optical modes E_u in the zone center Γ is observed, which is caused by long-range Coulomb forces (shown by the splitting of the E_u-mode in the region $= 11 - 14$ THz in Fig. 6.2a and 6.4c). On the transition to the metallic state this splitting disappears due to the screening of Coulomb forces in the CuO_2 planes by free charge carriers

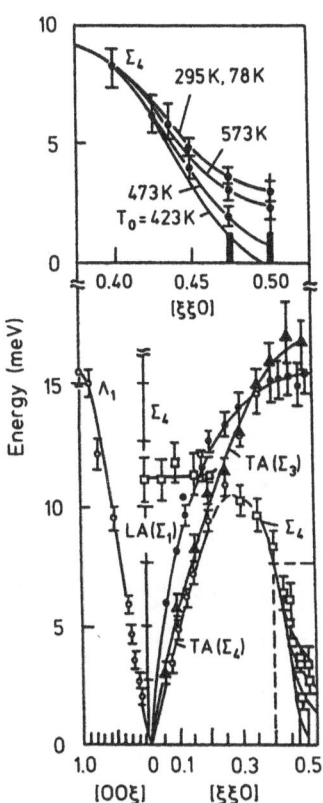

Fig. 6.3. Phonon dispersion curves and temperature dependence (*insert*) of the soft mode in LSCO having the temperature of the structural phase transition $T_0 = 423\,K$ [6.5]

(see Fig. 6.2b and 6.4d). On the other hand, vibrations of ions perpendicular to the CuO_2 planes preserve their ionic nature in view of small screening of these displacements. In this case, a sufficiently good theoretical description of the phonon dispersion curves (see Fig. 6.4a,b) may be obtained within the framework of the conventional shell model. In the metallic phase anisotropic screening of Coulomb forces are also introduced. The model, however, provides a poor description of a strong coupling of the vibrations A_g of apical oxygen along the c-axis and the vibrations E_u of oxygen ions in the basal plane, which shows itself in a strong splitting of A_g and E_u modes in the frequency range $\nu \simeq 12\,THz$ both in LSCO and in YBCO (see Fig. 6.2 and 6.4).

The behavior of high frequency "breathing" type modes both in LSCO ($\nu \simeq 21\,THz$) and in YBCO ($\nu \simeq 18\,THz$) in metallic phase is unusual. In this mode a symmetric stretching of Cu – O bonds occurs, which has a large matrix element of electron–phonon interaction. According to the conventional non-orthogonal tight binding theory of lattice dynamics [6.7], electron–phonon interaction should strongly renormalize the frequency of these vibrations. In experiments one observes only nonessential renormalization of these frequencies. At the same time, similar modes with stretching only two in-plane Cu – O bonds experience an essential softening at doping (mode Δ_1 at the point $(0.5, 0, 0)$ at $\nu \simeq 19\,THz$

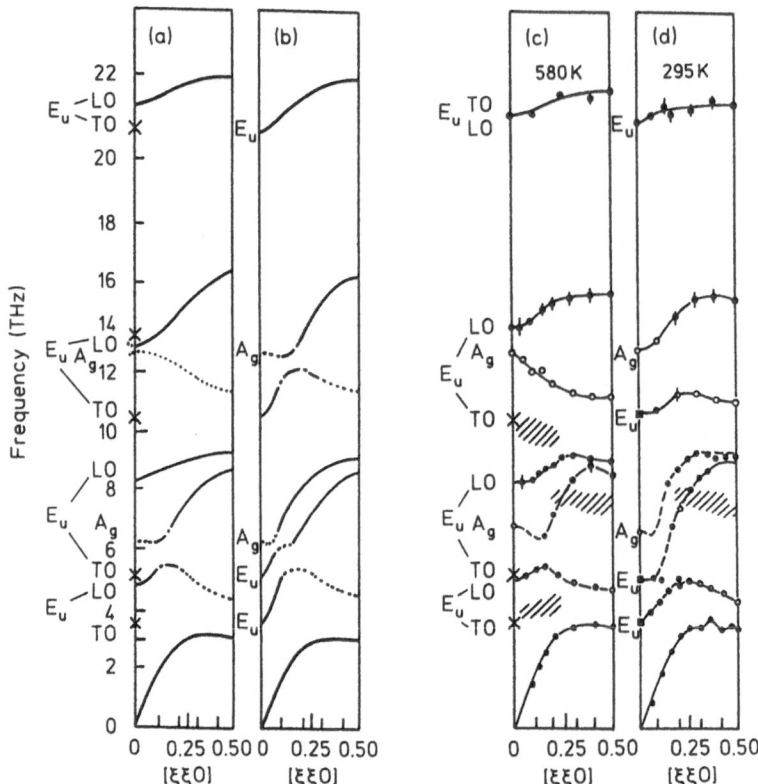

Fig. 6.4. Theoretical (**a, b**) and experimental (**c, d**) phonon dispersion curves in La$_{2-x}$Sr$_x$ CuO$_4$ in dielectric (a, c, $x = 0$) and metallic (b, d, $x = 0.1$) phases. The light and dark dots correspond to the longitudinal (*in ab plane*) and transverse (*along c-axis*) vibrations [6.3]

in LSCO [6.3]). Such a difference in the behavior of these modes may be explained by the peculiarities of band spectra of copper-oxide compounds. Strong Coulomb correlations on copper sites hinder the charge transfer from oxygen to copper (Sect. 5.3.1). In the "breathing" mode the renormalization (6.2) yields a small contribution because of the large denominator – the gap between the Cu 3d and O 2p bands. At a shift of only two oxygen atoms the charge transfer occurs between non-equivalent oxygen positions, and the contribution of (6.2) is large since the denominator is now small – there is no gap in the O 2p band. Such a strong difference in electron–phonon interaction is also observed for the soft mode at a rotation of the CuO$_6$ octahedrons in LSCO: a reconstruction of the electronic spectrum for a rotation round the axis [100] (or [010]) is much larger than for the rotations round the axes like [110] (Sect. 2.2).

Anomalous behavior of scattering intensities is also observed in the mid frequency range, shown hatched in Fig. 6.4c,d. In the first experiments it was supposed that in this frequency range there is an additional excitation branch, probably of an electronic nature [6.2]. However, further studies [6.3] demonstrated that anomalous scattering intensities in this range may be related to a splitting of modes in the orthorhombic phase which is partially preserved at transition to the tetragonal phase. Strong anharmonic effects in this frequency range also hinder an unambiguous determination of the phonon dispersion curves.

Measuring phonon dispersion curves in the dielectric phase of Nd_2CuO_4 confirmed an ionic nature of bonding in copper-oxide superconductors [6.3]. The phonon dispersion curves obtained were described in the frame of the shell ionic model accounting for only the nearest neighbor interaction. Measuring the phonon dispersion curves in doped Nd_2CuO_4 samples, as well as in more complicated compounds of Bi and Tl, requires synthesis of large high quality single crystals.

Thus, investigations of phonon spectra by inelastic neutron scattering reveal an ionic nature of bonding in copper-oxide compounds. A transition to the metallic state with doping leads to screening of the interaction only in the CuO_2 plane, preserving the ionic nature of the bonds for the vibrations perpendicular to these planes. Only a small number of the phonon dispersion curves demonstrate an anomalous behavior which may be related to strong electron–phonon interaction.

6.2 Optical Investigations

Investigations of lattice dynamics by Raman scattering and infrared absorption have played an important role in the assignment of certain modes in copper-oxide superconductors. A review of the original studies in this field is available [6.8, 9], and the later investigations have been discussed [6.10]. Unlike neutron scattering, optical investigations allow one to observe phonon modes only in the center of the Brillouin zone, at $q = 0$. However, the existence of definite symmetry selection rules for phonon modes, by an order of magnitude higher precision of measurements, and the possibility of carrying out experiments on samples of small size, render the optic methods unique for rigorous quantitative investigations. In this section the results of only those experiments which are a probe of the magnitude of electron–phonon interaction in copper-oxide compounds will be discussed. These are Fano effect, temperature dependence of the phonon frequency shift and width at the superconducting transition and polaronic effects in light scattering. A general symmetry analysis of optically active phonons in LSCO, YBCO and compounds on the basis of Bi and Tl, as well as the results of their experimental studies, have been considered [6.8–10].[2]

Certain indications regarding a possibility of strong electron–phonon interaction for some optical modes have been obtained in observations of the Fano effect [6.10]. It appears whenever discrete and continuous excitations may interfere coherently, and shows itself in an asymmetry of the lines for a discrete scattering spectrum. In this case, the scattering intensity as a function of frequency ω is proportional to (Fano profile)

$$f(\omega) = \pi \varrho(\varepsilon) T_e^2 \frac{(q + \varepsilon)^2}{1 + \varepsilon^2}. \tag{6.5}$$

---

[2] See also [6.38].

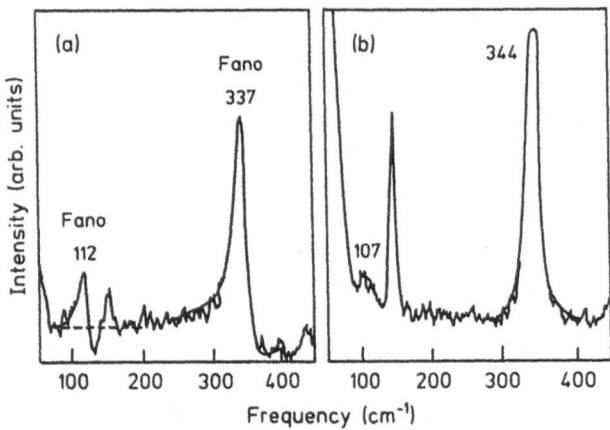

Fig. 6.5. Asymmetric (Fano effect) (**a**), and symmetric (**b**), Raman spectra for (**a**) super-conducting, and (**b**) semiconducting samples of YBCO [6.10]

In the approximation of a constant density of states in the continuous spectrum $\varrho(\varepsilon) = \varrho$, the dimensionless frequency $\varepsilon = (\omega - \omega_p)/\Gamma$ and the lineshape parameter $q = T_p/\pi\varrho V T_e$ define the profile of the line. Here T_e and T_p are the scattering matrix elements for the continuous (electronic) spectrum and a discrete excitation (phonon of frequency ω_p), V is their coherent interaction, Γ is the total linewidth. In the non-interacting case, $V \to 0$, we have $q \to \infty$ and the line shape becomes Lorentzian. At $q = 1$ the line shape is its most asymmetric. At the frequency $\omega_{min} = \omega_p - \Gamma q$ an antiresonance is observed when $f(\omega_{min}) = 0$.

The Fano effect is most pronounced for the B_{1g} type mode at frequency $\nu \simeq 340 \, \text{cm}^{-1}$ in YBCO, where it was discovered [6.10]. In this mode antiphase vibrations of in-plane oxygen ions O2, O3 occur along the c-axis of the crystal with respect to the copper ions Cu2. Figure 6.5 compares the intensity of Raman scattering for (a) superconducting at $y = 0$, and (b) semiconducting at $y = 1$, samples of YBa$_2$Cu$_3$O$_{7-y}$ [6.10]. In the latter case, the form of the scattering line is symmetrical, which corresponds to $q \to \infty$ in (6.5). For a finite coupling constant $VT_e \neq 0$ it is natural to assume that, in dielectric samples, the density of electronic states with a continuous spectrum $\varrho = 0$, while in superconducting samples $\varrho \neq 0$, and the Fano effect appears at a finite value of $q \simeq -4$. It is interesting to note antiresonance behavior of the A_g-mode of Ba vibrations ($\nu = 122 \, \text{cm}^{-1}$), which indicates a strong interaction of these ionic vibrations with in-plane charge carriers. The Fano effect remains for the A_g-mode when the temperature falls below T_c, which indicates a finite density of quasiparticle excitations in this energy range, below the energy 2Δ of the superconducting gap [6.10].

Quantitative estimates of the electron–phonon interaction constant were obtained in the course of studies of the effect of the superconducting transition on optical phonons. At the superconducting phase transition a gap Δ appears in the spectrum of quasiparticles which essentially alters the form of the phonon polarization operator (6.2). The calculations [6.13] show that for phonons of frequency

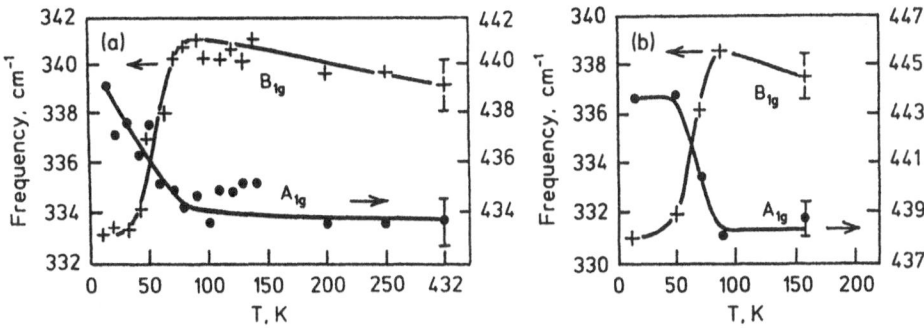

Fig. 6.6. Variation of the A_{1g} and B_{1g} phonon frequencies near T_c in YBCO (**a**) and TmBCO (**b**) [6.11]

$\hbar\omega > 2\Delta$ the renormalization $\Delta\omega_\nu$ due to electron–phonon interaction leads to an increase their frequency (hardening), and for phonons with $\hbar\omega < 2\Delta$ to a reduction of frequency (softening). An increase in the density of states at the superconducting transition in the energy range $\hbar\omega \geq 2\Delta$ also leads to an increase in the relaxation rate and linewidth γ_ν. These effects were investigated in a series of experiments [6.10–12] for RBCO compounds by Raman scattering. The temperature dependence of phonon frequency and linewidth for the modes of the symmetry A_{1g} ($\omega \simeq 440\,\mathrm{cm}^{-1}$) and B_{1g} ($\omega \simeq 340\,\mathrm{cm}^{-1}$), in which O2 and O3 ions vibrate along the c-axis in-phase or out-of-phase respectively were studied. Varying the composition, R=Eu, Dy, Er, Tm, Y and doing the isotopic exchange $^{16}\mathrm{O}\rightarrow{}^{18}\mathrm{O}$ one can vary in a certain range the frequencies of B_{1g} and A_{1g} modes, keeping the superconducting transition temperature T_c and the supposed gap in the spectrum of quasiparticles Δ constant. An external magnetic field H, which reduces T_c, simultaneously reduces the phonon softening temperature T_s, $dT_c/dH \propto dT_s/dH$, which confirms the relation of the phonon anomalies to the superconducting transition.

Figure 6.6 shows the temperature dependence of the frequency for B_{1g} and A_{1g} modes for crystals of YBCO (a) and TmBCO (b) [6.11]. A strong softening of the B_{1g}-mode and the hardening for the A_{1g}-mode at $T < T_c$ lead to the conclusion that a superconducting gap forms at T_c whose value may be estimated from the conditions $\tilde{\omega} = \hbar\omega/2\Delta < 1$ for $\omega \simeq 340\,\mathrm{cm}^{-1}$ and $\tilde{\omega} > 1$ for $\omega \simeq 440\,\mathrm{cm}^{-1}$.

Figure 6.7 shows the dependencies of the frequency shift $\Delta\omega_\nu/\omega_\nu$ (a) [6.11] and the linewidth $-\Delta\gamma_\nu/\omega_\nu$ (b) [6.12] at the superconducting transition for different samples of RBCO. The solid line shows the theoretical curves for these parameters as obtained in the strong coupling approximation [6.13]

$$\Delta\omega_\nu/\omega_\nu = \lambda_\nu f(\omega_\nu/2\Delta, T/T_c). \tag{6.6}$$

Here f $(\tilde{\omega}, \tilde{T})$ is a universal function, $|\,f\,| \sim 1$, and λ_ν is a dimensionless coupling constant

$$\lambda_\nu = 2N(0)\langle|g_\nu(\boldsymbol{k}, k)|^2\rangle_{FS}/\omega_\nu, \tag{6.7}$$

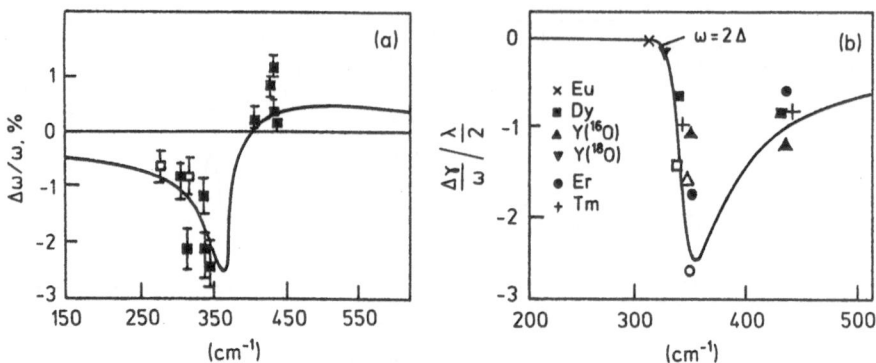

Fig. 6.7. Frequency shift $\Delta\omega_\nu/\omega_\nu$ (**a**) and the change of linewidth ($\Delta\gamma_\nu/\omega_\nu$) (**b**) at the superconducting transition in RBCO [6.12]

where $\langle\cdots\rangle_{FS}$ stands for the averaging of the matrix element of electron-phonon interaction (in one band model) over the Fermi surface. A good agreement of the theoretical results [6.13] with experimental data was obtained for the coupling constants $\lambda_\nu(B_{1g}) = 0.02$ and $\lambda_\nu(A_{1g}) = 0.01$ at $2\Delta(0) \simeq 316\,\mathrm{cm}^{-1}$, $2\Delta/kT_c \simeq 5$ [6.12].

Similar results were obtained for other vibration modes: the A_{1g}-mode for apical oxygen O4 ($\omega \simeq 500\,\mathrm{cm}^{-1}$) which is related to stretching of Cu – O bonds along the c-axis of the crystal, and the B_{2g}, B_{3g} modes related to bending of these bonds [6.14]. In this case, the electron–phonon interaction coupling constants also appear small: $\lambda_\nu(A_g) \simeq 0.01$, $\lambda_\nu(B_{3g}) < 0.003$, $\lambda_\nu(B_{2g}) < 0.02$. Observe that in some models high values of T_c in YBCO compounds are explained by a strong electron–phonon interaction for vibrations of apical oxygen, which is not confirmed for these optical modes at $q = 0$. The modes of symmetry $B_{1u}(z)$ which are active in infrared spectra were investigated [6.15], and small coupling constants $\lambda_\nu = 0.01 - 0.02$ were obtained for them too.

Doping with impurities like Au and Pr has an essential effect on the anomalies in frequency and phonon linewidth at the superconducting transition [6.16]. To explain the effect of impurities [6.16] a model of a gap, strongly anisotropic in the a, b plane, was proposed with the minimal value $2\Delta/kT_c \simeq 5$ which is observed in clean YBCO crystals. Presence of impurities averages the gap which raises its experimentally observed minimal value and changes the temperature dependence of the anomalies at T_c. The anisotropy of the gap and its impurity dependence may explain the large scatter of the $2\Delta/kT_c$ values observed in different experiments (see Sect. 5.6). With the average value $2\Delta/kT_c \simeq 3.5$, the deviations of the gap from the conventional BCS value may be related to its anisotropy, without using the hypothesis of a strong electron–phonon interaction.

A certain confirmation of the anisotropic gap (or two gaps) model was obtained in the investigations of Raman scattering in untwinned crystals of $YBa_2Cu_4O_8$ [6.17]. Electronic Raman scattering finds two gaps: for in-plane polarization of light one observes a large gap $2\Delta_2/kT_c \simeq 6.5$, and, for the polarization (Y, Y)

along the chains additionally a small gap $2\Delta_1/kT_c \simeq 2.3$. Investigations of the temperature dependence of the phonon frequency shift and their linewidth at the superconducting transition have confirmed the existence of two gaps, generally, anisotropic in k-space. At the same time the formation of the superconducting gap (or two gaps) affects not only the Cu2, O2, O3 ion vibrations in the basal plane (modes A_{1g} and B_{1g} in Figs. 6.6, 7), but also the vibrations of other ions, in particular Ba and in-chain Cu1, O1. This observation confirms a strong electron–phonon interaction for the A_{1g}-mode of Ba found in the Fano effect (Fig. 6.5).

Summarizing the investigations of phonon spectra by light scattering, the following are the most important results.

1. Observations of the Fano effect – interference of scattering on phonons and a continuous excitation spectrum suggest a sufficiently strong interaction of certain optical modes with doped charge carriers (see Fig. 6.5).

2. Elaborate investigations of the phonon frequency shift and its broadening at the transition to the superconducting state prove that a sufficiently sharp gap $2\Delta/kT_c \simeq 5$ appears in the spectrum of quasiparticles at T_c. Experimental curves agree quite well with the theoretical ones obtained within the limit of strong coupling. However, the constant of electron–phonon interaction for optical modes $\lambda_\nu \leq 0.02$ is too small to explain the high-T_c in RBCO compounds [3].

As noted in Chap. 2, local structural distortions in the oxygen sub-lattice due to doping or the superconducting transition are due to a strong coupling of the charge carriers to the lattice. For doping of YBCO, the charge transfer from chains to planes leads to a contraction of the Cu2 – O4 bonds (see Fig. 2.15), which is accompanied by an increase in the frequency of the A_g-vibrations of apical oxygen ($\omega \simeq 480 - 500\,\mathrm{cm^{-1}}$) [6.9], while a series of other frequencies experience some softening due to screening of Coulomb interaction by charge carriers. In the course of the investigations of the fine structure of X-ray absorption (EXAFS) on apical oxygen, a model was proposed for vibrations of O4 along the Cu1 – O4 – Cu2 bond in a double-well potential whose parameters change at the superconducting transition [6.18].

Confirmation of the strongly anharmonic nature of the apical oxygen vibrations was obtained in studies of photoinduced Raman scattering [6.19]. Exciting a small number of carriers with the aid of laser radiation leads to an increase of the frequency of A_g-vibrations of apical oxygen in semiconducting samples of $YBa_2Cu_3O_{6.3}$ and $Tl_2Ba_2Ca_{1-x}Gd_xCu_2O_8$ to the same extent as doping. The analysis of the frequency dependence of photoinduced conductivity [6.20] demonstrates that the carriers form polaronic states due to a strong coupling to the lattice. A hypothesis was also put forward that the conductivity in the mid-infrared range (see Sect. 5.4.3) may be related to multiphonon excitation processes which arise when the free carriers loose their phonon polarization cloud. A strong electron–

---

[3] Assuming equal electron–phonon interaction for all the $3r = 39$ phonon branches in YBCO one gets for an effective coupling constant (6.4) $\lambda = \sum_{\nu=1}^{3r} \lambda_\nu \cong 0.8$ where r is the number of atoms in the unit cell of the crystal.

phonon interaction for vibrations of O4 ions was also suggested in the studies of photoinduced infrared absorption which becomes possible at local displacements of O4 at the charge transfer Cu1 – O4. A resonant nature of the absorption indicates a formation of electronic states inside the charge transfer gap [6.21].

An anomalous behavior at the superconducting transition was found for the infrared mode B_{1u} at $570\,cm^{-1}$ which is also related to vibrations of Cu1 – O4 ions [6.22]. The frequency of this mode decreased at the superconducting transition, although according to the strong coupling model (6.5) at $\hbar\omega > 2\Delta$ it should increase. Impurity Co substitution for Cu1 confirmed the assigning of this mode to vibrations of Cu1 – O4 and established a correlation between the superconducting transition and a certain local structural reconstruction, probably due to formation of bipolarons. Within the framework of the anisotropic gap model [6.16] one has to assume that infrared-active phonons are coupled to the maximum of the gap at the Fermi surface, i.e. $2\Delta_{max} \geq 570\,cm^{-1}$.

Another model of polaron formation in CuO_2 planes was proposed in the investigations of Raman spectra in $La_{2-x}Sr_xCuO_4$ [6.23]. In a certain concentration range of Sr ($x \simeq 0.12$), infrared-active modes appear in the Raman spectrum related to transversal optical vibrations of symmetry B_{2u} or B_{3u}, which should not be observed in a crystal having a center of inversion. Their appearance is explained [6.23] as the result of a local symmetry distortion at a formation of polaron states on holes in CuO_2 planes. These experiments indicate a sufficiently strong electron–phonon interaction of a deformation type in CuO_2 planes.

Thus, although the estimates of the electron–phonon interaction constant on the basis of the strong coupling theory (6.5) for optical modes yield small values of $\lambda_\nu \leq 0.02$, a series of experiments demonstrate a strong coupling of charge carriers to the lattice with formation of polaronic type local distortions. This contradiction may be explained by a weak coupling of carriers with uniform (optical) deformations of the lattice which appear for optical phonons with $q \simeq 0$, and a sufficiently strong coupling to local distortions of the lattice which corresponds to phonons with large q.[4]

6.3 Theoretical Models

Symmetry analysis of lattice vibrations allow one to classify possible phonon modes with respect to the irreducible representations of the space group for the wavevector k, and thus to determine polarization vectors for the phonon modes under study [6.2, 3]. However, to compute the vibration frequencies and polarization vectors for separate atoms one must construct a model of atomic interaction in the crystal. The simplest one is the force constant method, in which one specifies the harmonic coupling constants of the interatomic interaction accounting for the

---

[4] Superconductivity-induced phonon shifts in $YBa_2Cu_3O_x$ observed by inelastic neutron scattering [6.37] were found to decrease with increasing q ($x = 7.0$) or exhibit a maximum at finite q ($x = 6.92$).

interaction in several coordination spheres. The fixing of constants is performed on the basis of a best fit to experimental data, and therefore this method is a phenomenological one and does not have a predictive power. An approach based on a model of atom-atom potentials seems to be more consistent. For ionic crystals, sufficiently simple is the shell model. Into it, besides the direct Coulomb interaction of rigid ion cores, one introduces a coupling of the valence electronic shell with the ionic core and accounts for their short range repulsive Born-Mayer interaction. On the basis of this model, the lattice dynamics of all the main types of copper-oxide superconductors has been computed (see, e.g. [6.24]). The same shell model, with an inclusion of anisotropic screening, was used to analyze the phonon dispersion curves measured by inelastic neutron scattering [6.2, 3] (see Fig. 6.4). A rather good agreement between the experimental and theoretical results for these calculations, with a small number of adjustable parameters in ionic shell model, have confirmed the ionic nature of interatomic forces in copper-oxide compounds.

In view of this fact, we can begin our construction of the theory of the electron–phonon interaction within the framework of the ionic model [6.25]. To describe the electronic system in copper-oxide superconductors one can use the effective Hamiltonian of the p-d model (5.6). In this case, accounting for lattice vibrations leads to fluctuations of the model parameters – i.e., of the hopping integral t_{ij} and the site energies of p, d states in the crystalline field [6.25]:

$$\delta t_{ij} \simeq t_{ij}(u_{ij}/d) + \dots , \qquad (6.8)$$

$$\delta \varepsilon_i \simeq \alpha_i V(u_i/d) + \beta_i V(u_i/d)^2 + \dots , \qquad (6.9)$$

where $u_{ij} = u_i - u_j$, u_i is a displacement of the type i ion, d is the interatomic distance. In conventional metals with broad bands the main contribution is due to the first term in (6.8), since due to strong screening the Coulomb fields $V \simeq e^2/\varepsilon d$ are small, and small is the interaction (6.9). In copper-oxide compounds, due to strong localization of the wavefunctions of valence electrons, their overlap is small (see Sect. 5.3.2) and therefore the transfer integral t_{ij} is small. At the same time, due to small concentration of carriers and poor screening (especially for directions out of CuO_2 planes), the Coulomb energy is large, $V \gg t_{ij}$, and the second contribution (6.9) is greater than the first (6.8). Estimates of the dimensionless parameters α_i and β_i for several typical displacements of ions Cu2, Ba, O2, O3, O4 in YBCO crystals in Raman-active modes, and for the ions Cu and O for the soft mode in La_2CuO_4 have been made [6.25]. In YBCO, due to local symmetry, there is a linear electron–phonon interaction with $\alpha_A \sim 2$ in (6.9) for A_g type modes. In LSCO we have for oxygen displacements in the soft mode $\alpha_{SM} = 0$, while $\beta_{SM} \cong 1$. As a result, for the Coulomb interaction $V_{pd} \simeq 1 - 2\,\text{eV}$, for the deformation potential we obtain the estimates $g_A \sim (\alpha_A V/d) \sim 1 - 2\,(\text{eV/Å})$, $g_{SM} \sim \beta V(u/d^2) \sim 0.1\,(\text{eV/Å})$ at $d \simeq 2 - 3\,\text{Å}$ and equilibrium displacements of oxygen ions in soft mode $u \sim 0.2 - 0.3\,\text{Å}$.

The renormalization of phonon frequencies according to (6.1) and (6.2) is determined by the relations

$$\Delta\omega_\nu \sim |g_\nu|^2 \chi_\nu , \qquad\qquad\qquad\qquad (6.10)$$

where χ_ν is the electronic susceptibility for a given vibration mode ν. As has been noted [6.25], for the p-d model of the type (5.6) the susceptibility χ_{pd} related to charge transfer through the gap Δ_{pd} appears much less than the susceptibility χ_{pp} describing charge transfer between oxygen ions inside the primitive cell CuO_2. Therefore one should anticipate a rather strong renormalization of frequencies for the vibration modes in which the equivalence of oxygen ion positions is violated and the shift (6.10) is determined by the susceptibility χ_{pp}. The renormalization of frequencies related to p-d transitions and the susceptibility χ_{pd}, e.g. for the "breathing" mode, will be small. An essential ionic contribution to electron–phonon interaction has also been found in [6.26]. Using the frozen-phonon approach in the frame of LMTO (linearized muffin tin orbital) method the electron–phonon interaction coupling constants were computed for some displacements of ions. Especially large coupling constants were obtained for vibrations of apical oxygen when accounting for charge transfer.

A detailed study of lattice structure and dynamics for LSCO and YBCO on the basis of the linear augmented plane wave method (LAPW) has been performed [6.27, 28, 5.27]. It proves possible to perform sufficiently precise calculations of the ground state energy of the crystal, and, using the frozen-phonon approach, to find the frequency and the polarization vector for certain modes in high symmetry points of the Brillouin zone. In particular, it was shown that the La_2CuO_4 lattice (in the metallic phase) is stable with respect to the condensation of the "breathing" mode – its frequency is high at the X-point of the Brillouin zone, but it is unstable with respect to the tilting soft mode [6.27]. In Fig. 2.19 the dependence of the crystal energy on rotations of octahedra describing this mode was shown. Similar calculations of a series of optical frequencies for YBCO have confirmed high reliability of this method [6.28]. For some optical A_g-mode sufficiently large coupling constants $\lambda_\nu \sim 1 - 2$ were obtained, which significantly exceed the corresponding values computed in the rigid muffin-tin approximation [5]. The latter disregards the ionic contribution due to changes of the Madelung potential at the displacement of the atom considered. These first-principles calculations confirm an essential influence of Coulomb forces on lattice dynamics and electron–phonon interaction.

An extensive band structure calculation of YBCO compounds by the LMTO method was performed by O. Anderson et al. [6.29] with special emphasis on the electron–phonon coupling constants. The latter, in particular, were used [6.11, 12] to calculate the phonon shift and linewidth in the strong coupling approximation (6.6). An analysis of the dependence of the coupling constants on the wavevector on the basis of the relation (6.4), $\lambda_{q\nu} \sim \gamma_{q\nu}/\omega_{q\nu}^2$, is available [6.29]. The estimate of the averaged constants leads [6.29] to the conclusion that the magnitude of the coupling constant is moderate in YBCO compounds, $\lambda \le 1$ (see also Sect. 7.4.2). There was remarked [6.28, 29] an irregular dependence of the susceptibility $\chi(q, \omega)$

---

[5] An effective coupling constant (6.4) in [6.27, 28] is defined as $\lambda = (1/3r) \sum_{\nu=1}^{3r} \lambda_\nu$. Compare with the previous definition, in footnote [3].

and linewidth $\gamma(q, \omega) \sim g^2 \operatorname{Im} \chi(q, \omega)$ on the wavevector q inside the Brillouin zone. In particular, the existence of saddle points on the Fermi surface $15-25$ meV away from the Fermi level leads to a strong mixing of lattice vibrations with quasiparticle excitations and to a weaker energy dependence of their linewidth $\gamma_\nu \sim (\varepsilon_k - \mu_F)^{3/2}$ than in the conventional Fermi liquid. These non-adiabatic effects may also lead to a violation of the Migdal approximation in the derivation of the Eliashberg equation.

As noted in Sect. 5.3.1, in electronic band structure calculations in the one-particle local density approximation the Coulomb correlations on copper sites are not properly taken into account, and therefore the picture of an electronic spectrum obtained in these calculations may differ from the real one-particle spectrum of excitations in copper-oxide compounds. The same problem of strong Coulomb correlations arises in investigations of electron–phonon interaction. One method [6.30] accounts for strong Coulomb correlations in electron–phonon interaction, based on the slave boson method [6.31]. In the case of the charge transfer semiconductor model, $U_d > \Delta = \varepsilon_p - \varepsilon_d > t_{pd}$ (see Table 5.1), strong Coulomb interaction U_d leads to a renormalization of the hopping integral $t_{pd} \to \tilde{t}_{pd} \cong t_{pd}\sqrt{\delta n}$, a sharp decrease of the $pd\sigma$ band width, $W \sim \tilde{t}_{pd}^2 \sim \delta n$, and a weakening of the electron–phonon interaction (6.8) at small hole concentrations δn. There also arises a strong renormalization of the ε_d energy level, but the positions of the ε_p levels and the electron–phonon interaction (6.9) on oxygen sites remain unchanged. Thus, accounting for strong Coulomb correlations U_d in the mean field approximation for slave bosons should not weaken the electron–phonon interaction on oxygen sites.

In the t-J model (5.11) describing hole motion as singlet states, the electron–phonon interaction leads to modulation of the hopping integral $t_{ij} \sim t_{pd}^2/\Delta$ and the exchange interaction (5.13) $J \sim t_{pd}^4/\Delta^3$. In both cases one should anticipate only a weak effect of the electron–phonon interaction on the spectrum of quasiparticles. A significantly larger effect arises from antiferromagnetic fluctuations leading to formation of magnetic polarons (see Sect. 7.1). As mentioned above, a more essential role is played by phonon modes corresponding to relative displacements of oxygen ions. The latter lead to mixing of singlet states with oxygen bands of a different symmetry ($a_1 -$ see (5.9)), which cannot be represented within the framework of the t-J model.

The effects of Coulomb interaction and electron–phonon interaction for apical oxygen in YBCO were investigated [6.32] by an exact diagonalization for a cluster of O4 – Cu1 – O4. This study accounts for an interaction both with Raman-active (A_g type) and infrared-active (B_{1u} type) vibrations already discussed in Sect. 6.2. Direct calculation of the wave functions for hole states on oxygen and the excitation spectrum demonstrated that for an infrared-active phonon one observes a correlated motion of a hole and ions in the form of a polaron, which are observed in infrared absorption spectra (and in the EXAFS method) as a double-well potential for O4. In Raman spectra, even for strong coupling, no anomalies are observed and the bound states are featured by a one-well potential. These investigations explain the discrepancies in the interpretation of the results for Raman scattering on the

A_{1g}-mode of apical oxygen, and infrared absorption on the B_{1u} mode, discussed at the end of Sect. 6.2.

In a number of works it was supposed that a strong electron–phonon interaction may arise in the case of anharmonic vibrations of ions in a double-well potential. A model of anharmonic vibrations of apical oxygen interacting with carriers in CuO_2 planes was proposed [6.33]:

$$H = -\Omega \sum_i S_i^z + g \sum_{i\delta} n_{i+\delta} S_i^x + H_0 \,, \tag{6.11}$$

where H_0 is the Hubbard Hamiltonian (see Sect. 7.1). Pseudo-spin operators S_i^α represent the ground E_0 ($S_i^z = 1/2$) and the first excited E_1 ($S_i^z = -1/2$) states of oxygen in a double-well at a site i having the energy difference $\Omega = E_1 - E_0$. The second term in (6.10) determines the electron–phonon interaction of these pseudo-spin excitations, i.e. strongly anharmonic phonons, with the carrier density $n_{i+\delta}$ on four oxygen sites δ inside the cell i. The effect of the electron–phonon interaction of this type on superconducting pairing is discussed in Sect. 7.4.

Another model to describe strongly anharmonic vibrations of oxygen ions in the tilting soft mode and their interaction with the deformations has been developed [6.34, 35]. On the basis of this model a microscopic theory of structural phase transitions in LMCO-compounds was developed (see Sect. 2.2.2). The Hamiltonian of the model reads

$$H = H_R + H_{R-\varepsilon} + H_\varepsilon \,, \tag{6.12}$$

$$H_R = H_0 + (B_0/4) \sum_{l,k=1,2} u_z^4(l,k) + \sum_{l,k=3,4} \{(B_1/4)[u_x^4(l,k)$$
$$+ u_y^4(l,k)] + (B_2/2)u_x^2(l,k)u_y^2(l,k)\} \,, \tag{6.13}$$

where H_0 is the harmonic part of the soft mode Hamiltonian H_R, B_0, B_1, B_2 are the anharmonic constants, $u_\alpha(l,k)$ is the α-component of the in-plane ($k = 1,2$) and the apical ($k = 3,4$) oxygen atom displacement in the l-th primitive cell (see Fig. 2.3). The interaction of atomic displacements with local deformations $\varepsilon_{\alpha\beta}$ is of the form

$$H_{R-\varepsilon} = \sum_{l,k} g_{\alpha\beta,\delta\gamma}^k \varepsilon_{\alpha\beta}(l) u_\delta(l,k) u_\gamma(l,k) \,, \tag{6.14}$$

where the coupling constants are given by the tensor $g_{\mu\nu}^k$ (in Voigt notation: $\alpha\beta = \mu$, $\delta\gamma = \nu$) having only seven independent components: g_{12}, g_{23}, g_{33} ($k = 1,2$) and g_{31}, g_{11}, g_{12}, g_{66} ($k = 3,4$). H_ε in (6.12) determines the deformation energy of the crystal in the tetragonal phase (see (2.4)).

To describe oxygen atom displacements in the soft mode which are rigid rotations of CuO_6 octahedra (see Sect. 6.1), requires the local normal coordinates $R_\lambda(l)$ – rotations of an octahedron round the axes $\lambda = x, y$ in a primitive cell l. Then the individual displacements of each oxygen atom may be represented in the form

$$u_z(l, k = 1, 2) = (1/2\sqrt{2m})[R_\lambda(l + a_k) - R_\lambda(l)], \qquad (6.15)$$

$$u_\alpha(l, k = 3) = -u_\alpha(l, k = 4) = (1/\sqrt{2\tilde{m}})R_\lambda(l), \qquad (6.16)$$

where $\lambda = x(y)$ for $k = 2(1)$ or $\alpha = y(x)$. The effective mass of the apical atom is $\tilde{m} = m(d_x/d_z)^2$, where $d_x = a_x/2$ and d_z are the Cu – O distances in the plane and along the z-axis respectively (Sect. 2.2). The tilting soft mode is determined by the linear combination of the rotations $R_{1,2}(l)$ according to (2.1):

$$R_{1,2}(l) = (1/\sqrt{2})(R_x(l) \mp R_y(l)). \qquad (6.17)$$

Its average value below the structural transition temperature T_0 is related to the macroscopic order parameter $C_i = (C_1, C_2)$ via the relation

$$\langle R_i(l) \rangle = C_i \exp(iq_i l). \qquad (6.18)$$

where $q_{1,2} = k_x(1, 2)$ is the wavevector (2.2) in the X-point of the Brillouin zone.

Substituting the representation (6.15) and (6.16) in (6.12), the Hamiltonian of the model in terms of the local normal coordinates $R_\lambda(l)$ or $R_i(l)$ (6.17) is obtained. The free energy of this model has been computed [6.35] and the temperature dependence of the order parameter C_i and the soft mode frequency Ω_s^2 investigated. Near the structural transition temperature T_0 the free energy is written in the form of the expansions (2.3–2.5) with the parameters depending on the interaction constants in the model (6.12): $(u, v) \propto (B_0, B_1, B_2)$, $(\alpha, \beta, \gamma) \propto (g_{\mu\nu})$. Comparing the results of the model calculation with the ground state energy calculation by the density functional method (see Fig. 2.9) one also gets the absolute values of the parameters: $u = 560\,(\mathrm{meV})^3$, $v = 1440\,(\mathrm{meV})^3$. The coupling constants of the soft mode to deformation (α, β, γ) are determined by comparing the calculated changes of the elastic coefficients (2.6) under the structural phase transition with experimental values. In particular, the interaction with the deformations ε_{xy} is sufficiently strong: $\gamma^2/C_{66} \simeq 180\,(\mathrm{meV})^3$. Thus a small change of the parameters u and v in the expansion (2.3) leads to a reversion of sign in the inequality $v - 4\gamma^2/C_{66} > u$ and to the transition HTT→LTT changing to HTT→LTO (see Sect. 2.2.2). In the microscopic model (6.13) the parameters (u, v) are expressed in terms of the anharmonic constants B_0, B_1, B_2 whose value depend on the ionic radius of the impurity M in $\mathrm{La}_{2-x}\mathrm{M}_x\mathrm{CuO}_4$. This dependence of the parameters of the model (6.13) on the concentration of impurities qualitatively explains the $T - x$ phase diagram for isovalent substitutions M=Sr, Ba and Ca or M=RE=La, Nd, Sm, and Gd (Sect. 2.2.2).

With a strong anharmonicity when the potential barrier height in the double-well potential for the soft mode is comparable to the phonon frequencies the model (6.12) may be written in a pseudo-spin representation. Accounting like in the model (6.11) only for the two lowest energy states in the double-well potential, E_0 and E_1, the Hamiltonian (6.13) may be represented in the form [6.36]

$$H_s = -\Omega \sum_{l,\lambda} S_{l\lambda}^z - \frac{1}{2} \sum_{lm\lambda\mu} J_{\lambda\mu}(l - m)S_{l\lambda}^x S_{m\mu}^x, \qquad (6.19)$$

where $\Omega = E_1 - E_0$ and the pseudo-spin operator $S_{l\lambda}^\alpha$ are related to the local normal mode coordinates: $R_\lambda(l) \propto 2\langle 0 \mid R_\lambda(l) \mid 1\rangle S_{l\lambda}^x$. The interaction of the rotations in different cells is described by the function $J_{\lambda\mu}(l - m)$. The phase transition in the model (6.19) according to (6.18) is due to an appearance of a nonzero average of the pseudo-spin operator: $\langle S_{l\lambda}^x\rangle \propto C_i \exp(iq_i l)$, which corresponds to a certain ordering of octahedra rotations. In this case the structural phase transition is viewed as an order-disorder transition in a crystal where in each cell there are four equilibrium positions (two for each rotation). These are determined by the four minima in the potential energy of the crystal in Fig. 2.9 [2.18].

To consider the interaction of doped holes in the LMCO crystal with the soft tilting mode we employ the ionic lattice model with the electron–phonon interaction given by (6.9). In the ordered low-temperature phases of the LMCO (where superconductivity occurs) with a nonzero value of the order parameter (6.18) (see Sect. 2.2.2) a linear electron–phonon coupling is possible. It can be written in the pseudo-spin representation for the model (6.19) in the form

$$H_{el-ph} = \sum_{l\lambda} g_\lambda(l) S_{l\lambda}^x n_l, , \qquad (6.20)$$

where n_l is the hole number operator for the unit cell l in the LMCO crystal. The coupling constant according to (6.9) $g_\lambda(l) \cong \beta V_c (u/d)^2 \cong (10^{-1} - 10^{-2}) V_c$ may be of moderate value only for the strong Coulomb interaction $V_c \cong 10\,\mathrm{eV}$. But, as will be shown in Sect. 7.4, even for a small value of the coupling constant $g_\lambda(l)$ the effective electron – electron coupling λ due to soft phonons may bring about a strong coupling and high value of T_c in the anharmonic model for superconductor (see (7.63)).

To summarize the studies of lattice dynamics of copper-oxide compounds we should point out the following results:

1. The Coulomb forces play the major role in the lattice dynamics of copper-oxide compounds both in insulating and metallic phases. In the latter case a highly anisotropic screening occurs which is weak in the c-direction and strong in the basal, CuO_2 plane.

2. "First principles" band structure calculations in the local density approximation point to a strong or moderate electron–phonon coupling for some phonon modes. But to allow for the strong Coulomb correlations on copper sites the ionic lattice model may be used to estimate the electron–phonon coupling due to the long-range Coulomb forces. The latter are much more important in copper-oxides in comparison to conventional metals with strong screening effects.

3. Highly anharmonic lattice vibrations for oxygen ions (e.g. in the A_g-mode for apex oxygen in YBCO or in the soft tilting mode in LSCO) are typical for oxide superconductors. Due to the large amplitude of vibrations in the double-well potential these modes may bring about strong electron–phonon coupling for doped holes and enhance the superconducting pairing.

7. Theoretical Models
of High-Temperature Superconductivity

In developing the theory of high-temperature superconductivity it is necessary to solve two problems which are of foremost importance and which are definitely interrelated: namely, what is the nature of the normal state for the electrons in the oxide compounds and what is the mechanism of the formation of the superconducting phase? While in conventional superconductors the picture of the Fermi liquid with a properly determined spectrum of quasi-particles near the Fermi surface is sufficiently well established, another mechanism for the formation of the ground state and of the spectrum of low-energy excitations is possible in copper-oxide compounds due to strong Coulomb correlations [7.1, 2]. Therefore, the Bardeen-Cooper-Schrieffer (BCS) theory of pairing which works perfectly well for the system of weakly bounded quasi-particles in conventional metals, can be of no use for the system of electrons with strong correlation, the interactions of which are not described by a quasi-particle picture.

In this connection, in the present chapter we shall first discuss the problem of the ground state and spectrum of one-electron excitations in the normal phase of copper-oxide compounds and then we shall discuss the most general mechanisms of a superconducting pairing.

7.1 Quasi-Particles in Models with Strong Correlation

To determine the spectrum of quasi-particles it is necessary to calculate the one-particle Green's function

$$G_\sigma(k, \omega) = \langle \Psi_{0N} | c_{k\sigma} \frac{1}{\omega - (H - E_0) + i\eta} c_{k\sigma}^+ | \Psi_{0N} \rangle, \tag{7.1}$$

where $c_{k\sigma}^+ (c_{k\sigma})$ is a creation (annihilation) operator of the electron or that of the hole with the wave vector k and spin σ above the ground state with the wave function Ψ_{0N} and the energy E_0 for a system of N particles. Singling out the energy of one-particle states (band states) $\varepsilon_\sigma(k)$ in the Hamiltonian of the system H, the Green's function can be represented in the general form

$$G_\sigma(k, \omega) = \frac{1}{\omega - \varepsilon_\sigma(k) - \Sigma_\sigma(k, \omega)}. \tag{7.2}$$

The self-energy operator $\Sigma_\sigma(k, \omega + i\eta)$ describes many-particle corrections of the spectrum $\varepsilon_\sigma(k)$. Its real part, ReΣ, and imaginary part, Im Σ, determine the renormalization and damping of the spectrum of one-particle states, respectively.

If the self-energy operator depends smoothly on the energy ω near the Fermi surface, it can be expanded in powers of $[\omega - \xi(k)]$ and the causal Green's function (7.2) can be written in the form of two contributions

$$G(k, \omega) = G_{QP}(k, \omega) + G_{\text{inc}}(k, \omega), \tag{7.3}$$

$$G_{QP}(k, \omega) = \frac{Z(k)}{\omega - \xi(k) + i \operatorname{sgn} \xi(k)}, \tag{7.4}$$

$$Z(k) = [1 - \frac{d}{d\omega} \operatorname{Re} \Sigma(k, \omega)]^{-1}_{\omega \to \xi(k)}. \tag{7.5}$$

Here the quasi-particle spectrum is determined from the equation

$$\xi(k) = \varepsilon(k) + \operatorname{Re} \Sigma(k, \omega = \xi(k)),$$

where $\xi(k_F) = 0$ on the Fermi surface. The first pole term in (7.3) describes a quasi-particle contribution with the weight $Z(k)$ and the second one describes a incoherent contribution which varies smoothly on the Fermi surface. The quasi-particle weight, $Z(k)$, (7.5) determines the degree of overlap of one-particle and many-particle states in the system and is connected with a jump of the distribution function on the Fermi surface

$$[n(k_F - \delta) - n(k_F + \delta)]_{\delta \to 0} = Z(k_F). \tag{7.6}$$

Coherent propagation with a fixed value of the wave vector k is possible only for a finite value of $Z(k)$.

For the standard model of the Fermi liquid for $T \to 0$ we have Im $\Sigma(k, \omega) \propto \omega^2$ and Re $\Sigma(k, \omega) \propto \omega$ which give a finite value for the quasi-particle weight $Z(k_F)$. On the other hand, in the one-dimensional Hubbard model for any finite value of the Coulomb repulsion U, the distribution function $n(k)$ varies smoothly near the Fermi surface

$$n(k) = n(k_F) - C|k - k_F|^\alpha \operatorname{sgn}(k - k_F),$$

where $\alpha < 1$ (see, for example, [7.3]). In such a system, which is called the Luttinger, or Thomonaga-Luttinger, liquid, the spectrum of excitations consists only of collective branches which describe fluctuations of spin and charge. According to Anderson [7.2], in the two-dimensional Hubbard model, electrons behave as a Luttinger liquid with a separation of spin and charge degrees of freedom. The absence of quasi-particle excitations in the Luttinger liquid and a non-standard Fermi surface does not permit the application of the usual ideas superconducting pairing from BCS theory.

The intermediate quasi-particle picture was proposed in the theory of 'marginal' Fermi liquid [7.4]. The theory is based on the assumption of an existence of spin and charge fluctuations with a wide spectrum of excitations which does not depend on the wave number q:

$$-\operatorname{Im} P(q, \omega) \propto \text{const (for } T < | \omega | < \omega_c)$$

or

$$-\operatorname{Im} P(q, \omega) \propto \omega / T \text{ (for } \omega \ll T) .$$

The interaction of electrons (holes) with these fluctuations leads to the self-energy operator

$$\Sigma(k, \omega) \propto \omega \ln(x/\omega_c) - \mathrm{i}(\pi/2)x$$

where $x = \max (| \omega |, T)$. In the ground state, we obtain

$$Z(k) \propto \ln | \xi(k)/\omega_c |^{-1} \to 0$$

for $T \to 0$ on the Fermi surface $\xi(k_F)$. Thus, the quasi-particle weight logarithmically tends to zero, disturbing minimally the picture of quasi-particles in the standard theory of the Fermi liquid.

Experimental investigations of the electronic structure of the copper-oxide compounds (Sect. 5.4) have proven the existence of a sufficiently large Fermi surface but they are not exact enough to find the jump in the distribution function on the Fermi surface and to prove the existence of quasi-particles according to (7.6).

In the following we shall study the basic models for the system of electrons with strong correlation and the results of the calculations of the quasi-particle spectrum.

7.1.1 Model Hamiltonians

As was noted in Sect. 5.3, the electronic properties of copper-oxide compounds are determined by charge carriers in the CuO_2 plane. Here the most important interactions are described by the effective p-d Hamiltonian (5.6). Here we shall consider the p-d model with a minimum number of parameters

$$
\begin{aligned}
H = \varepsilon_d \sum_{\iota,\sigma} n_{i\sigma}^d + \varepsilon_p \sum_{m,\sigma} n_{m\sigma}^p \\
+ U_d \sum_i n_{i\uparrow}^d n_{i\downarrow}^d + t_{pd} \sum_{im\sigma} S_{im}(d_{i\sigma}^+ p_{m\sigma} + \text{H.c.}),
\end{aligned}
\tag{7.7}
$$

where $d_{i\sigma}^+$ ($d_{i\sigma}$), $p_{m\sigma}^+$ ($p_{m\sigma}$) are the creation (annihilation) operators of the holes with spin σ in the copper sites i and in the oxygen sites m on the square lattice CuO_2, $n_{i\sigma}^d = d_{i\sigma}^+ d_{i\sigma}, n_{m\sigma}^p = p_{m\sigma}^+ p_{m\sigma}$, ε_d and ε_p are the energies of $3d(x^2 - y^2)$ or $O2p_\sigma(x, y)$ holes measured from the chemical potential μ, U_d is the Coulomb repulsion and t_{pd} is the integral of the overlap of the p-d states. The sign factor S_{im} is taken in the form

$$-S_{ii+x} = S_{ii+y} = S_{ii-x} = -S_{ii-y} = 1 ,
\tag{7.8}$$

where $m = i \pm x, i \pm y$ are the oxygen sites m nearest to the copper site i along the axes x and y, respectively, on the plane CuO_2 (see Fig. 5.13).

The parameters of the Hamiltonian (7.7) can be evaluated with the help of the data of Tab. 5.1. The reduction of their number in comparison with the total Hamiltonian (5.6) should result in the renormalization of the parameters. According to [7.5], the closest description of experimental data with the help of the Hamiltonian (7.7) can be obtained if the following relations for the parameters are accepted: $\Delta = \varepsilon_p - \varepsilon_d = 4t_{pd}$ and $U_d = 6t_{pd}$ which agrees with the data of Tab. 5.1. For these parameters, the band of Cu3d-states splits into two Hubbard subbands of one- and two-hole states, and bands of O2p-states lie between the Hubbard Cu3d-subbands as it is shown in Fig. 5.2a.

Under doping of the plane CuO$_2$ by more than one 3d-hole per unit cell, a complicated redistribution of the charges due to the hybridization t_{pd} and the Coulomb correlations U_d arises. The creation of singlet states of 3d- and 2p-holes the energy of which is inside the gap Δ is of foremost importance here. The creation of a band of singlet states permits us to carry out the further reduction of the Hamiltonian (7.7) to the t-J model (5.11) [5.44, 45]. Let us carry out this transformation according to [5.45, 7.6].

Consider the linear combination of O2p-states of the $p(b_1)$ type (5.9)

$$p_{i\sigma} = \frac{1}{2} \sum_m S_{im} p_{im} . \tag{7.9}$$

Another combination of the $p(a_1)$ type in (5.9) does not hybridize with the Cu3d-states and creates a non-bonding O2p-band without dispersion (in the approximation $t_{pp} = 0$). However, the states (7.9) are not orthogonal and, therefore, it is more convenient to pass to the Wannier operators $a_{j\sigma}$

$$p_{i\sigma} = \sum_j \lambda_{ij} a_{j\sigma} , \qquad \lambda_{ij} = \frac{1}{2N} \sum_k \gamma_k e^{ik(i-j)} , \tag{7.10}$$

where $\gamma_k = 2[\sin^2(k_x/2) + \sin^2(k_y/2)]^{1/2}$. Here N is the number of the copper sites in the lattice and the lattice constant a is taken to be equal to unity. If one substitutes (7.9) and (7.10) into the Hamiltonian (7.7), this yields an effective Hamiltonian depending only on lattice sites i, j, with the hybridization parameter $V_{ij} = 2t_{pd}\lambda_{ij}$ where $\lambda_0 = \lambda_{ii} \simeq 0.96$; $\lambda_1 = -0.14$ for $(i-j) = \pm(a_x, a_y)$; $\lambda_2 = -0.02$ for $(i-j) = \pm(a_x \pm a_y)$ etc.

Neglecting further the two-hole Cu3d^8−states in the limit $U_d \to \infty$, we obtain the representation

$$H = \sum_i \sum_\sigma [\varepsilon_p a_{i\sigma}^+ a_{i\sigma} + \varepsilon_d X_i^{\sigma\sigma} + V_0(X_i^{\sigma 0} a_{i\sigma} + \text{H.c.})]$$

$$+ \sum_{i \neq j} \sum_\sigma (V_{ij} X_i^{\sigma 0} a_{j\sigma} + \text{H.c.}) \tag{7.11}$$

for the Hamiltonian (7.7). Here the Hubbard operators act in the space of 3d-states having a single particle occupation

$$X_i^{\sigma\sigma} = n_{i\sigma}(1 - n_{i-\sigma}) = X_i^{\sigma 0} X_i^{0\sigma} ,$$
$$X_i^{\sigma 0} = d_{i\sigma}^+(1 - n_{i-\sigma}) . \tag{7.12}$$

The first single-site part of the Hamiltonian (7.11) can be reduced to the diagonal form in the space of one- and two-hole states and the second one can be considered as a small perturbation for $i \neq j$, since $V_{i \neq j} = 2t_{pd}\lambda_{ij} \ll V_0$. Let us consider this transformation.

We shall introduce one-hole p-d-states taking into account their hybridization in one cell. The lowest state of the d-type is determined by the function

$$|f_\sigma\rangle = \cos\theta_1 d_\sigma^+ |0\rangle - \sin\theta_1 a_\sigma^+ |0\rangle , \tag{7.13}$$

where $|0\rangle$ is a vacuum state without holes. The angle θ is found from the equation $\tan2\theta_1 = 2V_0/\Delta$ and the state energy (7.13) is equal to

$$E_d = (\varepsilon_p + \varepsilon_d)/2 - \sqrt{\Delta^2/4 + V_0^2} . \tag{7.14}$$

The hybridization results in lowering of the energy of the d-state by a value of the order $V_0^2/\Delta \simeq 4t_{pd}^2/\Delta$ and to an increase of the energy of the p-states by the same amount $E_p \simeq \varepsilon_p + V_0^2/\Delta$ (for $V_0 \ll \Delta$). Besides, owing to (7.13), a redistribution of charge between the d- and p-states takes place.

Two-hole states can be built from the singlet

$$|\phi(1,1)\rangle = \frac{1}{\sqrt{2}}(d_\uparrow^+ a_\downarrow^+ - d_\downarrow^+ a_\uparrow^+)|0\rangle ,$$

$$|\phi(2,0)\rangle = a_\uparrow^+ a_\downarrow^+ |0\rangle \tag{7.15}$$

and triplet states

$$\tau_0 = \frac{1}{\sqrt{2}}(d_\uparrow^+ a_\downarrow^+ + d_\downarrow^+ a_\uparrow^+)|0\rangle ,$$

$$|\tau_{2\sigma}\rangle = d_\sigma^+ a_\sigma^+ |0\rangle . \tag{7.16}$$

Taking into account the interaction V_0 in (7.11) does not lead to the hybridization of triplet states of the $d - p$-type the energy of which remains equal to $E_\tau = \varepsilon_d + \varepsilon_p$. The singlet states (7.15) turns out to be bounded. The lowest level of the $d - p$-type is described by the function

$$|\phi\rangle = \cos\theta_2 |\phi(1,1)\rangle - \sin\theta_2 |\phi(2,0)\rangle , \tag{7.17}$$

where $\tan2\theta_2 = 2\sqrt{2}V_0/\Delta$. Its energy decreases with respect to the initial level $|\phi(1,1)\rangle$

$$E_\phi = [(\varepsilon_d + \varepsilon_p) + 2\varepsilon_p]/2 - \sqrt{\Delta^2/4 + 2V_0^2} . \tag{7.18}$$

On the other hand, the excitation energy of the singlet state of the $p - p$-type $|\phi(2,0)\rangle$ increases. $E_{p-p} \simeq 2\varepsilon_p + 2V_0^2/\Delta$.

Thus, taking into account the hybridization V_0 in one cell and excluding $3d$ states with double occupation ($U_d \to \infty$) leads to a splitting of the two-hole p-d-states into a degenerate triplet level $E_\tau = \varepsilon_d + \varepsilon_p$ and a singlet level with lower energy E_ϕ (7.18). For $V_0 \ll \Delta$ their difference is given by the value $E_\tau - E_\phi \simeq 2V_0^2/\Delta \simeq 8t_{pd}^2/\Delta$ which agrees with the cluster calculations considered in Sect. 5.3. When there are no correlations, $U_d = 0$, the split of the singlet and triplet p-d-bands disappears.

Consequently, in the p-d-system, excitations of two types occur: low energy spin excitations when the singlet-triplet transition takes place and high-energy charge excitations when the d-hole is transferred to the p-orbital with the energy $E_{p-p} - E_\phi \simeq \Delta \gg E_\tau - E_\phi$. This separation of the excitations into ones mainly connected with spin and with charge degrees of freedom is a specific property of the systems with strong electron correlation [7.7].

To simplify further the Hamiltonian (7.11) one should project the initial operators of the p- and d-holes onto new states with one hole (7.13) and with two holes in the triplet (7.16) or in the singlet (7.17) states. As a result, we obtain new Hubbard operators acting in the space of states (7.13), (7.16), and (7.17)

$$
\begin{aligned}
Y^{\phi\sigma} &= | \phi\rangle\langle f_\sigma |, \\
Y^{\tau\sigma} &= (1/\sqrt{3}) | \tau_0\rangle\langle f_\sigma| + \sqrt{2/3} | \tau_{2\bar\sigma}\rangle\langle f_{\bar\sigma} | .
\end{aligned}
\tag{7.19}
$$

With the help of these operators and taking into account only states in the singlet-triplet band, the Hamiltonian (7.11) can be written in the form [7.6]

$$
\begin{aligned}
H_{\phi\tau} = {}& E_f \sum_{i\sigma} Y_i^{\sigma\sigma} + E_\phi \sum_i Y_i^{\phi\phi} + E_\tau \frac{3}{2} \sum_i Y_i^{\tau\tau} \\
&+ \sum_{i\neq j\sigma} V_{ij}[K_{\phi\phi} Y_i^{\phi\sigma} Y_j^{\sigma\phi} + K_{\tau\tau} Y_i^{\tau\sigma} Y_j^{\sigma\tau}],
\end{aligned}
\tag{7.20}
$$

where the contribution $K_{\phi\tau}$ describing the weak interaction of singlet-triplet states is omitted. The operators of the particle number in the σ-, ϕ-, and τ-states are determined in the form

$$
\begin{aligned}
Y_i^{\sigma\sigma} &= Y_i^{\sigma\phi} Y_i^{\phi\sigma}, \\
Y_i^{\phi\phi} &= Y_i^{\phi\sigma} Y_i^{\sigma\phi}, \qquad Y_i^{\tau\tau} = Y_i^{\tau\sigma} Y_i^{\sigma\tau}.
\end{aligned}
$$

The dispersion of the $\phi-$ and $\tau-$bands is connected with the coefficients $K_{\phi\phi}$ and $K_{\tau\tau}$ which depend on the hybridization parameters θ_1 and θ_2 in (7.13), (7.17) [7.6]. In the limit of small p-d-hybridization, $t_{pd} \ll \Delta$, one gets the usual value for the hopping integral in the t-J model (5.11). But as it has already been shown, the successive contribution of the strong hybridization in one cell when new states (7.13) and (7.17) are introduced also permits one to study the region of charge fluctuations, $V_0 \simeq 2t_{pd} \sim \Delta$, and to construct an effective one band t-J model under the condition for only a small overlap in the neighboring cells, $V_{i\neq j} \simeq 2\lambda_1 t_{pd} \ll \Delta$. Similar results were obtained in [7.8].

The one-band Hubbard model

$$
H = \sum_{ij\sigma} t_{ij} a_{i\sigma}^+ a_{j\sigma} + U \sum_i n_{i\uparrow} n_{i\downarrow},
\tag{7.21}
$$

where t_{ij} is an effective transfer integral and U is the Coulomb one-site energy, is often used for numerical calculations. Formally, the Hamiltonian (7.21) can be obtained from the p-d model (7.7) if the limit $\Delta = \varepsilon_p - \varepsilon_d \gg U_d$ is taken. In this case, the p-states are situated higher than the two-hole d-states and so can be excluded from the treatment. The model (7.21) permits one to consider both the case of weak correlations, $U \ll W$, and the case of strong correlations, $U \gg W$, where $W \propto | t_{ij} |$ is the bandwidth. In the case of strong correlations, the Hamiltonian (7.21) is also reduced to the t-J model (5.11) with an exchange integral $J = 4t^2/U$ where t is the transfer integral for the nearest neighbors. Some solutions for the model (7.21) are discussed, for example, in [7.7].

7.1.2 The One-Hole Quasi-Particle Spectrum

The study of the spectrum of quasi-particle excitations for a hole in an antiferromagnetic crystal revealed a rather complicated picture of its motion, and nowadays this is a separate field of research (see the reviews [7.9, 10]). Below, we shall discuss some results obtained for the t-J and p-d models, the account of which is necessary in the discussion of the theory of the superconducting pairing in the copper-oxide compounds.

The first time that the principle role played by spin correlations for holes in antiferromagnetic insulators was recognized was in [7.11–13]. In [7.11], it was shown that when the hole moves in a two-dimensional lattice with antiferromagnetic spin ordering (Néel state), it generates a chain of flipped spins – a string – the creation of which costs an energy proportional to the length of the path. As a result, the hole turns out to be trapped in a potential well in which the potential depends linearly on the distance. The spectrum of excitations for this quasi-oscillatory state turns out to be discrete [7.11]. The contribution of the interaction $J_\perp$ for the transversal spin components leads to string relaxation, as a result of which coherent motion of the hole is possible. The first evaluation of its effective mass was given in [7.12]. The spectrum of incoherent excitations of the hole in the limit $J \to 0$ was obtained in [7.13].

The picture of hole motion in the Heisenberg antiferromagnet can be described in the most simple form if a spin-polaron model is used [7.14, 15]. In this model for the two-sublattice Néel antiferromagnet, the creation operators of the hole are introduced with the help of the relations $c_{i\uparrow} = h_i^+$ for one sublattice and $c_{i\downarrow} = h_i^+ S_i^+$ for the other, where S_i^+ is the spin lowering operator. Using further the linear-spin-wave approximation, the Hamiltonian of the t-J model (5.11) is written in the form

$$H = \sum_q \omega_q \alpha_q^+ \alpha_q + \frac{zt}{\sqrt{N}} \sum_{kq} M(k,q) h_k^+ h_{k-q} \alpha_q + \text{H.c.} . \tag{7.22}$$

Here $\omega_q = \frac{1}{2} z J_z \nu_q$ is the spectrum of spin waves where $\nu_q = [1 - (\alpha \gamma_q)^2]^{1/2}$, $\gamma_q = (1/2)(\cos q_x + \cos q_y)$, $\alpha = J_\perp/J_z$ is the ratio of the exchange constants for transversal ($S_i^\pm$) and longitudinal (S_i^z) spin components, z is the coordination number ($z = 4$ for a square lattice). The creation operators α_q^+ and annihilation operators α_q of spin waves are connected with the Bose-operators $b_i \simeq S_i^+$, $b_i^+ \simeq S_i^-$ by the Bogolubov transformation

$$\alpha_q = u_q b_q - v_q b_{-q}^+$$

where $u_q = [(1 + \nu_q)/2\nu_q]^{1/2}$, $v_q = -\text{sgn}(\gamma_q)[(1 - \nu_q)/2\nu_q]^{1/2}$.

The matrix element of the spin-hole interaction is equal to $M(k,q)=(u_q\gamma_{k-q}+v_q\gamma_k)$. Unlike the well-known electron–phonon polaron model, the Hamiltonian (7.22) does not contain a term with the kinetic energy of the hole which describes its free propagation.

In this case, the one-particle function (7.2) for the hole takes the form

$$G_h(k,\omega) = \frac{1}{\omega - \Sigma(k,\omega)} . \tag{7.23}$$

In the self-consistent Born approximation [7.14–18] where the renormalization of the vertex is not taken into account, for the self-energy operator we obtain

$$\Sigma(k,\omega) = \frac{z^2 t^2}{N} \sum_q \frac{M^2(k,q)}{\omega - \omega_q - \Sigma(k - q, \omega - \omega_q)} . \tag{7.24}$$

The spectrum of the quasi-particle hole excitations $\xi(k) = \Sigma(k, \omega = \xi(k))$ and the quasi-particle weight $Z(k)$ (7.5) is determined by the self-consistent solution of the equations (7.23, 24).

In the case of the Ising model, $J_\perp = 0$, $\alpha = 0$, the system of equations (7.23, 24) is easy to solve, since the self-energy operator (7.24) does not depend on the wave number

$$\Sigma(\omega) = \frac{zt^2}{\omega - \omega_0 - \Sigma(\omega - \omega_0)} \tag{7.25}$$

for $\omega_q = \omega_0 = 2J_z$ and $M(k,q)= \gamma_{k-q}$. The iterative solution of (7.25) leads to a spectrum of the ladder type with the distance between the levels of the order $t(J/t)^{2/3}$ [7.11]. For $J_z = 0$, the solution (7.25) is of the form $E(\omega) = (1/2)(\omega + (\omega^2 - 4zt^2)^{1/2})$ and the Green's function (7.23) determines the spectrum of incoherent excitations in the band $\pm 4t\sqrt{z}$. In a narrower band, $\pm 4t(z - 1)^{1/2}$, such a spectrum was obtained by using the retraceable-path approximation in [7.13]. Thus, in the Ising limit, $\alpha = 0$, the approximation (7.24) reproduces the main conclusion of the studies [7.11, 13] concerning the absence of a quasi-particle contribution in the Green's function (7.23).

The numerical solution [7.16–18] of the system of equations (7.23, 24) reveals the appearance of a quasi-particle peak near the bottom of the band $\omega = -zt$. The width of the quasi-particle band is proportional to J^β, $\beta \sim 0.8$, and the quasi-particle weight $Z(k) \propto J^\varepsilon$, $\varepsilon \sim 0.7$ for $J < t$. The analysis of the dispersion of the quasi-particle spectrum shows that it can be approximated by the function

$$\xi(k) \simeq t_{\text{eff}}(\cos k_x + \cos k_y)^2 , \tag{7.26}$$

where $t_{eff} \simeq J/2$. Such spectrum corresponds to the tight-binding approximation with a transfer integral t_{eff} between next-nearest neighbors, i.e., to the motion of the hole on one sublattice of the Néel antiferromagnet. For a small hole concentration, we can expect the appearance of the Fermi surface in the form of four pockets near the points $(\pm\pi/2, \pm\pi/2)$ of the Brillouin zone where the quasi-particle energy (7.26) has its minimum.

The general picture of the quasi-particle spectrum for the hole obtained within the framework of the polaron model (7.22) was verified in the numerical calculations by the method of the exact diagonalization of the Hamiltonian of the t-J model for small clusters and by the quantum Monte Carlo method [7.9, 10].

Here we discuss, for example, the results obtained by the Lanczos method of the exact diagonalization for the t-J model in [7.19]. The spectral density of excitations for one hole in the cluster consisting of 16 lattice sites is shown in Fig. 7.1. For the Néel ground state in the limit $J \to 0$, only an incoherent spectrum of excitations (Fig. 7.1a) obtained earlier in [7.13] occurs. To obtain a quasi-particle peak, it is necessary to introduce a finite Ising interaction J_z which forms a quasi-oscillator state (string). If one chooses the wave function for the Heisenberg quantum model (singlet state) to be the ground state Ψ_{ON} for the calculation of Green's function of the hole (7.1), there arises a quasi-particle peak (at the bottom edge of the band) and an incoherent spectrum for higher energies at a finite value of the exchange interaction $J_\perp$ or J_z (Fig. 7.1b, where $J/t = 0.2$).

It is obvious that, for a small excitation energy $\omega \leq J$, the hole can move in a coherent way since the presence of quantum spin fluctuations in the ground state Ψ_{ON} and dynamic fluctuations due to the interaction of the transverse spin components, $J_\perp$, permits the retention of a 'polaron dressing' for the quasi-particle. For large energies, $\omega \simeq t \gg J$, the hole loses this spin deformation cloud and the spectrum becomes incoherent. The coherent contribution can arise only if the hopping of the hole over the sites of one sublattice due to the overlap of the second neighbors is taken into account [7.6].

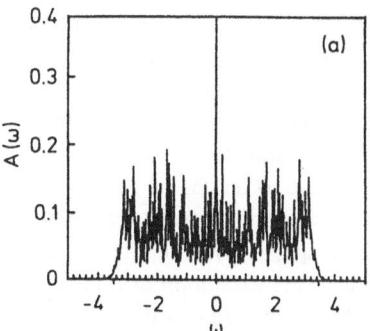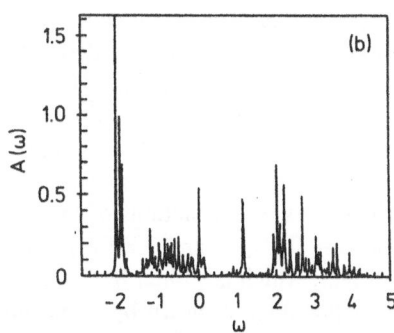

Fig. 7.1. The density of states for a single hole in the two-dimensional t-J model; in the Néel state for $J = 0$ (**a**) and for the exact singlet state in the quantum Heisenberg model for $J/t = 0.2$ (**b**) calculated by the method of exact diagonalization [7.19]

Quite a different picture of the quasi-particle spectrum appears in the t-J model for a finite hole concentration, $n_h \gtrsim 0.1$ [7.21, 22]. For clusters of 16–20 lattice sites with two holes, the spectrum of quasi-particles can be approximated by the function $\xi(\mathbf{k}) = -2t_{\text{eff}}(\cos k_x + \cos k_y)$ which corresponds to the spectrum in the tight-binding approximation for the nearest neighbors with the effective transfer integral $t_{\text{eff}}/t \simeq m/m^* \simeq J/1.5t$. In this case, a large electron Fermi surface arises, the form of which is similar to that of non-interacting particles (Fig. 5.12a). At the same time, the incoherent part of the spectrum with the bandwidth $(5 - 7)t$ is preserved and it indicates the presence of strong correlations between the charge and spin degrees of freedom. The analytic investigation of the excitation spectrum in the region of a moderate hole concentration is complicated, since the antiferromagnetic correlation length becomes comparable with the correlation length for two holes, and a simple polaron model (7.21) turns out to be inapplicable [7.23].

The change in the topology of the Fermi surfaces from a hole type to an electronic one with increasing concentration of holes was also discovered during the study of the motion of a polaron with a small radius in an antiferromagnetic matrix [7.24]. Supposing that the increase of the hole concentration leads to the frustration of antiferromagnetic couplings in the lattice, the authors have calculated the dependence of the quasi-particle spectrum of p-holes on the degree of frustration on the basis of the variational method. It turned out to be that with an increase of the frustration, the hole pockets in the neighborhood of the points $(\pm\pi/2, \pm\pi/2)$ in the Brillouin zone transform into one large electronic Fermi surface which is close to the Fermi surface of non-interacting particles (see Fig. 5.12a).

Besides the one-band t-J model, the spectrum of one-particle excitations in the multiband p-d model (7.7) has been studied in detail. Together with numerical methods for finite clusters (by exact diagonalization [7.20, 25, 26], quantum Monte Carlo method [7.27]) analytical methods, based on the projection technique, were also used [7.28, 29]. Let us consider the results obtained for a three-band model (7.7) with due regard for a direct hybridization of p-holes [7.29].

In this work [7.29] direct (PES) and inverse (IPES) photoemission spectra for p- and d-holes in the antiferromagnetic matrix were studied. To calculate them the Mori-Zwanzig projection technique for one-particle Green's functions (7.1) was used. In this method, by choosing a corresponding set of basic operators, the Green's function can be represented in the form of a matrix depending on a limited number of correlation functions [7.7]. To take into account strong correlations in the p-d model (7.7), besides one-particle operators, dynamical variables, including spin operators and charge fluctuations, were also considered as basic in [7.29]. This enabled photoemission spectra in the CuO$_2$ plane to be described in detail .

The results of calculations of the direct and inverse photoemission spectra are presented in Figs. 7.2 and 7.3 for p- and d-holes, respectively. The density of valence states, which are filled by electrons and where the excitation of a hole state (PES) is possible, is shown by a thin line; the density of states in the conduction band, where hole annihilation occurs, corresponding to the inverse photoemission process for the hole (IPES) is shown by a thick line. Here the excitation energy (in the units of t_{pd} for $\varepsilon_p = 0$, $\Delta = \varepsilon_p - \varepsilon_d = 4t_{pd}, t_{pp} = 0.6t_{pd}$) is shown on

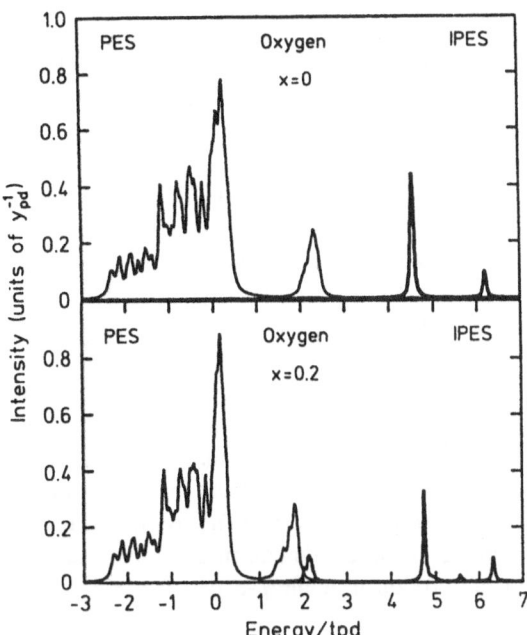

Fig. 7.2. Photoemission (PES, *thin lines*) and inverse photoemission (IPES, *thick lines*) spectra for p-holes for undoped ($x = 0$) and p-hole doped ($x = 0.2$) CuO_2 plane [7.29]

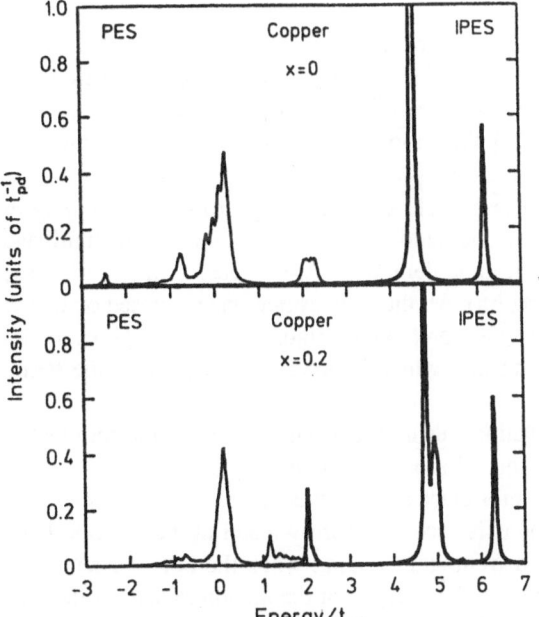

Fig. 7.3. PES and IPES spectra for d-holes (the notation is the same as on Fig. 7.2) [7.29]

the electronic scale where $\varepsilon_d > \varepsilon_p$. This scale is usually used in experimental investigations of PES, IPES, EELS (see Sect. 5.4.1).

Comparing p- and d- spectra we come to the following conclusions. There is a correlation gap between the singlet band (at $\omega \simeq 2$) and the upper Hubbard conduction band (at $\omega \simeq 4.5$) in the spectrum of one-particle excitations at half-filling ($x = 0$). The main contribution to the density of valence states (at $\omega \simeq 0$) comes from the non-bonding p-band and triplet p-d-states. Under doping there arise two main effects: the states from the conduction bands (IPES) appear in the neighborhood of the filled singlet band (PES) both in p- and d-spectra, which leads to the transition from an insulating to metallic state with a narrow band. The appearance of the conduction band inside the p-d gap is directly observed in O1s-spectra EELS in LSCO (Fig. 5.15), YBCO, and Bi/2212 (Fig. 5.17) compounds under doping.

The second important effect which appears under doping is the increase of the density of states in the singlet band due to decrease of the weight of the upper Hubbard band. This redistribution of the intensities and the small shifts of these bands are a consequence of strong correlation effects in the p-d model and are clearly observed in the optical absorption spectra. (The results for the t-J model are given in [7.22]). The increase of the Drude contribution under doping is accompanied by the decrease of absorption near the p-d gap $\hbar\omega \simeq 1 - 2\,\text{eV}$ (Figs. 5.21 and 5.22).

On the whole, the position of the peaks of PES and IPES as well as their dependence on doping, are in good agreement with the results of the exact diagonalization for the clusters [7.20, 25, 26]. Results for the density of quasi-particle excitations, close to those of [7.29], were obtained in the singlet-triplet model (7.20) [7.6] and in the two-band p-d model (7.7) where the Green's function was calculated using the composite operator technique [7.30]. The latter operators, as well as the basic operators in [7.29], take into account spin and charge fluctuations in p-hole motion.

The above results for the quasi-particle spectrum, appearing under doping of the antiferromagnetic matrix in the one-band t-J model, as well as in the multiband p-d model, shows the existence of quite definite quasi-particles with a finite value of the quasi-particle weight $Z(k)$ (7.5). All these investigations are based on one or another perturbation theory and, therefore, cannot rigorously prove the existence of quasi-particles on the Fermi surface, which is necessary for proving the Fermi liquid picture.

In this connection, the calculations, using the Monte Carlo method for clusters in the p-d model [7.27] and in the t-J model [7.31] are of foremost importance. For the first case, the calculations for clusters with dimensions of up to 4×4 CuO$_2$ elementary cells have proved the existence of a narrow band in the neighborhood of the Fermi surface, constructed from singlet states when the hole concentration is small ($x = 0.25$). At the same time, the analysis of the imaginary part of the self-energy operator of the Green's function (7.2) indicates a finite value of damping on the Fermi surface, which favors the Luttinger liquid model with $Z(k_F) = 0$.

Analysis of the jump in the distribution function $n(k)$ for one hole in the antiferromagnetic matrix for half-filling in the t-J model enables the dependence of $Z(k)$ on the dimensions of the cluster L to be calculated [7.31]. It turns out that

$Z(k) \propto L^{-\Theta}$, where the parameter Θ ranges from 0.4 to 0.7 depending on the value of J. According to the discussion in [7.31], such a strong dimensional effect (and vanishing of $Z(k)$ for $L \to \infty$) can be explained on the basis of "orthogonal catastrophe", considered for the two-dimensional Hubbard model with strong correlations by Anderson [7.2]. The essence of this phenomenon is connected with the non-trivial dependence of the scattering phase on the number of particles in the ground state in the process of the hole creation. Taking into account that the quasi-particle weight is determined by the overlap of a single-particle wave function of the hole $| \Psi_k \rangle$ and of the antiferromagnrtic ground state with one hole, $c_{k\sigma} | \Psi_{ON} \rangle$, we arrive at

$$Z(k) = |\langle \Psi_k | c_{k\sigma} | \Psi_{0N} \rangle|^2 \propto \exp(-\Theta \ln L) \propto L^{-\Theta},$$

if the long-range character (of a dipolar type) of perturbation, brought by a hole into magnetically ordered lattice, is taken into account [7.31]. It is obvious that a dependence of this kind cannot be obtained with the help of perturbation theory. Therefore, the final conclusion as to whether or not there are quasi-particles in the system with strong correlations (the Fermi liquid or the Luttinger liquid) can be made on the basis of non-perturbative methods.

In conclusion we note that the study of the dynamic properties of a two-dimensional single-band Hubbard model (7.21), e.g., a single-particle excitation spectrum and optical conductivity, made with the help of the exact diagonalization and Monte Carlo methods, leads to results which agree with those obtained from the t-J model in the limit of strong correlations (see, for example, [7.22, 32–34]).

7.1.3 Spin Fluctuation Spectrum

In describing the normal state of copper-oxide compounds, the other important problem is the study of the spectrum of antiferromagnetic spin fluctuations. As was noted in Chap. 3, strong dynamical spin fluctuations are observed for a wide region of concentration of charge carriers and this can lead to the superconducting pairing of the latter. In this connection, let us consider some results in this field.

The theoretical approaches for calculating dynamical spin susceptibilities or dynamical structure form-factors (3.17) can be divided into two main groups. In the first group, the Fermi liquid with a weak Coulomb interaction ($U \leq t$ in the model (7.21)) is studied. The Millis-Monien-Pines phenomenological model of an antiferromagnetic Fermi liquid [3.27] (see (3.32) in Sect. 3.3.2), the nested-Fermi liquid model with "nesting" of the Fermi surface [7.35], numerical calculations of the spin susceptibility in the two-dimensional Hubbard model with lattice-structure effects taken into account [7.36, 37], the self-consistent renormalization theory of spin fluctuations [7.38], and a number of others all belong to this group.

In the second group, the strong-correlation limit, $U \gg t$, which is usually studied on the basis of the one-band t-J model, is considered. We mention here the calculations performed with the help of the slave-boson method [7.39] and the diagram technique in the t-J model [7.40]. The quantum Monte Carlo method is used in the investigation of the region of intermediate coupling $t < U$ for the Hubbard model (see [7.32] and the literature quoted there).

In the weak coupling limit, $U < t$, calculations are usually performed on the basis of the formula for the dynamic spin susceptibility in the random phase approximation [7.41]

$$\chi(\boldsymbol{q}, \omega) = \chi_0(\boldsymbol{q}, \omega) [1 - U\chi_0(\boldsymbol{q}, \omega)]^{-1}, \tag{7.27}$$

where $\chi_0(\boldsymbol{q}, \omega)$ is the susceptibility of a non-interacting electron gas (3.23) with the spectrum $\varepsilon(\boldsymbol{k})$ for the lattice of a definite symmetry.

Let us consider, for example, the results obtained in [7.37] for the two-dimensional lattice with an electronic spectrum of the form

$$\varepsilon(\boldsymbol{k}) = 2t(2 - \cos ak_x - \cos ak_y - \mu)$$

where the chemical potential varies in the range $0 < \mu < 4$. The dependence of the imaginary part of the susceptibility $\chi''(\boldsymbol{q}, \omega) \cdot t$ on the dimensionless energy ω/t for a given temperature kT/t is presented in Fig. 7.4. The chemical potential $\mu = 1.8$ which corresponds to a filling of the band $n = 0.81$ (for a half filling $n = 1$, the chemical potential is $\mu = 2$). The solid lines refer to the antiferromagnetic (AF) wave vector $\boldsymbol{Q} = (\pi/a, \pi/a)$. With increasing U/t, the intensity of AF fluctuations increases and the Stoner instability ($U\chi_0(\boldsymbol{Q}, \omega = 0) = 1$), where the approximation (7.27) cannot be used, appears for $U/t \simeq 2.42$. The results of the calculations for $U/t = 2$ with other values of the wave vector are shown by the dashed ($q = 0.98\boldsymbol{Q}$; $0.94\boldsymbol{Q}$) and by the dotted lines ($q_x = \pi/a, q_y = 0.95$; $0.8(\pi/a)$). It is seen that, for wave vectors $\boldsymbol{q} = \boldsymbol{Q} - (\delta q_x, \delta y_y)$, a sharp suppression of spin fluctuations for $\omega \leq \omega_0$ arises, where the spin gap is estimated as $\omega_0 = 4t(2 - \mu - \delta q_x - \delta q_y)$. The appearance of the gap in the spectrum is connected with the fulfillment of the nesting condition for the electronic spectrum: $\varepsilon(\boldsymbol{k} + \boldsymbol{Q}) = -\varepsilon(\boldsymbol{k})$ for $\mu = 2$.

In order to study the influence of AF spin fluctuations on the renormalization of the spectrum of quasi-particles, i.e., the self-energy operator of the Green's function (7.2), the exchange paramagnon interaction

$$P(\boldsymbol{q}, \omega) = \frac{3}{2}U^2\frac{1}{\pi}\chi''(\boldsymbol{q}, \omega), \tag{7.28}$$

and its first moments were also calculated by averaging over the wave vector. It turned out that both the zero moment $P_0(\omega)$ and the first moment $P_1(\omega)$ (the q-integrated (7.28) with $\phi_1 = -(\cos q_x a + \cos q_y a)$) increase linearly with ω for $\omega/t \leq 0.5$ and then attain a saturation value in accordance with the model of a "marginal" Fermi liquid [7.4]. The calculation of the damping of quasi-particles on the basis of averaged values (7.28) has shown that it is compatible with the dependence $\hbar/\tau^* \simeq 0.6(\pi kT + \hbar\omega)$ which is observed in the infra-red absorption experiments (see Sect. 5.4.3, Fig. 5.24). The mass renormalization $m^*/m = Z_s(\omega = 0) \leq 2$ also appears in low temperature region.

Similar results have been obtained by the self-consistent solution of the system of equations for $\chi(\boldsymbol{q}, \omega)$ (7.27) and the self-energy operator $\Sigma_s(\boldsymbol{k}, \omega)$ for the model

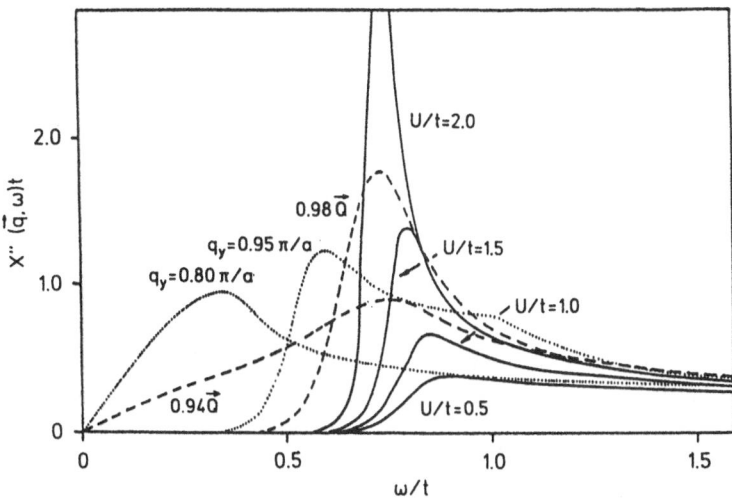

Fig. 7.4. Spin fluctuation spectrum for the temperature $kT/t = 0.02$ and $n = 0.81$ for wave vectors $q = Q$ (*solid line*), $q = 0.98Q$; $0.94Q$ (*dashed line*), $q = (q_x = \pi/a, q_y = 0.95; 0.8\pi/a)$ (*dotted line*) [7.37]

of the Fermi liquid with nesting [7.35] ; and in the detailed calculations [7.36] of the influence of spin fluctuations on electronic parameters of the Fermi liquid for the Hubbard model using approximation (7.27).

Thus, a number of anomalous properties of copper-oxide compounds in the normal phase can already be explained within the framework of the Hubbard model (7.21) for intermediate couplings $U \leq t$ by a self-consistent account of spin fluctuations and lattice-structure effects.

The real relation of the parameters ($U \simeq 4t$), however, makes its necessary to study the strong coupling limit which is usually modelled using the t-J Hamiltonian (5.11). We discuss here the results obtained for the t-J model in the slave-boson method [7.39] [1].

In this method, the hole creation and annihilation operators in the lower Hubbard subband $c_{i\sigma}^+, c_{i\sigma}$ (5.12) are represented in the form of the product of the Fermi operators $f_{i\sigma}$ (spinons) and the Bose operators b_i (holons), $c_{i\sigma}^+ = f_{i\sigma}^+ b_i$. Here the local constraint on the number of the particles on one site

$$b_i^+ b_i + \sum_\sigma f_{i\sigma}^+ f_{i\sigma} = 1 \tag{7.29}$$

should be taken into account. Using this representation, the Hamiltonian of the t-J model (5.11) can be written in the form

---

[1] Effect of the Fermi-surface geometry on spin dynamics in YBCO and LSCO compounds in the framework of the slave-boson technique for the p-d model are considered [7.140].

$$H = -t \sum_{ij} (b_j^+ b_i \chi_{ij} + \text{H.c.}) - \frac{1}{2} J \sum_{ij} \chi_{ij}^+ \chi_{ij}$$

$$- \mu \sum_{i\sigma} f_{i\sigma}^+ f_{i\sigma} - \sum_i \lambda_i (b_i^+ b_i + \sum_\sigma f_{i\sigma}^+ f_{i\sigma} - 1), \,, \tag{7.30}$$

where $\chi_{ij} = \sum_\sigma f_{i\sigma}^+ f_{j\sigma}$ is the spinon operator at a bond $i - j$ for the nearest neighbours; the Lagrange multiplier λ_i takes into account the local constraint (7.29).

In calculating the spin susceptibility, the approximation of a spatially uniform field for spinons and holons at a bond $\chi_{ij} \rightarrow \langle \chi_{ij} \rangle, b_i^+ b_j \rightarrow \langle b_i^+ b_j \rangle$, where the averages do not depend on the sites was used. Likewise, the local constraint (7.29) was replaced by the global one, $\lambda_i \rightarrow \lambda$; thus, an effective Hamiltonian in the self-consistent field was obtained. Its spin susceptibility is of the form

$$\chi(q, \omega) = \chi_0(q, \omega)[1 + J(q)\chi_0(q, \omega)]^{-1} , \tag{7.31}$$

where $J(q) = J(\cos aq_x + \cos aq_y)$ and $\chi_0(q, \omega)$ is the susceptibility of non-interacting fermions (3.23) with the quasi-particle spectrum $\varepsilon(k) = -2F(\cos ak_x + \cos ak_y) - \lambda_F$, $F = t\langle b_i^+ b_j \rangle + J/2\langle \chi_{ij} \rangle$ is the effective transfer integral. The chemical potential for spinons $\lambda_F = \mu + \lambda + J(1 - \delta)$ can be obtained from

$$1 - \delta = n = (1/N) \sum_{i\sigma} \langle f_{i\sigma}^+ f_{i\sigma} \rangle$$

according to (7.29). The calculations in [7.39] show that the parameters λ_F and F depend weakly on the temperature in the region $kT/J \leq 0.6$ but they are sensitive to the variation of the hole concentration δ. When there are no holes, $\delta = 0$, the Fermi surface of spinons is the same as for non-interacting electrons for $n = 1$ (see (5.4) and Fig. 5.12a), and it decreases with increasing δ under doping.

The detailed numerical calculations performed in [7.39] revealed a number of interesting properties of the dynamical spin susceptibility (7.31). The most important is the observation of a sharp decrease of the spin fluctuation intensity in the low temperature region for $q = Q$ and for energies $\omega < 2 | \lambda_F |$. The frequency dependence of the imaginary part of the spin susceptibility $\text{Im}\,\chi(Q, \omega)$ is shown in Fig. 7.5 for several temperatures for $\delta = 0.15$ and $t = 4J$. As the temperature decreases, we observe the sharp suppression of the spin fluctuation intensity below a definite energy as was the case in the Hubbard model with the susceptibility (7.27) (compare with Fig. 7.4).

The suppression of the spin fluctuation spectrum near the wave-vector $q = Q = (\pi/a, \pi/a)$ is easy to explain if we take into account the "nesting" condition for the quasi-particle spectrum $\varepsilon(k + Q) = -\varepsilon(k)$. In this case calculating the imaginary part of spin susceptibility (3.23) we find that

$$\text{Im}\,\chi_0(Q, \omega) = \frac{\pi}{2} N \left(\frac{\omega}{2}\right) [f(| \mu | - \frac{\omega}{2}) - f(| \mu | + \frac{\omega}{2})] , \tag{7.32}$$

where $N(\omega)$ is the density of states near the Fermi surface and $f(\varepsilon)$ is the Fermi function. At low temperatures, $kT \ll \mu$, suppression of the susceptibility (7.32)

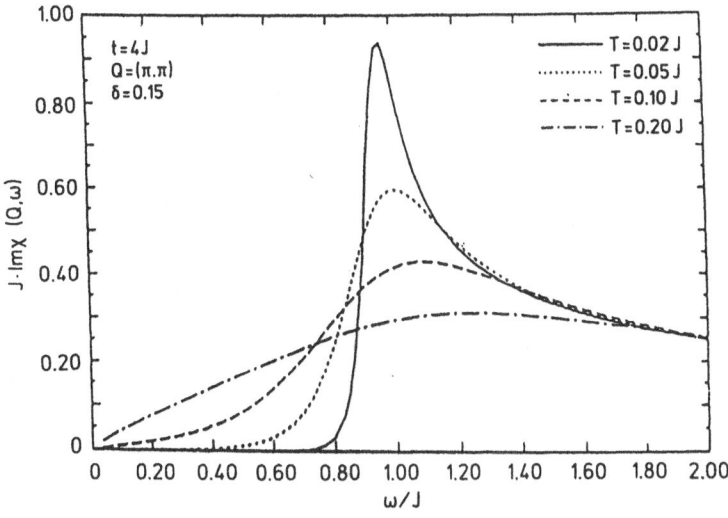

Fig. 7.5. The spectrum of spin fluctuations for $Q = (\pi/a, \pi/a)$ and the hole concentration $\delta = 0.15$ for various temperatures [7.39]

takes place for frequencies $\omega < 2|\mu|$ together with a suppression of the total susceptibility (7.27) or (7.31). Thus, the appearance of a "magnetic gap" at low temperatures is the result of the "nesting" condition for the fermion spectrum. Similar results were obtained in [7.36].

Another important result of the numerical calculations of susceptibility is the observation of an incommensurate structure with increasing hole concentration δ. In Fig. 7.6, the q dependence of the function Im $\chi(q, \omega)/\omega$ for $\omega \to 0$ and for two temperatures $kT/J = 0.01$ (a); 0.1 (b) for $\delta = 0.15$ is shown. The maximum of the function which was situated for $\delta = 0$ at $q = Q = (\pi, \pi)$ splits into four peaks for an increase of δ at $q = (\pi \pm q_x, \pi)$, $(\pi, \pi \pm q_y)$ where $q_x = q_y \propto \delta$. At the same time the minimum, which increases with a lowering of the temperature, appears at $q = Q$ and leads to a "magnetic gap" in the fluctuation spectrum.

The results obtained in [7.39] were used for the calculation of the spin-lattice relaxation rate for the nuclei ^{17}O and ^{63}Cu (Sect. 3.3.2). Integrating Im $\chi(q, \omega)$ with the corresponding structural factors $A_\alpha(q)$ (3.31), the authors obtained a reasonable agreement with experiment. A certain difficulty, however, appears in describing the temperature dependence of a static uniform susceptibility, $\chi(q \to 0, \omega = 0)$, for a small hole concentration (see the curve for $y = 0.37$ in Fig. 3.8). In the theory described above [7.39] (as well as in [7.37]) the uniform susceptibility does not depend on the temperature for $kT \ll \mu(\lambda_F)$.

In concluding this section, let us consider the results for the spin susceptibility obtained in [7.40] with the help of the diagram technique for the Hubbard operators in the t-J model. Since the Hubbard operators (see (7.12)) are subject to more complicated commutation relations than the Fermi operators, there arise, besides the vertex describing the exchange interaction J (or the Coulomb interaction U in

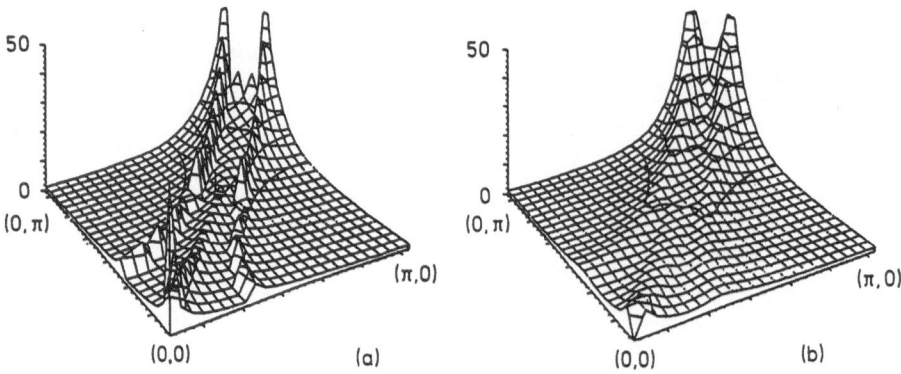

Fig. 7.6. The function $\mathrm{Im}\,\chi(q,\omega)/\omega$ for $\omega \to 0$ for $\delta = 0.15$ and the temperature $kT/J = 0.01$ (**a**); 0.1 (**b**) [7.39]

the Hubbard model), additional vertices (which describe "kinematic" interaction). In this connection, three loop diagrams which are constructed on these vertices appear additionally. The summation of all these diagrams leads to the following expression for the dynamical spin susceptibility in a generalized random-phase approximation (GRPA):

$$\chi(k) = \chi_0(k)\left\{[1 - \Lambda(k)][1 - Q(k)] + \chi_0(k)[J(k) + \Phi(k)]\right\}^{-1} . \qquad (7.33)$$

Here $k = (\mathbf{k}, i\omega_n)$ is a four-momentum, including the Matsubara frequency $i\omega_n$, $\Lambda(k)$, $Q(k)$, and $\Phi(k)$ are the additional loop diagrams (see [7.40]), and the "bare" susceptibility is

$$\chi_0(k) = \frac{1}{kT}nn_0\delta_{\omega_n,0} - \Pi(\mathbf{k}, i\omega_n) . \qquad (7.34)$$

Besides the usual hole-particle diagram $\Pi(\mathbf{k},\omega)$, the latter contains a localized contribution at zero frequency $\omega_n = 0$. It differs from zero only for $n > n_c$ where n_c is the critical hole concentration, for which localized magnetic moments appear. The chemical potential changes its sign at this point, so that $\mu(n_c) = 0$. The parameter $n_0 = 2e^{\mu/kT}[1 + 2e^{\mu/kT}]^{-1}$ changes abruptly from $n_0 = 0$ for $n < n_c$, to $n_0 = 1$ for $n > n_c$, in the limit $T \to 0$.

Thus, in the theory developed in [7.40], in comparison with the above approaches there appears a new physical element: the system behaves as an itinerant magnet for $n < n_c$ and for $n > n_c$ the contribution of localized moment with the Curie-Weiss susceptibility n/kT arises in (7.33). The evaluation of the critical concentration in the mean-field (Hubbard-I, see [5.6]) approximation gives $\mu = 0$ for $n = 2/3$, i.e., $n_c = 2/3$. The considerable temperature dependence of the static susceptibility χ_0 ($k \to 0$, $\omega = 0$) in (7.33) for $n > n_c$ (in the region of a small hole concentration $\delta = 1 - n < \delta_c$) can explain the corresponding temperature dependence of a uniform susceptibility which was discussed above in connection with Figure 3.8. The considerable frequency dependence $\mathrm{Im}\,\chi(\mathbf{Q},\omega)$ in (7.32) also arises for $n \leq n_c \simeq 2/3$.

Summarizing the discussion in this section we can outline the qualitative co-incidence of the results for the spectrum of spin excitations both in the case of a weak coupling $U \leq t$ for the Hubbard model, and, in the case of a strong coupling $U \geq t$ for the t-J model. The corresponding expressions for the dynamical spin susceptibility in the random phase approximation (7.27) and (7.31), differ only in the relative contribution of Coulomb correlation: in the first case in the form of a local, independent on q, potential U, and in the second case, with the help of exchange interaction for the nearest neighbors $J(q)$. This leads to the considerable increase of the q-dependence in (7.31), especially in the region $q = Q$. At the same time, the frequency and temperature dependences in (7.27) and (7.31) are mainly connected with the susceptibility of non-interacting particles $\chi_0(q, \omega)$. The latter determines the "magnetic gap" which appears according to (7.32) near $q = Q$ for a small hole concentration when the "nesting" conditions for the quasi-two-dimensional spectrum of quasi-particles are exactly fulfilled. To clarify the results obtained above it is necessary to go beyond the approximation of self-consistent field in the slave-boson technique following for a self-consistent renormalization of quasi-particle spectrum due to spin fluctuations. We believe that the self-consistent calculation of the spectrum of spin fluctuations and quasi-particle hole spectrum on the basis of a diagram technique in the t-J model seems to be very promising [7.40].

The influence of AF spin fluctuations on the superconducting pairing is discussed in Sect. 7.3.

7.2 Superconductivity in Systems with Strong Correlations

The search for new mechanisms of superconductivity due to strong Coulomb correlation of the electrons in copper-oxide compounds with a quasi-two-dimensional electronic structure, has narrowed, first of all, to the search for new unconventional ground states. This problem, which was briefly mentioned in Sect. 7.1, will be discussed below. We shall also discuss more traditional approaches to the study of superconducting correlations within the framework of the BCS theory of pairing for the t-J and p-d models.

7.2.1 Unconventional Ground States

Anderson was the first to note the possibility of an unconventional ground state and of a new mechanism of superconductivity in copper-oxide compounds connected with it [7.1]. He investigated the t-J model (5.11) which was obtained from the one-band Hubbard model (7.21) in the strong-coupling limit $U \gg t$. In the absence of holes, the antiferromagnetic (AF) exchange usually leads to the Néel ground state which is characterized by a long-range AF order in the spin alignment on the lattice sites. Anderson supposed that a disordered state in the form of resonating valence bonds (RVB) occurs instead of the Néel ground state in the two-dimensional system. The RVB state is characterized by a system of singlet states for a pair of electrons on the lattice sites (i, j)

$$b_{ij}^+ = \frac{1}{\sqrt{2}} \left(c_{i\uparrow}^+ c_{j\downarrow}^+ - c_{i\downarrow}^+ c_{j\uparrow}^+ \right) . \tag{7.35}$$

With the help of these operators, the exchange interaction in the t-J model is written in the form

$$H_J = -J \sum_{ij} b_{ij}^+ b_{ij} . \tag{7.36}$$

This very interaction is responsible for the creation of bounded RVB states $\langle b_{ij}^+ \rangle \neq 0$.

The breaking of the singlet bond (7.35) leads to the appearance of two Fermi excitations with spin 1/2 which have no charge. These neutral fermions are called "spinons". Upon doping, a new type of excitation, a "holon", namely the hole with a positive charge which has no spin, i.e., a charged Bose quasi-particle, arises. Thus, in the first model of the RVB state, the concept of the separation of spin and charge degrees of freedom in the two-dimensional Hubbard model was introduced by Anderson similarly to the one-dimensional system which represents the Luttinger liquid [7.2, 3] (see Sect. 7.1).

It is convenient to take into account the separation of spin and charge degrees of freedom with the help of auxiliary fields as was described above in the derivation of the Hamiltonian (7.30). In this case the RVB state is characterized by the order parameters for bosons $B_{ij} = \langle b_i^+ b_j \rangle$ and those for spinons $\Delta_{ij} = \langle f_{i\uparrow}^+ f_{j\downarrow}^+ - f_{i\downarrow}^+ f_{j\uparrow}^+ \rangle$ according to (7.35). At first, superconductivity in the RVB system was thought to be connected with the formation of the Bose condensate of holons $\langle b_i \rangle \neq 0$ [7.1]. Later, another mechanism of superconductivity was proposed which was based on the Josephson tunneling of a pair of electrons between the layers under the condition $\Delta_{ij} \neq 0$ (see [5.90]).

To go beyond the self-consistent field approximation in the method of auxiliary fields, it is necessary to take into account the fluctuations of the RVB order parameters and the Lagrange factor λ_i in (7.30) with the help of which the local constraint (7.29) on the number of particles at one site is taken into account. As was proposed in [7.42], this can be performed by the introduction of gauge fields. According to [7.43, 44], a time component of the gauge field is connected with the fluctuations of λ_i and spatial components are connected with the phase of the RVB order parameters. In this method, the investigation of two-dimensional systems having strong correlations within the framework (7.30) is reduced to the study of the gauge fields in the space with dimensionality (2+1).

Nowadays, the investigation of two-dimensional systems with a strong correlation in the theory of gauge fields is an independent branch of the statistical physics and the lattice gauge fields (see the reviews [7.10, 45]). We note here only the general features of the proposed models of the ground state of these systems.

The most interesting properties refer to flux phases [7.44, 45], which are characterized by a non-zero flow of the flux field in going around a closed contour on the two-dimensional lattice. An alternative description of the flux phases is possible with the help of particles with fractional statistics, the so-called anyons. Under the permutation of two anyons, their wave function acquires the phase factor $\exp(i\Theta)$. In the case of Fermi and Bose particles, $\Theta = \pi$ and $\Theta = 0$, respectively. The angle Θ is arbitrary for anyons, and these particles are called semions for $\Theta = \pm\pi/2$. One of the main properties of the flux phases is the violation of the time (T) and spatial (P) parity. The chiral spin liquid [7.44] which can be characterized by the order parameter $a_{ijk} = \langle S_i[S_j \times S_k]\rangle$ for non-complanar spins S_l on the sites of a two-dimensional lattice i, j, k is a simple example of a ground state which violates T- and P- invariance. The superconductivity in a system of anyons can be connected with the transition to the T- and P-violating state at some temperature T_{TP}, supposing that $T_c = T_{TP}$. Experimental investigations do not, however, reveal any violation of T- and P-invariance in the transition into the superconducting state for the copper-oxide compound YBCO [7.46].

A more detailed study of the ground-state energy of the commensurate flux phase [7.47] has shown that it is unstable for realistic values of the parameters of the t-J model [7.48].

The calculations for the normal properties of copper-oxide compounds which are closest to experimental results have been obtained for the uniform fluxless phase in [7.43, 49]. In the representation of auxiliary bosons (see (7.30)) this phase is characterized by RVB order parameters which do not depend on the coordinates. In this model, it is possible to reconcile a large Fermi surface, which is determined by spinons, $\langle \chi_{ij}\rangle$, with a low concentration of carriers which is proportional to the number of holons, $\langle B_{i,j}\rangle$. The free energy for the uniform RVB phase was calculated and it was compared with the results of high-temperature expansions [7.50]. Similar values for these energies can be obtained if one takes into account the gauge-field fluctuations connected with the order parameters. In this case, however, transverse fluctuations lead to the instability of the homogeneous phase with respect to the formation of the flux density wave for $T < T_c^{FDW}$. The phase is unstable for temperatures $T_c^{FDW} \geq 0.3J \sim 500\,\mathrm{K}$ and for $\delta \leq 0.2$. This fact casts doubt on the results of studies where the uniform RVB phase has been used for the description of normal properties of high-temperature superconductors for $T < T_c^{FDW}$.

We should also mention a large group of studies devoted to the investigation of the ground state of the quantum Heisenberg model for the antiferromagnet on the square lattice with frustration [7.10]. The latter appears when the competing spin interaction for the second (and the third) neighbors, tending to destroy the long-range antiferromagnetic (AF) order which is set due to the interaction of the nearest neighbors, is taken into account. This model is directly connected to the t-J model (5.11). Indeed, when there are no doped holes, the t-terms make no contribution and the AF exchange for spins $S = 1/2$ on the nearest sites of the square lattice leads to the Néel ground state for $T = 0$ [7.51]. The doping destroys the AF bonds on the nearest sites which is equivalent to the introduction of competing interactions (see, for example, [7.52]). Although the contribution from the kinetic energy of the hole, i.e, the t-term, is not directly taken into account the Heisenberg model with frustration, can be a convenient subject for the study of non-classical ground states. In particular, within the framework of this model, the chiral state (with the order parameter $a_{ixy} = \langle S_i[S_{i+x} \times S_{i+y}]\rangle$), the RVB state, and the dimer state close to it, and a number of others were studied (see [7.10]).

Besides a great number of calculations in the mean-field approximation which usually gives poor results in the case of the competing interactions, numerical calculations with the help of exact diagonalization for small clusters as well as high-temperature expansions have also been performed. However, the results obtained so far are of a contradictory character which makes a unique conclusion concerning the nature of the ground state in the frustrated Heisenberg model impossible [7.10].

For certain ranges of hole concentration $\delta = 1 - n$ and values of the ratio J/t in the t-J model, we can expect a phase-separation (see, for example, [7.53]). If $J \gg t$, there is an energetically favored phase separation into a region with AF order, having no holes, and the region, filled by holes [7.48]. The high-temperature expansion, carried out in [7.54] for the spin susceptibility in the two-dimensional t-J model, showed that phase-separation for a range of small electron concentrations $n \ll 1$ takes place for $J/t \geq 3.5$, and in the limit $n \to 1$ it takes place for $J/t \geq 1$. Therefore, it is hardly probable that phase separation can take place for realistic values of parameters for the t-J model in the CuO_2 plane ($J/t \simeq 0.35$, see [5.33]). The appearance of the ferromagnetic phase is possible for $n \to 1$ at a small value of the exchange interaction, $J/t \leq 0.1$.

So summarizing our discussion of work devoted to the study of the ground state in systems with strong correlation, we can say that the solution to this problem has not yet been found. A number of unusual properties, rigorously proved for the one-dimensional t-J model (or the Hubbard model), cannot be strictly established for the two-dimensional models. Physical properties, predicted in the theory of the gauge fields, require experimental verification and comparison with the results of the standard Fermi-liquid approach. Therefore, further investigations of models with strong correlation using standard approaches are still of importance.

7.2.2 Superconductivity in the t-J Model

In this section, we shall discuss the results obtained for calculations of the electronic Green's functions of the Hubbard operators without introducing auxiliary fields. Since the commutation relations for the Hubbard operators (7.12) automatically take into account the restrictions placed on the filling of quantum states, additional conditions of the type (7.29) are necessarily observed for them. However, the calculations with the help of the Hubbard operators require the use of a complicated diagrammatic technique (see, for example, [7.55]).

Below we shall discuss the results of the calculations of Green's functions in [7.56] using a simpler method based on the equations of motion. The results obtained in this method with the help of a projection technique of the Mori type are in agreement with the calculations of the self-energy operator in the diagrammatic technique up to the first skeleton diagrams.

Let us consider the t-J model (5.11) written in terms of the Hubbard operators (7.12) in the form

$$H - \mu N = t \sum_{\langle ij \rangle \sigma} X_i^{\sigma 0} X_j^{0\sigma} - \mu \sum_{i\sigma} X_i^{\sigma\sigma}$$

$$+ \frac{J}{2} \sum_{ij\sigma} \left(X_i^{\sigma\bar\sigma} X_j^{\bar\sigma\sigma} - X_i^{\sigma\sigma} X_j^{\bar\sigma\bar\sigma} \right), \tag{7.37}$$

where $\bar\sigma = -\sigma$ and μ is the chemical potential. Further, we will consider two-time anticommutator Green's functions [7.57]

$$G_{ij}^\sigma(t - t') = \langle\langle X_i^\sigma(t); X_j^\sigma(t')^+ \rangle\rangle. \tag{7.38}$$

where the Nambu operators

$$X_i^\sigma = \begin{pmatrix} X_i^{0\sigma} \\ X_i^{\bar\sigma 0} \end{pmatrix} \qquad (X_i^\sigma)^+ = (X_i^{\sigma 0} X_i^{0\bar\sigma}) \tag{7.39}$$

are introduced.

The equations of motion for the Nambu operators (7.39) can be projected onto the subspace of the one-particle operators with the help of the equation

$$i\frac{d}{dt} X_i^\sigma(t) = [X_i^\sigma, H] = \sum_j A_{ij}^\sigma X_j^\sigma + Z_i^\sigma. \tag{7.40}$$

Here the first term which is linear with respect to the set of operators (7.39) describes the time dependence in the mean-field approximation for the effective forces, and the last term Z_i^σ determines inelastic scattering processes. The system of equations

$$\sum_{l\beta} \left(A_{il}^\sigma \right)_{\alpha\beta} \langle\{X_l^\sigma, (X_j^\sigma)^+\}\rangle_{\beta\gamma} = \langle\{[X_i^\sigma, H], (X_j^\sigma)^+\}\rangle_{\alpha\gamma},$$

for the matrix $(A_{ij}^\sigma)_{\alpha\beta}$ are found from the orthogonality conditions for the linear set of the operators (7.39) $\langle\{Z_i^\sigma, (X_j^\sigma)^+\}\rangle = 0$, where $\{A, B\} = AB + BA$. The solution of this system permits us to find explicit expressions for the normal $(A_{ij}^\sigma)_{11}$ and anomalous $(A_{ij}^\sigma)_{12}$ matrix components. The first of these determines the excitation spectrum Ω_q^σ in the normal phase and the second determines a superconducting gap Δ_q^σ

$$\Omega_q^\sigma = \sum_j (A_{ij}^\sigma)_{11} e^{-iq(i-j)},$$

$$\Delta_q^\sigma = \sum_j (A_{ij}^\sigma)_{12} e^{-iq(i-j)}. \tag{7.41}$$

Taking into account only the first term in (7.40), we obtain simple expressions

$$\langle\langle X_q^{0\sigma}|X_{-q}^{\sigma 0}\rangle\rangle_\omega = \langle Q_i^\sigma\rangle \frac{\omega + \Omega_q^\sigma - \mu}{\omega^2 - (E_q^\sigma)^2}, \tag{7.42}$$

$$\langle\langle X_q^{\bar\sigma 0}|X_{-q}^{\sigma 0}\rangle\rangle_\omega = -\langle Q_i^\sigma\rangle \frac{\Delta_q^\sigma}{\omega^2 - (E_q^\sigma)^2}, \tag{7.43}$$

where the spectrum of one-particle excitations is determined by the function

$$E_q^\sigma = \left[(\Omega_q^\sigma - \mu)^2 + |\Delta_q^\sigma|^2\right]^{1/2}, \tag{7.44}$$

for the normal and anomalous Fourier components of the Green's function (7.38). In the paramagnetic state $\Omega_q^\sigma = \Omega_q$ and $\Delta_q^\sigma = \sigma\Delta_q$, ($\sigma = \pm 1$) and

$$\langle Q_i^\sigma\rangle = \langle X_i^{00} + X_i^{\sigma\sigma}\rangle = \langle 1 - X_i^{\bar\sigma\bar\sigma}\rangle = 1 - n/2$$

where n is the average number of electrons at the site. Calculating $\langle X_i^{\sigma\sigma}\rangle = \langle X_i^{\sigma 0} X_i^{0\sigma}\rangle$ with the help of the Green's function (7.42) we obtain the equation

$$\frac{n}{1 - n/2} = \frac{1}{N}\sum_k \left(1 - \frac{\Omega_k - \mu}{E_k}\right)\tanh\frac{E_k}{2T} \tag{7.45}$$

which determins n as a function of μ. This equation automatically satisfies the condition $n \leq 1$. Under this condition, the chemical potential μ is equal to zero for a filling $n \simeq 2/3$, unlike the case of the slave-boson approximation where $\mu = 0$ for a half-filling ($\lambda_F = 0$ for $n = 1$ [7.39]).

To determine the temperature of the phase transition into the state $\Delta_q \neq 0$, it is necessary to solve the self-consistent equation for Δ_q in (7.44). Calculating anomalous correlation functions with the help of the Green's functions (7.43) in (7.41) we obtain

$$\Delta_q = \frac{1}{N}\sum_k \frac{\Delta_k}{E_k}[t\gamma(k) + J\gamma(k+q)]\tanh\frac{E_k}{2T}, \tag{7.46}$$

where $\gamma(k) = 2(\cos ak_x + \cos ak_y)$. The first term in (7.46), proportional to the transfer integral t in the t-J model, describes a so-called kinematic interaction which is due to the non-fermionic character of the commutation relations for the Hubbard operators. It was obtained for the first time in the equation for T_c in [7.58], where a new "kinematic" mechanism for superconductivity was proposed. In the papers [7.59–61], in which superconductivity in the t-J model was also studied, the kinematic contribution was lost, since the mean-field approximation for the t term was used. The approximation $tb_i^+ b_j \rightarrow t\langle b_i^+ b_j\rangle = \tilde{t} \propto t\delta$, where $\delta = 1 - n$ is the hole concentration, was used in the method of slave-bosons [7.59, 60]. As a result, only the J−term describing the pairing due to AF fluctuations in the mean-field approximation was taken into account in (7.46).

Taking into consideration the separable character of the interaction $J\gamma(k+q)$ in (7.46), the solution for the gap can be written in the form

$$\Delta_q = \Delta_0 + \Delta_s \psi_s(q) + \Delta_d \psi_d(q), \tag{7.47}$$

where the q-independent solution Δ_0 contributes only for $J = 0$ ($U \rightarrow \infty$) (kinematic mechanism) and the s- and d-wave pairings are described by the corresponding functions

$$\psi_s(q) = \cos a k_x + \cos a k_y,$$
$$\psi_d(q) = \cos a k_x - \cos a k_y, \tag{7.48}$$

(see (5.53) and (5.54)). The solution of the self-consistent equation (7.46) for the s- and d-wave pairings and an analysis of the energy of the ground state for the transition to the superconducting phase, Δ_α, shows that s-wave pairing is possible for small hole (or electron) concentrations; and d-wave pairing is energetically favorable for $n \simeq 2/3$ (see, for example, [7.56]).

Using the solution (7.43), we consider the equation for the one-site correlation function

$$\langle X_i^{\sigma 0} X_i^{\bar{\sigma} 0} \rangle = \frac{1 - n/2}{N} \sum_k \frac{\Delta_k^\sigma}{2 E_k^\sigma} \tanh \frac{E_k^\sigma}{2T} = 0, \tag{7.49}$$

which prohibits any pairing on a single site due to the relation $X_i^{\alpha\beta} X_i^{\gamma\beta} = \delta_{\beta\gamma} X_i^{\alpha\delta}$ for the Hubbard operators. Equation (7.49) is satisfied for the d-pairing but it is violated for Δ_0 and s-pairing. Consequently, in the limit of strong correlations described by the t-J model, only the d-wave pairing is permissible. In this case, however, due to the symmetry of the function $\psi_d(q)$ in (7.48), the first term (the kinematic one) in (7.46) does not make any contribution and only the exchange interaction is left (as in the papers [7.59–61]).

A more detailed analysis of the superconducting equations in the t-J model, carried out in [7.62] on the basis of the diagrammatic technique, shows that (7.46) represents the contribution of only the first term of expansion for the self-energy operator of the electronic Green's function. The contribution of the subsequent terms, describing inelastic scattering from the spin fluctuations (contribution Z_i^σ in (7.40)), leads to suppression of the transition temperature T_c^0 which was obtained from the solution of (7.46). Especially strong scattering appears for $n \geq n_c \simeq 2/3$ when, due to (7.33) and (7.34), localized magnetic moments arise in the system [7.40]. As a result of this "pair-breaking" mechanism, the high superconducting transition temperatures T_c, due to the spin-fluctuation exchange, are unlikely [7.62].

Owing to an absence of rigorous methods which can deal with strong correlations, a number of numerical calculations, mainly for the one-band Hubbard model (7.21), were carried out in order to find correlations of the superconducting type.

Investigations using the Monte Carlo method do not show superconducting correlations in one-band models which have Coulomb repulsion in one plane (see, for example, [7.63]). To obtain a superconducting pairing, phonon degrees of freedom have to be included in these models as, for example, in the model "with the apex oxygen" [7.64]. In the two-layer models, it is possible to observe the superconducting correlations [7.65] and the transition to a state with long-range off-diagonal order [7.66]. Pairing was also found in the one-dimensional t-J model and for the Hubbard model in an external magnetic field with an AF ordering (staggered field) [7.67]. Although the observed pairing effects are usually small, switching on an additional attractive interaction, for example, due to phonons, can lead to a stable superconducting ground state.

A small quasi-particle weight $Z(k)$ for systems having a strong correlation obviously plays an important role, as was discussed in Sect. 7.1. Therefore, a search for superconducting correlations between the pairs of quasi-particles (the intensity of which is proportional to the small quantity $| Z(k)|^2 \ll 1$), often leads to negative results. As was noted in [7.68], the search for superconducting correlations in systems with strong Coulomb repulsion is facilitated by numerical calculations for composite quasi-particles which allow for spin-polaron effects (see Sect. 7.1.2).

7.2.3 Superconductivity in the p-d Model

Due to the limited character of the one-band t-J model, the possibility of super-conducting correlations of the mixed p-d type in the multiband model (7.7) has also been discussed extensively (see, for example, [7.69]). In this model, charge excitations in the spin-fluctuation regime can be excluded with the help of the cor-responding canonical transformation as a result of which the exchange interaction between the spins of d- and p-holes

$$H_{\text{exc}} = 2 \sum_{im} J_{im} \left(S_i \hat{\sigma}_m - \frac{1}{4} n_i^d n_m^p \right) , \tag{7.50}$$

appears in the Hamiltonian, where $S_i = (1/2) \Sigma_{\sigma\sigma'} X_i^{\sigma 0} \hat{\sigma}_{\sigma\sigma'} X_i^{0\sigma'}$, is the spin of a d-hole and $\hat{\sigma}_m = (1/2) \Sigma_{\sigma\sigma'} p_{m\sigma}^+ \hat{\sigma}_{\sigma\sigma'} p_{m\sigma'}$ is the spin of a p-hole, $\hat{\sigma}_{\sigma\sigma'}$ is the Pauli matrix, $J_{im} \propto t_{pd}^2/\Delta$ (see, for example, [7.7, 52]). Introducing the operators of the singlet state of the $d - p$ pair

$$b_{im} = \frac{1}{\sqrt{2}} \left(X_i^{0\downarrow} p_{m\uparrow} - X_i^{0\uparrow} p_{m\downarrow} \right) ,$$

the exchange interaction (7.50) can be represented in the form

$$H_{exc} = - \sum_{im} J_{im} b_{im}^+ b_{im} , \tag{7.51}$$

as in the t-J model (7.36). Using again the self-consistent mean-field approximation $b_{im} \to \langle b_{im} \rangle$, it is easy to obtain an equation of the BCS type

$$\Delta_i = \sum_m t_{pd} S_{im} \langle b_{im} \rangle \tag{7.52}$$

for anomalous averages [7.69]. The solution of the self-consistent equation (7.52) determines the temperature T_c, below which a non-zero solution for the gap exists.

As for the t-J model, T_c is identified with the temperature of the transition to the superconducting state due to the exchange interaction (7.50) [7.69].

A more general approach, based on the equation-of-motion method for the Green's functions, was used for the p-d model in [7.70]. As for the t-J model (7.37), the system of equations for the matrix Green's function with the dimensionality 4×4 for the operators $X_i^{\sigma 0}$, p_{ms} and those conjugate to them were studied. As in (7.40), on the basis of the projection technique, a closed system of equations was derived and anomalous averages of all types $\langle X_{-q}^{\sigma 0} X_{-q}^{\bar\sigma 0} \rangle$, $\langle p_{q\sigma}^+ p_{-q\bar\sigma}^+ \rangle$, $\langle X_q^{\sigma 0} p_{-q\bar\sigma}^+ \rangle$ were obtained. In this case, it turned out that all the anomalous averages are proportional to the function (7.52) which does not depend on site indices. As a result, the local constraint on the pairing of two d-holes on one site

$$\langle X_i^{\sigma 0} X_i^{\bar\sigma 0} \rangle = \frac{1}{N} \sum_q \langle X_q^{\sigma 0} X_{-q}^{\bar\sigma 0} \rangle = 0 \tag{7.53}$$

cannot be satisfied for $\Delta \neq 0$. Consequently, as in the case of the t-J model, condition (7.53) prohibits the s-pairing described by the function (7.52).

In reference [7.69], this constraint was not taken into account, since in the mean-field approximation the substitution

$$t_{pd} X_i^{\sigma 0} p_{m\sigma} \rightarrow \tilde{t}_{pd} d_{i\sigma}^+ p_{m\sigma} , \tag{7.54}$$

was used. Here, as in the slave-boson method the non-fermionic character of the Hubbard operators was taken into account in the average by renormalizing the hybridization parameter $\tilde{t}_{pd} \propto t_{pd} \sqrt{\delta}$, $\delta = (1 - n)/(1 - n/2)$. In this approximation, the restriction on the appearance of two holes on one d-site does not arise since Fermi statistics does not prohibit pairing of the type $\langle d_{i\sigma}^+ d_{i\bar\sigma}^+ \rangle$.

In this connection, note that the approximation of the particle statistics when the Hubbard operators are approximated by the Fermi ones both in (7.54) and (7.30), when the local constraints for slave bosons (7.29) are replaced by global ones, $\lambda_i \rightarrow \lambda$, seems to be rather a rough approximation. It violates for instance the dependence of the chemical potential μ on the number of the particles n which is connected with it. No other approximations, commonly used for many-particle interactions in many-body theory distorts the physical properties of the systems with strong correlations, as much as this approximation for the kinematic interaction. The latter results from non-fermionic commutation relations.

We shall postpone until Sect. 7.4 a more general analysis of superconductivity in the basic p-d model (7.7) or (7.11) based on the method of Green's functions where all types of interactions responsible for pairing, i.e., including spin and charge fluctuations and phonons will be taken into account.

In concluding this section, let us discuss the results of the search, by means of computations, for superconducting correlations in the p-d model (7.7). These were performed for a small number of particles in the form of a cluster consisting of 16–20 cells of CuO_2. The contribution of the internal degrees of freedom in the cell of CuO_2 considerably limits the possible dimensions of the system in the computations.

Let us consider the results obtained in [7.5] by the quantum Monte Carlo method (for 16 cells of CuO_2) for a three-band p-d model (7.7). In this method, the correlation function

$$P_\alpha = \langle \Delta_\alpha \Delta_\alpha^+ \rangle, \tag{7.55}$$

where

$$\Delta_\alpha = \frac{1}{N} \sum_{m,l} g_\alpha(l) c_{m+l\downarrow}^+ c_{m\uparrow}^+ , \tag{7.56}$$

was calculated for pairs of particles. Here N is the number of cells in the cluster; $g_\alpha(l) = \pm 1$ determines the pairing symmetry (of s- and d-type) in the summation over the lattice sites lattice, l, nearest to the site m. The independent propagation of a pair of particles is described by the function

$$\bar{P}_\alpha = \frac{1}{N} \sum_{mm'} \sum_{ll'} g_\alpha(l) g_\alpha(l') G_{mm'}^\uparrow G_{m'+l'\ m+l}^\downarrow , \tag{7.57}$$

where $G_{m'm}^\sigma = \langle c_{m'\sigma}^+ c_{m\sigma} \rangle$. Superconducting correlations are indicated when the value of the pair-interaction vertex $S^* = P_\alpha - \bar{P}_\alpha$ is positive.

The dependence of the vertex S^* for the s-pairing on the concentration of doped holes δ for the 4×4 cluster of CuO_2 with $U_d = 6$ and $\Delta = 4$ (in the units of t_{pd}) for the model (7.7) is shown in Fig. 7.7. The calculations were carried out for a temperature $kT/t_{pd} = 0.1$ ($\beta = (kT/t_{pd})^{-1} = 10$). The investigation of the temperature dependence ($2 \le \beta \le 12$) confirms the enhancement of superconducting pairings as the temperature decreases (and β increases). Comparing the results for the vertices of s- and d-types shows that the s-pairing is more effective than the d-pairing. As was noted in [7.5], the curve for the dependence of S^* (δ) coincides qualitatively with the dependence of the superconducting transition temperatures, T_c, on the hole concentration for copper-oxide superconductors (Fig. 5.3).

Thus, in some cases, investigation of models with a strong Coulomb repulsion leads to conclusion that there exist pair correlations of the superconducting type. The most probable reason for their appearance in systems with repulsion is due to the exchange interaction, as indicated by analytical calculations in the

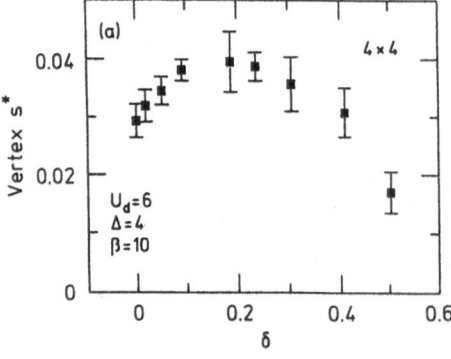

Fig. 7.7. The dependence of pair correlations on the hole concentration δ. The calculation was perfomed by the quantum Monte Carlo method [7.5]

self-consistent-field approximation. Though it is difficult to prove that the latter calculations are reliable for the case of strong coupling, $U \gg t$, calculations of the same type in the case of weak coupling, $U \ll t$, lead to similar results.

7.3 Antiferromagnetism and Superconductivity

As was shown in the previous sections, antiferromagnetic (AF) spin fluctuations are vitally important in explaining many of the anomalous properties of high-temperature superconductors in the normal phase. It is thus natural to suppose that these boson excitations are also responsible for superconducting pairing in copper-oxide compounds [7.71, 72, 38]. Earlier the magnetic mechanism of pairing was proposed as an explanation for superconductivity in systems with heavy fermions [7.73], where d-wave pairing is observed. This was then investigated with regard to high-temperature superconductors [7.74]. It was shown that, under the exchange of an AF paramagnon, an attraction appears in the d-channel and acts most effectively near the AF instability. In a more rigorous treatment, a self-consistent account of all three types of instability connected with the formation of a spin or charge density wave and with the transition into the superconducting state is required [7.75].

Actually, the AF exchange mechanism has already been discussed in Sect. 7.2. There we took the limit of strong correlations, $U \gg t$, and obtained the effective Hamiltonian with AF exchange (see (7.37), (7.50)), which determines the hole pairing. In this case, however, the retarded character of the exchange interaction is not taken into account. Therefore, we shall only discuss in this section investigations where the effective exchange interaction is directly connected with the dynamic spin susceptibility for copper-oxide compounds.

The dynamical spin susceptibility in the approximation of weak coupling, $U \ll t$, is described by the formula (7.27) and leads to an effective interaction for two holes (7.28). The solution of the corresponding Eliashberg equation for this interaction shows that the temperatures where the superconducting transition T_c occurs for the extended s- and d-channel are much lower than the temperatures observed in the copper-oxide compounds [7.37, 76]. An increase of T_c can be obtained by moving into the region of AF instability, where, however, the approximate formula (7.27) is not applicable. On the other hand, self-consistent calculations within the theory of spin fluctuations [7.38] lead to a sufficiently high T_c in the d-channel.

The retarded interaction between the quasi-particles on the two-dimensional square lattice under the exchange of the AF paramagnon was studied in [7.72]. Using the phenomenological expressions for the dynamical spin susceptibility (3.34) with parameters obtained from NMR experiments, the authors solved the equation for the gap in the d-wave channel. The results for the numerical calculation can be represented in the form of a formula of the BCS type

$$T_c = \alpha(\Gamma(T_c)/\pi^2)\exp[-1/\lambda(T_c)]\,, \tag{7.58}$$

where $\Gamma(T_c)/\pi^2$ is the characteristic energy of the spin fluctuations, $\lambda(T_c)=\eta g_{eff}^2(T_c)\chi_0(T_c)N(0)$ is the effective coupling constant ($\eta \sim 1$ and $\alpha \sim 1$ are constants). The formula (7.58) gives T_c which are experimentally observed for LSCO and YBCO compounds even for a sufficiently weak coupling, $\lambda \simeq 0.33 - 0.48$ due to the large value of the characteristic electron Fermi energy, $\Gamma(T_c) \simeq 0.4\,\text{eV}$ (see Sect. 3.3.2).

Studies of the temperature dependence of the gap $\Delta(T)$ show that near T_c the gap grows rapidly when the temperature decreases and reaches maximum value $2\Delta(0)/kT_c = 6 - 8$ which agrees with experiment (Sect. 5.6). In these calculations, the high T_c and the d-wave pairing are conditioned by the quasi-two-dimensional character of the electron spectrum and by a strongly anisotropic interaction due to the AF spin fluctuations. The finite life-time effects and the strong coupling limit were considered in the subsequent papers [7.72].

In connection with this latter problem, we will outline the ideas contained in [7.62] where the Eliashberg equations in the approximation of strong coupling were obtained, based on the diagram technique in the t-J model. In this theory, the effective attraction is determined by spin fluctuations which are described by the spin susceptibility (7.33) and by the charge fluctuations. In the itinerant electron regime ($n < n_c \sim 2/3$), d-wave pairing with high T_c near the AF instability is possible. In the range $n > n_c$, where local magnetic moments appear (see (7.34)), scattering of electron pairs from these "magnetic impurities" prevents the occurrence of a high T_c. Thus, according to the authors of [7.62], pair-breaking effects play an important role in the magnetic-pairing mechanism and, therefore, they should be taken into account in an explicit form in order to obtain quantitative results.

Above the paramagnetic ground state was considered. For the antiferromagnetic ground state, transverse fluctuations of the AF order parameter ensure that an effective attraction is more favorable for the manifestation of a magnetic pairing mechanism (see [7.71, 10]). In this connection, we consider the concept of the spin bag proposed by Schrieffer et al. [7.77]. Although in copper-oxide compounds superconductivity arises for hole concentration $n_h \geq 0.05$, where long-range AF order is already absent, due to the small superconducting correlation length ξ_0 we can use the theory in the region $\xi_N > \xi_0$ (where ξ_N is a correlation length of the AF spin fluctuations).

The theory [7.77] is based on the assumption of a local depression by a hole of the AF gap in the electron spectrum. As a result, a magnetic polaron – a spin bag which moves together with the cloud of spin deformation – appears. In this case, two holes, i. e., polarons, attract each other due to the overlap of their regions of deformation of the AF order. A more detailed analysis shows that the longitudal spin fluctuations lead to a singlet pairing with the maximum contribution coming from the d-wave channel. The transversal fluctuations here suppress the effective attraction.

In the paramagnetic phase, the contribution of spin fluctuations near the AF wave vector, $q = Q$, leads to the appearance of a pseudo-gap in the electron spectrum near half-filling. The additional exchange of the AF spin fluctuations for two quasi-particles – spin bags – decreases their energy, similar to the case of long-range AF order, and results in their mutual attraction. In this case, the effective interaction potential $V_{kk'}$ is attractve in the range of small momentum transfer $q = k - k,'$ and of a repulsive character for large $q \simeq Q$. Therefore, in this case, the symmetry of the superconducting order parameter, $\Delta(k)$, and the value of T_c strongly depend on the form of the Fermi surface for quasi-particles ("hole-pockets" or the large Fermi surface, see Sect. 7.1.2).

Thus, the pairing mechanism due to the two-magnon scattering processes proposed in [7.77] is of a universal character but requires complicated self-consistent calculations of the quasi-particle spectrum and spin fluctuations in order to yield quantitative results. These calculations were performed under the assumption of weak or intermediate couplings and within the framework of the one-band Hubbard model (7.21) (for $U \leq 4t$). Though the results for weak coupling and for the t-J model with strong coupling agree, the theory [7.77] should be generalized for the case of strong coupling and for the multiband p-d model to compare the conclusions of the theory with the experiment in the copper-oxide superconductors.

In discussions about the spin-fluctuation pairing mechanism, it is often pointed out that such a pairing does not correspond to the s-wave nature of the superconducting gap that is observed in some experiments (Sect. 5.6). Therefore, theoretical models in which the spin fluctuations ensure s-wave pairing are of special interest (see, however, the discussion in [7.72]). Thus, let us consider the results of the paper [7.78], where an s-wave pairing was obtained. As is noted by the authors, in the case of a small hole density, their Fermi momentum p_F is small and, therefore, one-magnon scattering with the transfer of the large AF momentum Q takes the hole far away from the Fermi surface, $2p_F \leq | Q |$. In this case, the two-magnon scattering which brings the hole back onto the Fermi surface turns out to be more effective in the Cooper channel. According to the evaluations in [7.78], the effective coupling constant in the s-channel for the two-magnon scattering is $\lambda_s \simeq 1$. If the energy of spin fluctuations ω_s is sufficiently large, a high superconducting transition temperature $T_c \sim \omega_s \exp(-1/\lambda_s)$ may be obtained. In this rather general model, it is also possible to obtain a reasonable dependence of T_c on the hole concentration x by allowing for the dependences $\lambda_s(x)$ and $\omega_s(x)$.

Thus, the electronic exchange interaction by means of AF spin fluctuations can lead to a superconducting pairing. The magnetic pairing mechanism is most effectively manifested in the d-channel, although under certain assumptions it can also lead to an attraction in the s-channel. For the numerical comparison of the theoretical predictions with experiment, it is necessary to take into account the pair-breaking effects which can lead to a considerable suppression of T_c. The theory of pairing due to AF spin fluctuations, in the limit of strong coupling when the spectrum of hole excitations has a coherent quasi-particle peak and a wide incoherent component, has not yet been worked out in full (Sect. 7.1.2). The investigation of superconducting pairing by allowing for both the spin-fluctuation effects and electron–phonon interaction is of considerable interest (see Sect. 7.4).

7.4 Electron–Phonon Mechanism

The electron–phonon pairing mechanism which is the main mechanism in conventional superconductors, surely plays some role in the formation of the superconducting state in the oxide superconductors; though according to many physicists, it is not a decisive one. The first theoretical calculation of T_c based on a theory of strong electron–phonon coupling [6.7] was not successful due to the fact that it contradict some experiments (see [7.79]). On the other hand, the discovery of strong Coulomb correlations led to the development of non-phononic models of high-temperature superconductivity, as discussed in Sects. 7.2 and 7.3. Subsequent experimental data and theoretical calculations, however, showed the urgent need to develop an electron–phonon model of superconductivity for copper oxides (Chap. 6).

7.4.1 Isotope Effect

The observation of the isotope effect in conventional superconductors, $\alpha \simeq 0.5$ in (1.3), was the direct proof of the existence of an electron–phonon pairing mechanism in these materials. Similar values were found in the three-dimensional oxide superconductors $BaPb_{1-x}Bi_xO_3$ and $K_xBa_{1-x}BiO_3$ which again proved the existence of an electron–phonon mechanism [5.104, 7.80]. The first measurements of α in copper-oxide superconductors revealed a slight change in the transition temperature, $\Delta T_c < 0.5\,\mathrm{K}$, for the isotopic replacement of ^{16}O by ^{18}O; this led to $\alpha \leq 0.2$ in the LSCO compounds, and $\alpha \leq 0.05$ in YBCO and in the Bi and Tl based compounds [7.81].

Later, a strong dependence of α on the composition was found: the index α increased with the suppression of T_c. Thus in $La_{2-x}Sr_xCuO_4$, a value $\alpha \simeq 0.6$ was obtained for $x \simeq 0.11$ and $T_c \simeq 30\,\mathrm{K}$ [7.82]; in $Y_{1-x}Pr_xBa_2Cu_3O_7$ a value $\alpha \simeq 0.4$ was observed for $T_c \simeq 30\,\mathrm{K}$ [7.83]; and in the compounds $La_{1.85}Sr_{0.15}Cu_{1-x}M_xO_4$ it was found that $\alpha = 0.2 - 1.3$ depending on the type and concentration x of impurities M=Ni, Zn, Fe [7.84]. It should, however, be noted that the isotope effect is small in the optimized compounds, with the maximum T_c. This was found e.g. for the compound $Nd_{1-x}Ce_xCuO_x$ with electron type conductivity [7.85].

Several explanations for the increase in α when T_c decreases have been proposed. The simplest is based on the assumption that there exist pair-breaking effects due to the scattering of Copper pairs from magnetic impurities [7.86]. These effects lead to the suppression of the initial transition temperature T_c^0 according to the relation

$$\ln\left(\frac{T_c^0}{T_c}\right) = \Psi\left(\varrho + \frac{1}{2}\right) - \Psi\left(\frac{1}{2}\right),\tag{7.59}$$

where $\Psi(x)$ is the digamma function and $\varrho = \hbar/[2\pi\tau_p(1+\lambda)kT_c]$, τ_p is the relaxation time for the scattering on paramagnetic impurities. Supposing that, when there is no scattering, $\alpha_0 = -d\ln T_c^0/d\ln M$ differs from zero, equation (7.59) yields

$$\alpha = \alpha_0/[1 - \varrho\Psi'(\varrho + 1/2)].\tag{7.60}$$

In the limit of weak scattering when $\tau_p \to \infty$, and $\varrho \to 0$, we have $\alpha = \alpha_0/$ $(1-0.7\tau_p^c/\tau_p)$ where τ_p^c is the critical value of the relaxation time at which $T_c \to 0$. In the latter limit, $\alpha \simeq 0.24\alpha_0(T_c^0/T_c)^2$ diverges. Relation (7.60) agrees sufficiently well with the experiments of [7.83] and [7.84] for small initial values $\alpha_0 \simeq 0.1$.

Thus, the analysis of the isotope effect in the copper-oxide superconductors proves that the electron–phonon interaction makes a contribution to the superconducting pairing ($\alpha_0 \neq 0$), though its role is not decisive ($\alpha_0 \ll 0.5$) [7.86].

7.4.2 Strong Electron–Phonon Coupling

In the theoretical discussion of the electron–phonon mechanism several models were proposed. To obtain a high-T_c within the framework of the standard BCS-Eliashberg theory, it is necessary to assume the existence of strong coupling, i.e., large values of the coupling constant λ (6.4). For example, in [6.13, 7.87] a model with strong coupling was proposed to describe superconductivity in the YBCO compounds. In this model a high-T_c was obtained for $\lambda \simeq 2.4$.

As was discussed in Chap. 6, large constants λ can be obtained for a sufficiently low density of states $N(0)$ if the specific features of the electronic structure of copper-oxide compounds are taken into account: the quasi-two-dimensional nature of the Fermi surface, the strong hybridization of Cu3d- and O2p-states, the ionic nature of the coupling between CuO_2–layers. The small density of charge carriers ensures a weak screening, and consequently a large contribution of long-range Coulomb forces to the electron–ion interaction [7.87, 6.25–29]. According to evaluations of the coupling constants, performed later on the basis of the density-functional method, their averaged values are equal to $\lambda \simeq 1.4$ in YBCO [7.88] and $\lambda \simeq 1.7$ in LSCO [7.89]. For these values of the coupling constants, and taking into account the contribution of high-frequency phonons in the region 50–60 meV, we obtain $T_c \simeq 40$ K. Though these calculations do not give $T_c \simeq 90$ K in YBCO, they indicate a large contribution of the electron–phonon interaction to the superconducting pairing. According to [7.90], the renormalization of the electron–phonon interaction due to the strong Coulomb correlations may suppress the interaction at low concentration of holes [2].

7.4.3 Anharmonic Model of a Superconductor

As was first pointed out in [7.91], strong anharmonicity in the oscillations of the lattice ions in a double-well potential can also lead to a large value of the coupling constant λ, even for a low density of states $N(0)$ and for a moderate constant of the electron–ion interaction $\langle g^2 \rangle$ in (1.2). This model was used in [7.92] and in a number of other papers (for example, [7.93–95]) to explain the existence of high temperatures T_c in oxide superconductors with a structurally unstable lattice.

---

[2] It has been found [7.141] that the on-site electron–phonon coupling is in general enhanced in forward scattering but dramatically suppressed for large momentum transfer scattering by electron correlations.

In cases of strong anharmonicity it is necessary to use a pseudo-spin representation (see (6.11) or (6.19, 20)) to describe the dynamics of the lattice. Composing the Eliashberg equations for this model, we come up with the standard representation for the coupling constant, wich has the form

$$\lambda_s \simeq N(0) \langle g_s^2 \rangle \chi_s , \tag{7.61}$$

where the static pseudo-spin susceptibility of the lattice in the model (6.19) is equal to

$$\chi_s = \frac{1}{N} \sum_{q,\lambda} \frac{\Omega \langle S^z \rangle}{\Omega_\lambda^2(q)} \equiv \frac{2\Omega \langle S^z \rangle}{\omega_s^2} . \tag{7.62}$$

Here $\Omega_\lambda^2(q) = \Omega^2 - \Omega \langle S^z \rangle J_{\lambda\lambda}(q)$ is the spectrum of the pseudo-spin excitations in the paraphase ($\langle S^x \rangle = 0$) and $\langle S^z \rangle = (1/2)\tanh(\Omega/2kT)$.

To compare the coupling constants in the pseudo-spin model (7.61) and in the standard electron–phonon model (1.2), we suppose that electron–ion contributions are equal in them and that

$$\langle g_s^2 \rangle = (2x_{01})^2 \langle (\nabla V_{e-i})^2 \rangle \simeq (2x_{01})^2 \langle g_{ph}^2 \rangle ,$$

where $x_{01} = \langle 0 \mid \hat{x} \mid 1 \rangle$ is the matrix element of the displacement operator $\hat{x}$ in the pseudo-spin model (6.19). Then the ratio of the constants (7.61) and (1.2) can be written in the form

$$\frac{\lambda_s}{\lambda_{ph}} \simeq \frac{(2x_{01})^2 \chi_s}{\chi_{ph}} \simeq \frac{(2x_{01})^2}{\langle u^2 \rangle} \frac{\omega_{ph}}{\omega_s} \gg 1 , \tag{7.63}$$

where χ_s is approximated by the expression $\chi_s \simeq 1/\omega_s \simeq 1/\Omega$ in (7.62); and the harmonic phonon susceptibility is approximated by the function $\chi_{ph} = \langle 1/M\omega^2 \rangle \simeq \langle u^2 \rangle / \hbar \omega_{ph}$ where $\langle u^2 \rangle \simeq \hbar/M\omega_{ph}$ is the mean square ion displacement for oscillations having a characteristic frequency ω_{ph}.

Consequently, due to a much larger amplitude for the ion oscillations in a double-well potential in comparison with the harmonic potential, $(2x_{01})^2 \gg \langle u^2 \rangle$, and due to the presence of low-frequency modes in "soft" ion configurations, $\omega_s \ll \omega_{ph}$, the effective coupling constant λ_s for the structurally unstable lattice can be much greater then the coupling constant λ_{ph} in the harmonic lattice. Direct evaluations for the tilting mode type in LSCO give $\lambda_s \simeq 2.3$ and $T_c \simeq 40\,\text{K}$ for a small density of electron states $N(0) \simeq 1\,\text{eV}^{-1}$, and for a weak coupling $g_s \simeq 0.5(\text{eV/Å}) \times 0.3\,\text{Å} \simeq 0.15\,\text{eV}$ and $\omega_s \simeq 10\,\text{meV}$.

In the anharmonic model of the superconductor, the coupling constant (7.61) depends on the ion mass, since the pseudo-spin susceptibility of the lattice (7.62) is connected in a rather complicated manner with the ion mass oscillating in the double-well. A numerical study of the dependence of T_c and of the isotopic exponent α on the degree of anharmonicity was carried out in [7.96]. A gradual increase in the anharmonic interaction leads to a softening of the oscillation frequency, $\omega_s \simeq \Omega$, in (7.62) and to the growth of T_c up to a certain maximum value

T_c^{max} for optimal value Ω_{max}. Under a further increase in the depth of the two minima of the potential, we have $\Omega \to 0$ and $T_c \propto \Omega^{2/3}$. (In the strong coupling limit $T_c \simeq 0.18 \omega_{ph} \sqrt{\lambda_{ph}} \simeq 0.18[N(0)\langle g_{ph}^2 \rangle /M]^{1/2}$ for $\omega_{ph} \to 0$ in the harmonic theory).

When Ω decreases, the exponent of the isotope effect, α, decreases at first and can become negative ($\alpha < 0$ is the inverse isotope effect), but as the anharmonic interaction continues to increase in the region of strong coupling, $\lambda \gg 1$ and $\Omega < \Omega_{max}$, grows rapidly and reaches a value $\alpha \simeq 0.8$ in the region $T_c \leq T_c^{max}$. If the value of the anharmonic interaction is assumed to be related to the concentration of Sr ions in $La_{2-x}Sr_xCuO_4$ which determines the temperature of the structural transition $T_0(x)$ (see Sect. 2.2.2), the dependence $\alpha(x)$ obtained in [7.96] permits us to obtain a qualitative description of the rapid growth of α when T_c decreases in the experiments [7.82].

Thus, the anharmonic model [7.92] permits us not only to obtain a high T_c but also to get a qualitative description of the anomalous isotope effect. We also note that, within a framework of the anharmonic model, we manage to establish the correlation of T_c with the structural transitions in $La_{2-y-x}RE_ySr_xCuO_4$ and $La_{2-x}Ba_xCuO_4$ (see Sect. 2.2.1) Recent experiments in $YBa_2Cu_3O_7$ have revealed the instability of the lattice near T_c with respect to the shift of the apex oxygen ions in the pyramid in the tilting type mode, as well as in the La_2CuO_4 crystals [7.97]. All these facts undoubtedly show a connection of the structural instability of the lattices of the copper-oxide compounds with the transition to the superconducting state.

7.4.4 Polarons and Bipolaron Superconductivity

A number of experiments have revealed the appearance of a local distortion of the lattice in doped oxide superconductors which is connected with the formation of polarons and bipolarons (see Section 6.2 and [7.98]). In this case, the adiabatic approximation is invalid and the Migdal-Eliashberg theory is not applicable [7.99]. In this connection, a polaron theory of superconductivity was developed [7.100, 101] in which small polaron states were considered.

In the process of the creation of a polaron with a small radius, the most important effect is a sharp reduction of the electronic band width, which is evaluated by the formula

$$W = D \exp(-g^2) \ll D, \tag{7.64}$$

where g is the dimensionless electron–phonon coupling constant and D is the initial width of the band. We suppose that the formation of polarons, i.e., localized states, takes place for a moderate value of the coupling constant, $\lambda = g^2 \omega / D \geq 1$, where ω is a characteristic phonon frequency. The temperature at which the transition to the superconducting state occurs in the polaron model is estimated from the value [7.101]

$$T_c \simeq W \exp(-1/\lambda_p), \tag{7.65}$$

where $\lambda_p \simeq g^2\omega/W \simeq \lambda\exp(g^2) \gg \lambda$. Due to the large energy W, the width of the polaron band, and to the strong coupling, $\lambda_p \gg \lambda$, the temperature T_c (7.65) reaches high values. According to [7.101], the maximum values for the transition temperature are $T_c \sim 0.5\omega \sim 500\,K$ for the interaction with high frequency oxygen modes.

In the region of strong electron–phonon coupling, $\lambda_p \gg 1$, the formation of polarons and the transition to the bipolaron superconductivity is possible, below the temperature for Bose-condensation,

$$T_B \sim \frac{n^{2/3}}{m^*}, \qquad m^* \sim m\exp(g^2). \tag{7.66}$$

Here m is the effective mass and n is the density of bipolarons (in the three-dimensional case). In this case, the properties of the resulting bipolaron super-conductivity are determined by the condensate of local pairs which have a small correlation length $\xi_p \ll n^{-1/3}$.

Discussing the general properties of the superconductors with local pairs we can proceed from the Hubbard model (7.21) with negative U describing the attraction of a pair of particles, $U < 0$. Besides single-site attraction, models with inter-site attraction, $V_{ij} < 0$, can also be considered. The properties of these models are discussed in detail in the review [7.102]. In the case of weak coupling, $|U| \ll D$, the transition temperature increases with an increase in the constant $\lambda = |U|/D$; as in the BCS model, $T_c \sim D\exp(-1/\lambda)$. In the strong coupling case we have $|U| \gg D \simeq zt$, where the Bose condensation of local pairs takes place, and $T_c \sim t^2/|U|$, where t is the transfer integral in the Hubbard model and z is the lattice coordination number.

The most important predictions of the polaron theory of superconductivity are the increase of the effective mass of carriers, $m*/m \sim \exp(g^2) \geq 30$ [7.101], and the existence of local pairs above T_c in the case of bipolarons. Neither of these predictions has yet been unambiguously experimentally verified.

At the present time, there exist several specific models for polaron superconductivity: polarons of a large radius [7.103], bipolarons on diatomic chains in the buffer layers of copper-oxide superconductors [7.104], spin-lattice polarons [7.105]. Also of some interest are more complicated models of two-component Fermi and Bose liquids consisting of itinerant electrons and local pairs [7.106, 107]. Bose condensation of local pairs leads to the simultaneous formation of the Cooper pairs in a system of itinerant electrons.

We would like also to outline the ideas of references [7.108, 109]. Here a considerable increase of the electron–phonon coupling constant λ due to strong antiferromagnetic correlations in the t-J model and in the Hubbard model for electron–phonon interactions, was found. In particular, in [7.109] it was shown that the assumption of strong Coulomb correlations and antiferromagnetic spin fluctuations for a small value of the initial electron–phonon coupling constant, $\lambda \sim 0.2-0.4$, leads to the self-localization of the hole for a large effective coupling constant, $\lambda \sim 20 - 30$. In this case, a strong anharmonicity of the double-well type appears. In the anharmonic model, this leads to a high value for T_c (see Sect. 7.4.3).

7.4.5 Weak Coupling in the Quasi-Two-Dimensional Lattice

In the standard BCS-Eliashberg theory, in deriving the equations of superconductivity, it is supposed that the Fermi surface and electron–phonon interaction are isotropic which permits one to perform an averaging of the matrix elements of this interaction and lattice susceptibility over the Fermi surface. As a result, we obtain a formula for T_c of the form (1.1); with the averaged coupling constant depending only on the density of electron states on the isotropic Fermi surface $N(0)$. In copper-oxide superconductors, the band spectrum is of a quasi-two-dimensional nature in the CuO_2 plane, which, therefore, requires the generalization of the Eliashberg theory to the strongly anisotropic case; see [7.110]. In this case, the existence of singularities of the van Hove type in the electronic spectrum can considerably change the value of T_c [7.111–114] [3].

Let us examine the standard BCS equation to determine the temperature of the superconducting transition T_c:

$$\frac{1}{V} = \int_{-\omega}^{\omega} N(\varepsilon) \frac{d\varepsilon}{\varepsilon} \tanh \frac{\varepsilon}{2kT_c} , \qquad (7.67)$$

where ω is a characteristic phonon frequency and V is the coupling constant. If the function $N(\varepsilon)$ is smooth, it can be replaced by $N(0)$, i.e., by the density of states on the Fermi surface and equation (7.67) leads to a formula of the form (1.1) with $\lambda = VN(0)$. For a quasi-two-dimensional Fermi surface crossing a van Hove anomaly (see Fig. 5.12), we have $N(\varepsilon) \simeq N_0 \ln| D/\varepsilon |$. Integrating (7.67) in this case we obtain [7.112, 113]

$$T_c = 1.36 D \exp \left\{ - \left[\frac{2}{\lambda} + \left(\ln \frac{D}{\omega} \right)^2 - 1 \right]^{1/2} \right\}$$
$$\sim D \exp(-\sqrt{2/\lambda}) , \qquad (7.68)$$

where $\lambda \simeq N_0 V$ and D are the characteristic width of the two-dimensional band. This expression is valid for $\ln (D/\omega) \ll \sqrt{2/\lambda}$, i.e., for weak coupling, $\lambda \ll 1$, provided that the main contribution to the integral in (7.67) is determined by the logarithmic singularity in the density of states and not by the region ω of the electron–phonon attraction. According to the evaluations [7.112], the transition temperature is $T_c \sim 100\,K$ for $\lambda \lesssim 0.1$ and $D \simeq 0.3\,eV$.

The existence of the van Hove anomaly in the electronic spectrum leads to an anomalous behaviour of the isotope effect. A direct calculation of dT_c/dM in

---

[3] Some consequences of a much stronger extended saddle point singularity, experimentally observed in ARPES of YBCO, on the superconducting transition are discussed by Abrikosov et al.[7.142].

(7.67), taking into account the dependence $\omega \propto \sqrt{1/M}$, gives a small isotope effect for the maximum T_c (when the Fermi surface crosses the van Hove anomaly) and a rapid growth of α along with a decrease in T_c as the carrier density decreases [7.113]. As was noted in [7.86], this theory does not however describe the dependence of α in $Pr_x Y_{1-x} Cu_3 O_7$ that was experimentally observed in [7.83]. In the case of the singular behavior in $N(\varepsilon)$, the self-consistent solution of the Eliashberg equations for $Z(\omega)$ and $\Delta(\omega)$ indicates a smoother dependence of α on the carrier concentration x or $N(\varepsilon, x)$ than expected from the simple formula (7.67).

As was discussed in Sect. 7.3, the quasi-two-dimensional character of the Fermi surface in copper-oxide superconductors, in view of the Coulomb correlations, favors the development of spin and charge fluctuations near half-filling. The proximity of the system to a phase transitions leading to the formation of a spin or charge density wave, also leads to an enhancement of the density of states and to an increase of T_c, as in the case of the van Hove anomaly. We note that this mechanism for the enhancement of T_c was first proposed by Kopaev et al. [7.115]. A self-consistent solution of the system of equations which describes the instability with respect to spin, charge, and superconducting fluctuations, is studied in [7.75].

7.4.6 Spin-Fluctuation and Electron–Phonon Interactions

The construction of realistic models for superconductivity in copper-oxide superconductors requires that the spin fluctuations and the electron–phonon interaction should simultaneously be taken into account. In the approximation of weak coupling for the Hubbard model, $U \leq t$, such calculations have been carried out; for example, in [7.76]. A calculation only for the spin-fluctuation interaction in [7.37] leads to a small value for the coupling constant, $\lambda_{sf} \leq 0.15$, and to low values for T_c. The additional contribution of the electron–phonon interaction with a coupling constant $\lambda_{ph} \geq 2$ allows higher values for T_c.

The electron–phonon interaction in the limit of strong correlations, $U \gg t$, was studied both in the one-band Hubbard model (see, for example, [7.116, 117]) and in the p-d-model [7.118]. Here we also mention numerical calculations performed by the Monte Carlo method [7.64]. In these papers, extended s-wave and d-wave couplings were obtained depending on the symmetry of the electron–phonon interaction. Taking into account strong correlations, one obtains a narrower band which results in an enhancement of the density of states and the transition temperature T_c [7.116, 118].

Taking into account strong correlations within the p-d-model (7.11), one also obtains a spin-fluctuation contribution and an interaction through charge fluctuations [7.118]. To clarify these processes, we consider the equation of motion for the Hubbard operator (7.12) in the model (7.11)

$$i\frac{d}{dt}X_i^{0\sigma} = [X_i^{0\sigma}, H] = (\varepsilon_d - \mu)X_i^{0\sigma}$$
$$+ \sum_j V_{ij}[X_i^{\bar\sigma\sigma}a_{j\bar\sigma} + (1 - n_{i\bar\sigma}^d)a_{j\sigma}], \tag{7.69}$$

where we have taken into consideration the anticommutation relations for the Hubbard operators $[X_i^{0\sigma}, X_i^{\sigma'0}] = X_i^{00}\delta_{\sigma\sigma'} + X_i^{\sigma'\sigma}$ and the relation $X_i^{00} + X_i^{\sigma\sigma} = 1 - X_i^{\sigma\bar\sigma} = 1 - n_{i\bar\sigma}^d$. In the second line appear the operators which describe spin flip $X_j^{\bar\sigma\sigma} a_{i\bar\sigma}$ and charge fluctuation on d-site, $\delta n_{i\sigma}^d = n_{i\sigma}^d - \langle n_{i\sigma}^d \rangle$, where $\langle n_{i\sigma}^d \rangle = n_d/2$, is the average number of d- holes in the paramagnetic state.

As was shown in [7.118], the composition of the equations of motion for the 4×4 matrix Green's functions from the Nambu operators (see Sect. 7.2.2) taking into account (7.69) leads to a system of equations for the self-energy operator and the Green's functions. Here the kernel of the integral equation for the self-energy operator (and for its anomalous component, i.e., the superconducting gap) contains contributions both from the electron–phonon interaction (introduced into the Hamiltonian (7.11)) and from the spin and charge fluctuations which are described by the following dynamical susceptibilities:

$$
\begin{aligned}
\chi_{sf}^{\sigma\bar\sigma}(q,\omega) &= \langle\langle X_q^{\sigma\bar\sigma} | X_{-q}^{\bar\sigma\sigma} \rangle\rangle_\omega , \\
\chi_{cf}^{\sigma}(q,\omega) &= \langle\langle \delta n_{q\sigma}^d | \delta n_{-q\sigma}^d \rangle\rangle_\omega ,
\end{aligned}
\tag{7.70}
$$

where the corresponding Fourier-components (q,ω) of the retarded Green's functions are introduced [7.57].

As follows from (7.69), the coupling constants for these two pairing channels are determined by the same hybridization parameter of the model $V_{ij} = 2t_{pd}\lambda_{ij}$ or $V(q) = t_{pd}\gamma_q$ (see (7.10)) and not by the fitting parameter as, for example, in the spin-fluctuation coupling model in [7.72].

Reference [7.118] presents a numerical solution of the Eliashberg equation with regard to the electron–phonon interaction and spin fluctuations with the susceptibility (7.70), chosen in the form (3.34) in the MMP model as in [7.72]. It tuns out, that in the case of weak electron–phonon coupling, $\lambda_{\text{ph}} \simeq 0.5$, and for the standard parameters of the p-d model, $\varepsilon_p - \varepsilon_d = 3.5 t_{pd}$, a high T_c can be obtained both for the extended s-wave coupling and for the d-coupling. The spin-fluctuation contribution greatly increases T_c in the case of the d-coupling, but gives also a considerable contribution in the s-coupling. A superconducting pairing in the singlet band of the model (7.20) has been considered in [7.119]. Since the maximum density of hole states in the singlet band occurs at the concentration of holes $n \simeq 1.24$ [7.6] the calculated $T_c(n)$ dependence shows a maximum at this concentration $n \simeq 1.24$ (which is characteristic for copper-oxides – see Fig. 5.3), both for electron–phonon s-wave pairing and spin-exchange d-wave pairing. Though the calculations carried out in [7.118, 119] are only qualitative (i.e., the quasiparticle damping is not taken into account, the electron–phonon interaction is of a model character, etc.), they undoubtedly show that the spin-fluctuation mechanism should be taken into account when studying the electron–phonon mechanism in the systems with strong correlation.

7.5 Exciton Models

The general formula for the superconducting transition temperature (1.1) shows that a large enhancement of T_c can be obtained if the boson energy which is responsible for the electron coupling is of the order of the electron energies, $\hbar\omega \sim 1\,\text{eV}$. Then, even in the case of weak coupling, $\lambda \ll 1$, transition temperatures around the room temperature are possible. This exciton model, in which the coupling is due to the exchange of electronic excitations, was first proposed by Little [7.120] for quasi-one-dimensional organic superconductors and by Ginsburg [7.121] for layered systems (see [1.3]).

To realize this model we must assume the existence of two groups of electrons: one of them is connected with the conductivity band where the superconducting pairing is due to the exchange of excitons, i.e., excitations in a second group of almost localized electrons, takes place. In the oxide superconductors, due to the multiple-band nature of the electron spectrum, layered structure and other unusual propertities of the electronic structure, this separation of the electron states is quite probable. It has been used as the basis of a great number of exciton models [7.122]. Let us consider some of them.

7.5.1 Plasmon Model

The effective interaction of two electrons in a crystal can be represented in the framework of the permittivity formalism [1.3] in the form

$$V_{\text{eff}}(\boldsymbol{q}, \omega) = \frac{v(\boldsymbol{q})}{\varepsilon(\boldsymbol{q}, \omega)}, \tag{7.71}$$

where $\boldsymbol{q} = \boldsymbol{p} - \boldsymbol{p}'$ and $\hbar\omega = \varepsilon_p - \varepsilon_{p'}$ are respectively the difference of momenta and energies of two electrons, $v(\boldsymbol{q})$ is the Coulomb potential and $\varepsilon(\boldsymbol{q}, \omega)$ is the longitudinal permittivity of the system. The latter satisfies the dispersion relation given by the Kramers-Kronig formula:

$$\frac{1}{\varepsilon(\boldsymbol{q}, \omega)} = 1 - \int_0^\infty \frac{dz^2 \varrho(\boldsymbol{q}, z)}{z^2 - \omega^2 - i\delta}, \tag{7.72}$$

where the non-negative spectral density

$$\varrho(\boldsymbol{q}, \omega) = -\frac{1}{\pi} \operatorname{Im} \frac{1}{\varepsilon(\boldsymbol{q}, \omega + i\eta)}, \tag{7.73}$$

is directly related to the scattering spectrum of fast electrons in the crystal, $S(\boldsymbol{q}, \omega) \propto \varrho(\boldsymbol{q}, \omega)$.

Writing the permittivity in the form of a sum over the electronic $\alpha_e(\boldsymbol{q}, \omega)$ and ionic $\alpha_i(\boldsymbol{q}, \omega)$, polarizabilities

$$\varepsilon(\boldsymbol{q}, \omega) = 1 + \alpha_e(\boldsymbol{q}, \omega) + \alpha_i(\boldsymbol{q}, \omega), \tag{7.74}$$

one can represent the effective interaction (7.71) in the form of the sum

$$V_{\text{eff}}(q, \omega) = \frac{v(q)}{\varepsilon_e(q, \omega)} - \frac{v(q)}{\varepsilon_e(q, \omega)} \frac{\tilde{\alpha}_i(q, \omega)}{1 + \tilde{\alpha}_i(q, \omega)}, \tag{7.75}$$

where the first term which is proportional to $\varepsilon_e^{-1}(q, \omega) = (1 + \alpha_e(q\omega))^{-1}$ determines the contribution to the interaction from electronic excitations (excitons) and the second determines the contribution from phonons with a lattice polarizability $\tilde{\alpha}_i = \alpha_i/\varepsilon_e$. In the BCS, theory, a low-frequency limit is considered (7.75) when the first term in the region q $\simeq 2k_F$ gives the Coulomb repulsion and the second term gives the attraction for frequencies $\omega < \omega_{\text{ph}}$ where ω_{ph} is a characteristic phonon frequency. In the case where low-frequency collective excitations – plasmons – occur, the condition $[1/\varepsilon_e(q, \omega) - 1] < 0$ can be fulfilled (for $\omega < \omega_0$, plasmon frequencies), and, therefore, an attraction appears due to the exchange of plasmon excitations. Several plasmon models have been proposed (see, for example, [7.123] and the literature cited therein).

The studies [7.124, 125] consider plasma excitations in layered crystals which have a quasi-two-dimensional electronic spectrum. These can give rise to the appearance of weakly damped acoustic plasmons with a spectrum $\Omega_q = (aq + bq^2)^{1/2}$, which for small q can lead to an effective attraction for a wide range of frequencies, $\omega < \Omega_{\text{max}}$. However, for large scattering momenta, $q \sim 2k_F$, which make the main contribution to Cooper pairing, the plasma modes undergo strong Landau damping and they become ineffective in the pairing channel. Therefore, in homogeneous systems, even in the case of strong anisotropy, the effective plasmon interaction averaged over the transfer momentum q leads to a repulsion or to a very weak attraction.

In this connection, multiband models in which low-frequency plasmons are formed in the two-dimensional band of heavy holes, seem to be more promising [7.123]. If the maximum plasmon frequency is larger than the width of the narrow band of heavy holes W_h, $\hbar\Omega_h > W_h$, then an effective exciton attraction appears for light holes in the broad band. In this case, the spectrum of low-frequency plasmons and the width of the Landau damping region are periodic functions of the longitudinal momentum $q = (q_x, q_y)$, which leads to wide permissible bands for plasma excitations without damping. Consequently, the region where the plasmon mechanism makes a contribution extends to the entire Brillouin zone, and the averaged exciton pairing interaction, i.e., the first term in (7.75) turns out to be positive.

In a number of papers, it has been noted that the interaction of optical phonons with plasma oscillations can lead to a considerable enhancement of the electron–phonon interaction in (7.75) (see, for example, [7.126]). Furthermore, in ionic crystals which have a large dielectric constant (for example, near the ferroelectric phase transition) due to the hybridization of plasmons with optical phonons, the region of attraction broadens up to a maximum frequency $\omega_{\text{max}} = (\omega_{\text{LO}}^2 + \Omega_h^2)^2$, where ω_{LO} is the frequency of longitudinal optical phonons [7.124].

In Ref. [7.127], a model of collective electron excitations connected with the charge transfer $Cu^{2+}O^{2-} \rightarrow Cu^{1+}O^{1-}$ was proposed. For a sufficiently strong Coulomb repulsion V_{pd} for the holes on the p- and d-sites, $V_{pd} \sim t_{pd}$ in (5.6), the frequency of the collective excitations has been estimated to be $\hbar\omega_0 \sim t_{pd}$

[7.127]. In a more detailed elaboration of this model, strong Coulomb correlations on the d-sites which considerably change both the nature of the one-particle p-d excitations and the fluctuations of the related spin and charge density, should be taken into account. The general phenomenological model for the spectrum of collective and one-particle excitations for the copper-oxide compounds, i.e., the model of the marginal Fermi-liquid [7.4] was discussed in Sect. 7.1. The solution of the Eliashberg equations and calculation of T_c in this model was considered in [7.128].

A detailed analysis of the role of the charge-transfer exitations in oxide superconductors was carried out in [7.129]. In the opinion of the authors, the strong electron–phonon interaction for certain phonon modes connected with charge transfer from the buffer layers into the CuO_2 plane could account for the high T_c and a number of anomalous properties of oxide compounds.

7.5.2 Jahn-Teller Excitations and the d-d Model

As was noted in Sect. 5.1, in the crystal electric field of the lattice, the splitting of the $3d$-levels of the copper ion leads to the appearance of low-energy electronic excitations of the quadrupole type (see Fig. 5.1). The latter can play the role of excitons which are responsible for the pairing of electrons in the $pd\sigma$ band for the CuO_2 plane. Such a pairing mechanism due to virtual d-d excitations was proposed in [7.130, 131].

The attraction mechanism arising in the d-d model can be described as follows [7.131]. According to the scheme for the electron levels in the crystal field shown in Fig. 5.1, the lowest energy belongs to the excitation of the Jahn-Teller type between the split e_g levels $d_x \rightarrow d_z$, where $d_x = d(x^2 - y^2)$, $d_z = d(3z^2 - r^2)$. Its value is estimated by $E_{JT} \simeq 0.5\,eV$. The Coulomb repulsion V_{pd} between the p-hole on the orbitals $p_\sigma(x, y)$ and the d_x–orbital, V_{pd}^x, is much larger than the Coulomb interaction with the d_z–orbital, V_{pd}^z, where $\Delta V = V_{pd}^x - V_{pd}^z \simeq 0.5\,eV$. Therefore, when a p-hole moves, the excitation of the d-system into the state d_z (which leads to a decrease of the Coulomb energy in the CuO_4–cluster) is energetically favorable. As a result, the other hole is attracted to the initial p-hole dressed by the d-excitations. This attraction mechanism seems to be similar to the electron–phonon one, where the attraction arises between two electrons dressed by phonon excitations.

In the simplest approximation where the spin degrees of freedom on the copper sites are neglected, only the system of spinless local d_x and d_z states with the Coulomb interaction is relevant. This "exciton" system is also connected by Coulomb interactions V_{pd}^x and V_{pd}^z to a broad conductivity p-band. The evaluation of the temperature for superconducting pairing in the p-band leads to the standard BCS expression (1.1)

$$T_c \simeq E_{JT} \cdot \exp(-1/\lambda_{ex}), \tag{7.76}$$

where the coupling constant is $\lambda_{ex} \simeq N(0)(\Delta V)^2/E_{JT}$. For the parameters characteristic for copper-oxide compounds, we obtain $\lambda_{ex} \simeq 0.1 - 0.3$, $E_{JT} \simeq 5000\,K$ and the transition temperature $T_c \sim 100\,\mathrm{K}$. For a more rigorous evaluation, we should take into account the spin degrees of freedom of the Cu3d-states, in particular, the formation of the singlet and triplet bands (see Sect. 7.1.1).

To experimentally verify the d-d model, the spectra of excitations of the Cu3d-states should be investigated. The existence of the hole $3d(3z^2-r^2)$-states, as a 10–15 % admixture to the $3d(x^2-y^2)$-states in the CuO$_2$-plane, is proved by EELS and XPS experiments (see Sect. 5.4.1). However, the observation of the multiplet d-d-excitations in the energy range $0.3-1.5\,\mathrm{eV}$ is a complicated problem, since they are not connected with dipole transitions and, therefore, contribute a small intensity in the optical absorption spectra. Special optical experiments are necessary to single out the d-d-excitation of the quadrupole type from the background of strong absorption in the mid-infrared frequency region (see Figs. 5.21, 22) [4].

The Jahn-Teller electronic excitations always reveal a strong interaction with a local deformation of the lattice of the corresponding symmetry. In this connection, an additional mechanism of the electron attraction due to phonon exchange, i.e., the fluctuation of the local Jahn-Teller type distortions of the lattice, is possible. Several specific models based on the dynamic Jahn-Teller effect were proposed: a model of off-center polarons with coupling to an odd optical mode [7.132], model of "double-well" potentials of degenerate $O(p_\pi)-O(p_\pi)$ bonds in the CuO$_2$ plane [7.133] and a number of others.

7.5.3 Coulomb Interaction in the Multiband Models

In concluding this section which has been devoted to electronic pairing mechanisms, we will discuss some models in which the direct Coulomb interaction leads to a superconducting gap due to the multiband nature of the electronic spectrum. The two-band model of a superconductor with interband Coulomb scattering was first proposed in [7.134]. We shall now discuss this electron-pairing mechanism for the multiband model studied in [7.135].

Suppose that there are several quantum states α for the charge carriers, i.e., separate bands or regions in q-space connected by the scattering potential $V_{\alpha\alpha'}(kk')$. Then the self-consistent equation for the order parameter $\Delta_\alpha(k)$ is of the form [7.135]

$$\Delta_\alpha(k) = -\sum_{\alpha'k'} V_{\alpha\alpha'}(k,k')\frac{\Delta_{\alpha'}(k')}{E_{\alpha'}(k')}\tanh\frac{E_{\alpha'}(k')}{2kT}. \tag{7.77}$$

This equation can have non-zero solutions even for a repulsive, $V_{\alpha\alpha'} > 0$, potential ; in particular, if the order parameter $\Delta_\alpha(k)$ has different signs for different quantum numbers α. In this case, "self-tuning" of the order parameter occurs, as a result of which the superconducting state with an alternating order parameter is energetically favorable. The simplest example is the d-wave pairing,

---

[4] Recently d-d excitations in the crystal field for a La$_2$CuO$_4$ crystal have been observed [7.143].

$\Delta(\mathbf{k}) = \Delta_d(\cos k_x - \cos k_y)$, which appears for the anisotropic (alternating in the $\mathbf{k}$-space) interaction $V(\mathbf{kk'})$ due to a one-magnon exchange.

As is outlined in [7.135], the Coulomb pairing mechanism in (7.77) is effective near the dielectric instability due to the interband Coulomb interaction. In this case, the intraband Coulomb repulsion of a pair of particles inside the band turns out to be smaller than the interband Coulomb repulsion describing the interaction of the order parameters Δ_α between different bands. Due to a sufficiently broad energy band where the interband interaction acts, high values of T_c can be expected in this model.

Several specific multiband models with a Coulomb interaction have also been proposed. For example, a three-band model where electron–hole pairing in two spatially separated bands leads to an excitonic insulating state having low-lying collective excitation modes, was examined in [7.136]. Due to the strong polariz-ability of such systems, the appearance of an effective attraction for the conduction electrons in a third band connected only by Coulomb interaction with the first two bands, is possible. In [7.137] the dependence of T_c on the hole concentration was calculated in the two-band models of LSCO and YBCO compounds with an effective interaction $V_{\alpha\alpha'}(\mathbf{k}, \mathbf{k'}) > 0$.

An original model of hole superconductivity with an energy-dependent gap was proposed by Hirsh [7.138]. Taking into account the strong polarization of oxygen ions, the author introduces an additional correlated hopping term in the Hubbard model (7.21)

$$H_1 = -\Delta t \sum_{ij} c_{i\sigma}^+ c_{j\sigma}(n_{i-\sigma} + n_{j-\sigma}) + \text{H.c.}, \tag{7.78}$$

which increases the possibility of the hopping of holes from sites which are already occupied by holes. The result is an effective Coulomb interaction that depends on the hole scattering momentum. For the two-dimensional lattice it is of the form

$$V_{\mathbf{kk'}} = U - K(\cos k_x + \cos k_x' + \cos k_y + \cos k_y'), \tag{7.79}$$

where U is the one-site repulsion in the Hubbard model (7.21) and $K = 8\Delta t$. In the approximation of weak coupling, the BCS equation with the interaction (7.79) leads to non-zero solutions for the extended s-wave pairing. In common with the general model (7.77), the model of hole superconductivity proposes that there is a mutual compensation of the interaction signs (7.79) and the order parameter $\Delta(\mathbf{k})$: in the repulsion region where $V_{\mathbf{kk'}} > 0$, the gap becomes negative ensuring a positive contribution to (7.77). It is interesting to note that if the electron–phonon interaction in the hole model (7.79) is additionally taken into account, an increase of T_c occurs only for small values of the parameter K [7.139].

Thus, for the multiband electron spectrum in oxide compounds, there is one more channel for carrier pairing due to interband Coulomb scattering. In describing realistic models based on this pairing mechanism, the strong single-site (intraband) Coulomb correlations, which usually lead to a suppression of the electronic mech-anisms of attraction, must be consistently taken into account.

8. Conclusion

Summarizing the results of the study of the physical properties and mechanisms of the superconductivity in oxide superconductors, we still cannot answer the principal question: what is the nature of their superconducting state; without this we are unable to answer another question of practical importance – whether it is possible to obtain compounds with even higher T_c of the order of room temperature. To answer these questions, many experimental investigations have to be carried out in order to clarify the problems mentioned in the previous chapters. A quick solution of these problems is hardly possible due to the complicated physical and chemical structure of the oxide materials. We recall that it took over 46 years to develop a microscopic theory of superconductivity from the discovery of this phenomenon in 1911.

Leaving aside the discussion of the main results of the investigation of the structure, antiferromagnetic correlations, thermodynamic, electronic and lattice properties of the oxide compounds which are listed at the end of the corresponding sections of the book, we can make only general statements concerning the importance of these properties in the formation of the superconducting state.

In the three-dimensional $BaBiO_3$-based oxide compounds, the superconducting pairing is apparently due to a strong electron–phonon interaction coming from certain vibrational modes connected with the Bi – O charge transfer. Therefore, the same pairing mechanism should also make a certain contribution in the copper-oxide superconductors, since it is difficult to think of reasons why it should be suppressed in them.

Of course, specific properties of copper-oxide compounds should be taken into account. The main one is the strong Coulomb correlations, leading to an electronic spectrum of the correlation type and to strong antiferromagnetic fluctuations. The latter apparently determines many anomalous properties of copper-oxide compounds and has a high probability of making a contribution to the superconducting pairing. The formation of the Zhang-Rice singlet states inside the p-d gap is specific to the CuO_2 plane. Under doping, their appearance can presumably explain the occurrence of superconductivity only in a narrow range of hole concentration, where these strongly correlated states – composite quasi-particles – are stable.

As a whole, the "synergetic" pairing mechanism seems to be the most probable one in copper-oxide compounds. This mechanism includes the electron–phonon interaction enhanced by charge fluctuations and the spin-fluctuation interaction in the CuO_2 plane. In the author's opinion, a number of anomalous properties of copper-oxide compounds in the normal and superconducting states can only be accounted for when the interactions of all the degrees of freedom – lattice, electronic, and spin – are taken into consideration. The elaboration of the corresponding theory requires not only the performance of complicated numerical calculations, but also the solution of a number of important problems in systems with strong electron correlations.

Experimental investigations can lead to the discovery of complitely new compounds with high T_c. A good example is the discovery of superconductivity with $T_c \simeq 30\,\mathrm{K}$ in various fullerene compounds $R_x C_{60}$, i.e., in the three-dimensional organic crystals built from the carbon moleculae C_{60} and alkaline metals R=Cs, Rb, K [8.1–3]. These compounds open up a new chapter in the study of high-temperature superconductivity, so far one of the most mysterious phenomena that Mother Nature has presented to us.

References

Chapter 1

1.1 J.G. Bednorz, K.A. Müller: Zs. Phys. B **64** (1986) 189
1.2 J.G. Bednorz, K.A. Müller: Rev. Mod. Phys. **60** (1988) 588
1.3 *High-Temperature Superconductivity*, ed. by V.L. Ginsburg and D.A. Kirzhnits (Consultant Bureau, New York, London 1982)
1.4 V.S. Vonsovsky, Yu.A. Izyumov, E.Z. Kurmaev: *Superconductivity of Transition Metals* (Springer, Berlin, Heidelberg 1982)
1.5 A.W. Sleight, J.L. Gillson, F.E. Bierstedt: Solid State Commun. **17** (1975) 27
1.6 S. Uchida, K. Kitazawa, S. Tanaka: Phase Transitions 8 (1987) 95
1.7 B.K. Chakraverty.: Phys. Lett. **40** (1979) L99; J. Phys. **42** 1351
1.8 J.G. Bednorz, M. Takashige, K.A. Müller: Europhys. Lett. **3** (1987) 379
1.9 M.K. Wu, J.R. Ashburn, C.J. Torng, P.H. Hor, R.L. Meng, L. Goa, Z.J. Huang, Y.Q. Wang, C.W. Chu: Phys. Rev. Lett. **58** (1987) 908; P.H. Hor, L. Gao, R.L. Meng, Z.J. Huang, Y.Q. Wang, K. Forster, J. Vassilious, C.W. Chu, M.K. Wu, J.R. Ashburn, C.J. Torng: Phys. Rev. Lett. **58** (1987) 911
1.10 H. Maeda, Y. Tanaka, M. Fukutomi, T. Asano: Jpn. J. Appl. Phys. **27** (1988) L209
1.11 Z.Z. Sheng, A.M. Hermann: Nature **332** (1988) 55; A.M. Hermann, Z.Z. Sheng, D.C. Vier, S. Schultz, S.B. Oseroff: Phys. Rev. B **37** (1988) 9742
1.12 L.F. Mattheiss, E.M. Gyorgy, D.W. Johnson, Jr.: Phys. Rev. B **37** (1988) 3745
1.13 R.J. Cava, B. Batlogg, J.J. Krajewski, R. Farrow, L.W. Rupp, Jr., A.E. White, K. Short, W.F. Peck, T. Kometani: Nature **332** (1988) 814
1.14 Y. Tokura, H. Takagi, S. Uchida: Nature **337** (1989) 345
1.15 H. Takagi, S. Uchida, Y. Tokura: Phys. Rev. Lett. **62** (1987) 1197
1.16 C.W. Chu, Y.Y. Xue, Y.Y. Sun, R.L. Meng, Y.K. Tao, L. Gao, Z.J. Huang, J. Bechtold, P.H. Hor, Y.C. Jean: Proc. Taiwan Intl. Symp. on Superconductivity (World Scientific, Singapore 1989)
1.17 B. Batlogg: Physica C **185–189** (1991) XVIII
1.18 Proc. Intl. Conf. High Temperature Superconductors and Materials and Mechanisms of Superconductivity (Interlaken, Switzerland 1988) Physica C, **153–155** (1988)
1.19 Proc. IId Intl. Conf. M²-HTSC (Stanford, California 1989) Physica C **162–164** (1989)
1.20 Proc. IIId Intl. Conf. M²-HTSC (Kanazawa, Japan 1991) Physica C **185–189** (1991)
1.21 *Physical Properties of High-Temperature Superconductors*, ed. by D.M. Ginsberg (World Scientific, Singapore 1989 v.1, 1990 v.2, 1992 v.3)
1.22 *Strong Correlation and Superconductivity*, ed. by H. Fukuyama, S. Maekawa, A.P. Malozemoff (Springer, Berlin, Heidelberg 1989)
1.23 *Earlier and Recent Aspects of Superconductivity*, ed. by J.G. Bednorz, K.A. Müller (Springer, Berlin, Heidelberg 1990)
1.24 *Physics of High-Temperature Superconductors*, ed. by S. Maekawa, M. Sato, (Springer, Berlin, Heidelberg 1992)

1.25 A. Schilling, M. Cantoni, J.D. Guo, H.R. Ott: Nature **363** (1993) 56
1.26 M. Lagues, X.M. Xie, H. Tebbji, X.Z. Xu, V. Mairet, C. Hatterer, C.F. Beuran, C. Deville-Cavellin: Science **262** (1993) 1850; J.-L. Tholence, B. Souletie, O. Laborde, J.-J. Capponi, C. Chaillout, M. Marezio: Phys. Lett. A. **184** (1994) 215

Chapter 2

2.1   R.M. Hazen: In *Physical Properties of High-Temperature Superconductors*, ed. by D.M. Ginsberg (World Scientific, Singapore 1990) v.2, p. 121
2.2   S. Pei, J.D. Jorgensen, B. Dabrowski, D.G. Hinks, D.R. Richards, A.W. Mitchell, J.M. Newsam, S.K. Sinha, D. Vaknin, A.J. Jacobson: Phys. Rev. B **41** (1990) 4126
2.3   R.J. Cava: Science **247** (1990) 656
2.4   J.B. Torrance, Y. Tokura, A. Nazzal, S.S.P. Parkin: Phys. Rev. Lett. **60** (1988) 542
2.5   R.J. Cava, R.B. Van Dover, B. Batlogg, J.J. Krajevski, L.F. Schneemeyer, T. Siegrist, B. Hessen, S.H. Chen, W.F. Peck, Jr., L.W. Rupp, Jr.,: Physica C **185–189** (1991) 180
2.6   Z. Hiroi, M. Takano, M. Azuma, Y. Takeda, Y. Bando: Physica C **185–189** (1991) 523; Nature **356** (1992) 775
2.7   M.G. Smith, A. Manthiran, J. Zhou, J.B. Goodenough, J.T. Markert:. Nature **351** (1991) 549
2.8   J.D. Jorgensen: Jpn. J. Appl. Phys. Suppl. **26–3** (1987) 2017
2.9   R.J. Birgeneau, G. Shirane: In *Physical Properties of High Temperature Superconductors*, ed. by D.M. Ginsberg (World Scientific, Singapore) **1** (1989) 151
2.10 C. Howard, R.J. Nelmes, C. Vettier: Solid State Commun. **69** (1989) 261
2.11 R. Moret, J.P. Pouget, C. Noguera, G. Collin: Physica C **153–155** (1988) 968; H.J. Kim, R. Moret: Physica C **156** (1988) 363
2.12 J.D. Axe, A.H. Moudden, D. Hohlwein, D.E. Cox, K.M. Mohanty, A.R. Moodenbaugh, Y. Xu: Phys. Rev. Lett. **62** (1989) 2751
2.13 T. Suzuki, T. Fujita: Physica C **159** (1989) 111; T. Suzuki, M. Tagawa, T. Fujita: Physica C **162–164** (1989) 983
2.14 M. Sato: In *Physics of High-Temperature Superconductors*, ed. by S. Maekawa, M. Sato (Springer, Berlin, Heidelberg 1992) p. 239
2.15 K.I. Kumagai, I. Watanabe, K. Kawano, H. Matoba, K. Nishiyama, K. Nagamine, N. Wada, M. Okaji, K. Nara: Physica C **185–189** (1991) 913
2.16 Y. Maeno, N. Kakehi, M. Kato, T. Fujita: Phys. Rev. B **44** (1991) 7753
2.17 S. Barisic, J. Zelenko: Solid State Commun. **74** (1989) 367
2.18 W.E. Pickett, R.E. Cohen, H. Krakauer: Physica B **169** (1991) 45; Phys. Rev. Lett. **67** (1991) 228
2.19 M.K. Crawford, R.L. Harlow, E.M. McCarron, W.E. Farneth, J.D. Axe, H. Chou, Q. Huang: Phys. Rev. B **44** (1991) 7749
2.20 M.K. Crawford, W.E. Farneth, R.L. Harlow, E.M. McCarron, R. Miao, H. Chou, Q. Huang: In Proc. Intl. Conf. Lattice effects in High-T_c Sureconductors (Santa Fe 1992) ed. by Y. Bar-Yam, T. Egami, J. Mustre-de Leon, A.R. Bishop (Wold Scientific, Singapore 1992) p. 531
2.21 B. Büchner, M. Braden, M. Cramm, W. Schlabiz, W. Schnelle, O. Hoffels, W. Braunisch, R. Müller, G. Heger, D. Wohlleben: Physica C **185–189** (1991) 903
2.22 Wu Ting, K. Fossheim, T. Laegreid: Solid State Commun. **75** (1990) 727
2.23 A. Migliori, M. Visscher, S. Wong, S.E. Brown, I. Tanaka, H. Kojima, P.B. Allen: Phys. Rev. Lett. **64** (1990) 2458
2.24 N.M. Plakida, V.S. Shakhmatov: Physica C **153–155** (1988) 233. Soviet Phys. Izvestia **53** (1989) 1236
2.25 K.S. Aleksandrov, T.I. Ivanova, T.I. Misyul, V.P. Sakhnenko, G.M. Chechin: Phase Transitions. **22** (1990) 245

2.26 A.D. Bruce, R.A. Cowley: *Structural phase transitions*, (Taylor and Francis Ltd., London 1981)

2.27 Y. Ishibashi: J. Phys. Soc. Jpn. **59** (1990) 800

2.28 Y. Tokura, H. Takagi, H. Watabe, H. Matsubara, S. Uchida, K. Hiraga, T. Oku, T. Mochiku, H. Asano: Phys. Rev. B **40** (1989) 2568

2.29 L.H. Greene, B.G. Bagley: In *Physical Properties of High Temperature Superconductors*, ed. by D.M. Ginsberg (World Scientific, Singapore 1990) v.2, p. 509

2.30 R. Sonntag, D. Hohewein, T. Bruckel, G. Kollin: Phys. Rev. Lett. **66** (1991) 1497

2.31 V.E. Zubkus, S. Lapinskas, E.E. Torneau: Physica C **166** (1990) 472

2.32 R.J. Cava, A.W. Hewat, E.A. Hewat, B. Batlogg, M. Marezio, K.M. Rabe, J.J. Krajewski, W.F. Peck, Jr., L.W. Rupp, Jr.: Physica C **165** (1990) 419

2.33 J.D. Jorgensen, B.W. Veal, A.P. Paulikas, L.J. Nowicki, G.W. Crabtree, H. Claus, W.K. Kwok: Phys. Rev. B **41** (1990 1863

2.34 B. Bücher, J. Karpinski, E. Kaldis, P. Wachter: J. Less Common. Met. **164–165** (1990) 20; E. Kaldis, J. Karpinski, S. Rusiecki, B. Bücher, K. Conder, E. Jilek: Physica C **185–189** (1991) 190

2.35 T. Graf, G. Triscone, A. Junod, J. Müller: Physica B **165–166** (1990) 1671

2.36 R.J. Cava, B. Batlogg, J.J. Krajewski, L.W. Rupp, L.F. Schneemeyer, T. Siegrist, R.B. van Dover, P. Marsh, W.F. Peck, Jr., P.K. Gallagher, S.H. Glarum, J.H. Marshall, R.C. Farrow, J.V. Waszczak, R. Hull, P. Trevor: Nature **336** (1988) 211

2.37 G. Roth, P. Adelmann, S. Heger, R. Knitter, Th. Wolf: J. de Phys. **1** (1991) 721

2.38 H. Maeda, Y. Tanaka, M. Fukutomi, T. Asano: Jpn. J. Appl. Phys. **27** (1988) L209

2.39 Z.Z. Sheng, A.M. Hermann: Nature **332** (1988) 55, 138

2.40 S.S.P. Parkin, V.Y. Lee, A.I. Nazzal, R. Savoy, T.C. Huang, G. Gorman, R. Beyers: Phys. Rev. B **38** (1988) 6531

2.41 A.W. Hewat, E.A. Hewat, P. Bordet, J.J. Capponi, C. Chaillout, J. Chenavas, J.L. Hodeau, M. Marezio, P. Strobel, M. Francois, K. Yvon, P. Fischer, J.L. Tholence: Physica B **156–157** (1989) 874

2.42 W. Dmowdski, B.H. Toby, T. Egami, M.A. Subramanian, J. Gopalakrishnan, A.W. Sleight: Phys. Rev. Lett. **61** (1988) 2608

2.43 B.H. Tobi, T. Egami, J.D. Jorgensen, M.A. Subramanian: Phys. Rev. Lett. **64** (1990) 2414

2.44 Y. Tokura, T. Arima: Jpn. J. Appl. Phys. **29** (1990) 2388; Y. Tokura: Physica C **185–189** (1991) 174

2.45 J.D. Jorgensen, D.G. Hinks, P.G. Radaelli, S. Pei, P. Lightfoot, B. Dabrowski, C.U. Segre, B.A. Hunter: Physica C **185–189** (1991) 184

Chapter 3

3.1 R.J. Birgeneau, G. Shirane: In *Physical Properties of High Temperature Superconductors*, ed. by D.M. Ginsberg (World Scientific, Singapore 1989) v.1, p. 151

3.2 T. Thio, T.R. Thurston, N.W. Preyer, P.J. Picone, M.A. Kastner, H.P. Jenssen, D.R. Gabbe, C.Y. Chen, R.J. Birgeneau, A. Aharoni: Phys. Rev. B **38** (1988) 905; T. Thio, C.Y. Chen, B.S. Freer, D.R. Gabbe, H.P. Jenssen, M.A. Kastner, P.J. Picone, N.W. Preyer: Phys. Rev. B **41** (1990) 231

3.3 M.A. Kastner, R.J. Birgeneau, T.R. Thurston, P.J. Picone, H.P. Jenssen, D.R. Gabbe, M. Sato, K. Fukuda, S. Shamoto, Y. Endoh, K. Yamada, G. Shirane: Phys. Rev. B **38** (1988) 6636

3.4 N.M. Plakida, V.S. Shakhmatov: Superconductivity: Phys., Chem., Techn. **6** (1993) 669

3.5   J.I. Budnick, M.E. Filipkowski, Z. Tan, B. Chamberland, Ch. Niedermayer, A. Wei-
      dinger, A. Golnik, R. Simon, M. Rauer, H. Gluckner, E. Recknagel, C. Baines: In
      Proc. Intl. Seminar High−T_c Superconductivity (Dubna 1989) ed. by N.N. Bogo-
      lubov, V.L. Aksenov, N.M. Plakida (World Scientific, Singapore 1990) p. 172
3.6   A. Aharony, R.T. Birgeneau, A. Coniglio, M.A. Kastner, H.E. Stanley: Phys. Rev.
      Lett. **60** (1988) 1330
3.7   K.B. Lyons, P.A. Fleury, J.P. Remeika, A.S. Cooper, T.J. Negran: Z. Phys. B **37**
      (1988) 2353
3.8   N. Pyka, L. Pintschovius, A. Yu. Rumiantsev: Z. Phys. B **82** (1991) 177
3.9   T.E. Mason, G. Aeppli, H.A. Mook: Phys. Rev. Lett. **68** (1992) 1414; T.E. Mason,
      G. Aeppli, S.M. Hayden, A.P. Ramirez, H.A. Mook: Phys. Rev. Lett. **71** (1993) 919
3.10 J. Rossat-Mignod, P. Burlet , M.J. Jurgens, J.Y. Henry, C. Vettier: Physica C **152**
      (1988) 19
3.11 J.M. Tranquada, D.E. Cox, W. Kunnmann, A.H. Moudden, G. Shirane, M. Suenaga,
      P. Zolliker, D. Vaknin, S.K. Sinha, M.S. Alvares, A.J. Jacobson, D.C. Johnston: Phys.
      Rev. Lett. **60** (1988) 156
3.12 J. Rossat-Mignod, J.X. Boucherle, P. Burlet, J.Y. Henry, J.M. Jurgens, G. Laperlot,
      L.P. Regnault, J. Schweizer, F. Tasset, C. Vettier: In Proc. Intl. Seminar High−T_c
      Superconductivity (Dubna 1989) ed. by N.N. Bogolubov, V.L. Aksenov, N.M.
      Plakida (World Scientific, Singapore 1990) p. 74
3.13 J. Rossat-Mignod, L.P. Regnault, C. Vettier, P. Bourges, Burlet P., J. Bossy, J.Y.
      Henry, G. Lapertot: Physica C **185−189** (1991) 86
3.14 G. Shirane: Physica C **185−189** (1991) 80; J.M. Tranquada, P.M. Gehring, G.
      Shirane, S. Shamoto, M. Sato: Phys. Rev. B **46** (1992) 5561; B.J. Sternlieb, G.
      Shirane, J.M. Tranquada, M. Sato, S. Shamoto: Phys. Rev. B **47** (1993) 5320
3.15 P.H. Hor, R.L. Meng, Y.Q. Wang, L. Gao, Z.J. Huang, J. Bechtold, K. Forster, C.W.
      Chu: Phys. Rev. Lett. **58** (1987) 1891
3.16 W.-H. Li, J.W. Lynn, S. Skanthakumar, T.W. Clinton, A. Kebede, C.-S. Lee, J.E.
      Crow, T. Mihalisin: Phys. Rev. B **40** (1989) 5300
3.17 Lynn J.W.: Physica B **163** (1990) 69
3.18 J.W. Lynn, W.-H. Li, H.A. Mook, B.C. Sales, Z. Fisk: Phys. Rev. Lett. **60** (1988)
      2781
3.19 C.H. Pennigton, C.P. Slichter: In *Physical Properties of High Temperature Super-
      conductors*, ed. by D.M. Ginsberg (World Scientific, Singapore 1990) v.2, p. 269
3.20 H. Alloul: Physica B **169** (1991) 51
3.21 H. Monien, D. Pines, M. Takigawa: Phys. Rev. B **43** (1991) 258
3.22 H. Monien, P. Monthoux, D. Pines: Phys. Rev. B **43** (1991) 275
3.23 M. Takigawa, A.P. Reyes, P.C. Hammel, J.D. Thompson, R.H. Heffner, Z. Fisk, K.C.
      Ott: Phys. Rev. B **43** (1991) 247
3.24 S.E. Barret, D.J. Durand, C.H. Pennington, C.P. Slichter, T.A. Friedmann, J.P. Rice,
      D.M. Ginsberg: Phys. Rev. B **41** (1990) 6283
3.25 M. Takigawa, P.C. Hammel, R.H. Heffner, Z. Fisk, K.C. Ott, J.D. Thompson: Phys-
      ica C **162−164** (1989) 853
3.26 F. Mila, T.M. Rice: Physica C **157** (1989) 561
3.27 A.J. Millis, H. Monien, D. Pines: Phys. Rev. B **42** (1990) 167
3.28 H. Alloul, T. Ohno, P. Mendels: Phys. Rev. Lett. **63** (1989) 1700; J. Less-Common
      Metals. **164, 165** (1990) 1022; M. Horvatic, T. Auler, C. Berthier, Y. Berthier,
      P. Butand, W.G. Clark, J.A. Gillet, P. Segransen: Phys. Rev. B **47** (1993) 3461
3.29 T. Ohno, T. Kanashiro, K. Mizuno: J. Phys. Soc. Jpn. **60** (1991) 2040
3.30 H. Yasuoka, T. Imai, T. Shimizu: In *Strong Correlation and Superconductivity*, ed.
      by H. Fukuyama, S. Maekawa and A.P. Malozemoff (Springer, Berlin, Heidelberg
      1989) p. 254
3.31 P.C. Hammel, M. Takigawa, R.H. Heffern, Z. Fisk, K.C. Ott, J.D. Thompson: Phys.
      Rev. Lett. **63** (1989) 1865

3.32 A.J. Millis, H. Monien: Phys. Rev. B **45** (1991) 3059
3.33 C.H. Pennigton, C.P. Slichter: Phys. Rev. Lett. **66** (1991) 381
3.34 R.J. Birgeneau, R.W. Erwin, P.M. Gehring, M.A. Kastner, B. Keimer, M. Sato, S. Shamoto, G. Shirane, J. Tranquada: Z.Phys. B **87** (1992) 15
3.35 R. Akis, C. Jiang, J.P. Carbotte: Physica C **176** (1991) 485
3.36 S.E. Barret, J.A. Martindale, D.J. Durand, C.H. Pennington, C.P. Slichter, T.A. Friedmann, J.P. Rice, D.M. Ginsberg: Phys. Rev. Lett. **66** (1991) 381; J.A. Martindale, S.E. Barret, C.A. Klug, K.E. O'Hara, S.M. Desoto, C.P. Slichter, T.A. Friedmann, D.M. Ginsberg: Physica C **185–189** (1991) 93
3.37 J.A. Martindale, S.E. Barret, K.E. O'Hara, C.P. Slichter, W.C. Lee, D.M. Ginsberg: Phys. Rev. B **47** (1993) 9151
3.38 D. Thelen, D. Pines, J.P. Lu: Phys. Rev. B **47** (1993) 9155

Chapter 4

4.1 E.M. Lifshitz, L.P. Pitaevskii: *Statistical Physics* ("Nauka", Moscow 1978) part 2
4.2 L.N. Bulaevskii: Intern. J. Mod. Phys. B **4** (1990) 1849
4.3 A. Umezawa, G.W. Crabtree, J.Z. Liu, T.J. Moran, S.K. Malik, L.H. Nunez, W.L. Kwok, C.G. Sowers: Phys. Rev. B **38** (1988) 2843
4.4 J.R. Clem: Physica C **162–164** (1989) 1137
4.5 E.Z. Meylikhov, V.G. Shapiro: Superconductivity: Phys., Chem., Techn. **4** (1991) 1437
4.6 A. Junod: In *Physical properties of high temperature superconductors,* ed. by D.M. Ginsberg (World Scientific, Singapore 1990) v.2, p. 13
4.7 F. Marsiglio, R. Akis, J.P. Carbotte: Phys. Rev. B **36** (1987) 5245; Physica C **153–155** (1988) 223
4.8 A. Junod, T. Graf, D. Sanchez, G. Triscone, J. Müller : Physica B **165–166** (1990) 1335; A. Junod, D. Eckert, T. Graf, E. Kaldis, J. Karpinski, S. Rusiecki, D. Sanchez, G. Triscone, J. Müller : Physica C **168** (1990) 47
4.9 A. Schilling, A. Bernasconi, H.R. Ott, F. Hulliger : Physica C **169** (1990) 237, A. Schilling, H.R. Ott, J.D. Guo, S. Rusiecki, J. Karpinski, E. Kaldis: Physica C **185–189** (1991) 1755
4.10 S.E. Inderhees, M.B. Salamon, J.P. Rice, D.M. Ginsberg : Phys. Rev. Lett. **66** (1991) 232
4.11 M.B. Salamon: In *Physical properties of high temperature superconductors,* ed. by D.M. Ginsberg (World Scientific, Singapore 1989) v.1, p. 39
4.12 N.E. Phillips, R.A. Fisher, J.E. Gordon, S. Kim, A.M. Stacy, M.K. Crawford, E.M. McCarron: Phys. Rev. Lett. **65** (1990) 357; N.E. Phillips, R.A. Fisher, J.E. Gordon: Prog. Low Temp. Phys. **13** (1992) 267
4.13 N. Wada, H. Muro-Oka, Y. Nakamura, K.-I. Kumagai: Physica C **157** (1989) 453
4.14 A.P. Malozemoff: In *Physical properties of high temperature superconductors,* ed. by D.M. Ginsberg (World Scientific, Singapore 1989) v.1, p. 71
4.15 K.A. Müller, M. Takashige, J.G. Bednorz : Phys. Rev. Lett. **58** (1987) 1143
4.16 E.H. Brandt: Physica B **169** (1991) 91; Int. J. Mod. Phys. B **5** (1991) 751
4.17 D.S. Fisher, M.P.A. Fisher, D.A. Huse : Phys. Rev. B **43** (1991) 130
4.18 A. Umezawa, G.W. Crabtree, U. Welp, W.K. Kwok, K.G. Vandervoort: Phys. Rev. B **42** (1990) 8744
4.19 Y. Yeshurun, A.P. Malozemoff, F. Holtzberg, T.R. Dinger : Phys. Rev. B **38** (1988) 11828
4.20 Wu Dong-Ho, S. Sridhar : Phys. Rev. Lett. **65** (1990) 2074
4.21 V.V. Moshchalkov, J.Y. Henry, C. Marin, J. Rossat-Mignod, J.F. Jacquot: Physica C **175** (1991) 407

4.22 U. Welp, W.K. Kwok, G.W. Crabtree, K.G. Vandervoort, J.Z. Liu: Phys. Rev. Lett. **62** (1989) 1908

4.23 U. Welp, M. Grimsditch, H. You, W.K. Kwok, M.M. Fang, G.W. Crabtree, J.Z. Liu: Physica C **161** (1989) 1

4.24 M. Hikita, Y. Tajima, H. Fuke, K. Sugiyama, A. Yamagishi, M. Date: J. Magn. Magn. Mat. **90–91** (1990) 681

4.25 J. Annet, N. Goldenfeld, S.R. Renn : Phys. Rev. B **43** (1991) 2778

4.26 A.T. Fiory, A.F. Hebard, P.M. Mankiewich, R.E. Howard : Phys. Rev. Lett. **61** (1988) 1419

4.27 L. Krusin-Elbaum, R.L. Greene, F. Holtzberg, A.P. Malozemoff, Y. Yeshurun: Phys. Rev. Lett. **62** (1989) 217

4.28 J. Rammer: Europhys. Lett. **5** (1988) 77

4.29 A. Schilling, F. Hulliger, H.R. Ott: Physica C **168** (1990) 272

4.30 A. Schilling, F. Hulliger, H.R. Ott : Z. Phys. B **82** (1991) 9

4.31 Y.J. Uemura, L.P. Le, G.M. Luke, B.J. Sternlieb, J.H. Brewer, R. Kadono, R.F. Kiefl, S.R. Kreitzman, T.M. Riseman: Physica C **162–164** (1989) 857

4.32 J.W. Loram, K.A. Mirza, J.R. Cooper, W.Y. Liang: Phys. Rev. Lett. **71** (1993) 1740; J.W. Loram et al.: In Proc. Intl. Conf. Low Temperature Physics LT-20 (Eugene, OR 1993) Physica B (1994).

4.33 A.P. Mackenzie, S.R. Julian, G.G. Lonzarich, A. Carrington, S.D. Huges, R.S. Liu, D.C. Sinclair: Phys. Rev. Lett. **71** (1993) 1238

4.34 M.S. Osofsky, R.J. Soulen, Jr., S.A. Wolf et al.: Phys. Rev. Lett. **71** (1993) 2315

4.35 W.N. Hardy, D.A. Bonn, D.C. Morgan, R. Liang, K. Zhang : Phys. Rev. Lett. **70** (1993) 3999; W.M. Hardy et al.: In Proc. Intl. Conf. Low Temperature Physics LT-20 (Eugene, OR 1993) Physica B (1994).

4.36 M. Zhengxiang, R.C. Taber, L.W. Lombardo, A. Kapitulnik, M.R. Beasley, P. Merchant, C.B. Eom, S.Y. Hou, J.M. Phillips: Phys. Rev. Lett. **71** (1993) 781

4.37 W. Braunisch, N. Knauf, V. Kataev, et al.: Phys. Rev. B **48** (1993) 4030; D. Khomskii: J. Low Temp. Phys. (1994)

4.38 D.A. Wollman, D.J. van Harlingen, W.C. Lee, D.M. Ginsberg, A.J. Legget: Phys. Rev. Lett. **71** (1993) 2134

4.39 Ch. Niedermayer, C. Bernhard, U. Binninger, H. Glückler, J.L. Tallon, E.J. Ansaldo, J.I. Budnick: Phys. Rev. Lett. **71** (1993) 1764

Chapter 5

5.1 A.W. Sleight: Science **242** (1988) 1519; Physica C **162–164** (1989) 3

5.2 G.A. Sawatzky: In *Earlier and Recent Aspects of Superconductivity*, ed. by J.G. Bednorz, K.A. Müller (Springer, Berlin, Heidelberg 1990) p. 345

5.3 J. Fink, N. Nücker, H. Romberg, M. Alexander, P. Adelmann, J. Mante, R. Classen, T. Buslaps, S. Harm, R. Manzke and M. Skibowski: In Proc. Intl. Seminar High−T_c Superconductivity (JINR E17-90-472, Dubna 1990) p. 8

5.4 W.A. Harrison: *Electronic Structure and the Properties of Solids* (W.H. Freeman and Company, San Francisko 1980)

5.5 N.F. Mott: *Metal-Insulator Transitions* (Taylor and Francis, London 1974)

5.6 J. Hubbard: Proc. Roy. Soc. A (London) **276** (1963) 238; Ibid. **281** (1964) 1386

5.7 P.W. Anderson: Phys. Rev. **115** (1959) 2; Solid State Phys. **14** (1963) 99

5.8 A. Fujimori, F. Minami, S. Sugano: Phys. Rev.B **29** (1984) 5225; A. Fujimori, F. Minami: Phys. Rev. B **30** (1984) 957

5.9 G.A. Sawatzky, J.W. Allen: Phys. Rev. Lett. **53** (1984) 2239; J. Zaanen, G.A. Sawatzky, J.W. Allen: Phys. Rev. Lett. **55** (1985) 418

5.10 J.T. Markert, Y. Dalichaouch, M.B. Maple: In *Physical Properties of High-Temperature Superconductors*, ed. by D.M. Ginsberg (World Scientific, Singapore 1989) v.1, p. 509

5.11 L.H. Greene, B.G. Bagley: Ibid, **2** (1990) 509

5.12 A.V. Narlikar, C.V.N. Rao, S.K. Agarwal: In *Studies of High Temperature Superconductors* , ed. by A. Navlikar (Nova Science Publisheres, New York 1989) v.1, p. 341

5.13 H. Takagi, T. Ido, S. Ishibashi, M. Uota, S. Uchida: Phys. Rev. B **40** (1989) 2254

5.14 H. Takagi, R.J. Cava, M. Marezio, B. Batlogg, J.J. Kraevski, W.F. Peck, Jr., P. Bordet, D.E. Cox: Phys. Rev. Lett. **68** (1992) 3777; B. Dabrowski, Z. Wang, J.D. Jorgensen, R.L. Hitterman, J.L. Wagner, B.A. Hunter, D.G. Hinks: Physica C **217** (1993) 455

5.15 Y. Tokura: In *Physics of High Temperature Superconductors*, ed. by S. Maekawa, M. Sato (Springer, Berlin, Heidelberg 1992), p. 191

5.16 C. Marayama, N. Mori, S. Yomo, H. Takagi, S. Uchida, Y. Tokura: Nature **339** (1989) 293

5.17 U. Welp, M. Grimsditch, S. Flesher, W. Nessler, J. Downey, G.W. Crabtree: Phys. Rev. Lett. **69** (1992) 2130

5.18 J.J. Neumeier, T. Bjoernholm, M.B. Maple, Iv.K. Schuller: Phys. Rev. Lett. **63** (1989) 2516

5.19 J.-M. Triskone, Ø. Fisher, O. Bruner, L. Antognazza, A.D. Kent, M.G. Karkut: Phys. Rev. Lett. **64** (1990) 804; Ø. Fisher, J.M. Triscone, L. Antognazza, O. Brunner, L. Mieville, I. Maggio-Aprile, A.D. Kent: In *Physics of High-Temperature Superconductors*, ed. by S. Maekawa, M. Sato (Springer, Berlin, Heidelberg 1992), p. 353

5.20 D. Ariosa, H. Beck: Phys.Rev. B **43** (1991) 344

5.21 G. Xiao, F.H. Streitz, A. Gavrin, Y.W. Du, C.L. Chien: Phys. Rev. B **35** (1987) 8782

5.22 Y. Oda, N. Kawaji, H. Fujita, H. Toyoda, K. Asayama: J. Phys. Soc. Jpn. **57** (1988) 4079

5.23 G. Xiao, M.Z. Cieplak, A. Gavrin, F.H. Streitz, A. Bakhshai, C.L. Chien: Phys. Rev. Lett. **60** (1988)1446; G. Xiao, M.Z. Cieplak, D. Musser, A. Gavrin, F.H. Streitz, C.I. Chien, J.J. Rhyne, J.A. Gotaas: Nature **332** (1988) 238

5.24 J.-M. Taraskon, P. Barboux, P.F. Miceli, L.H. Greene, G.W. Hull, M. Eibschutz, S.A. Sunshine: Phys. Rev. B **37** (1988) 7458

5.25 A.M. Balagarov, J. Piechota, A. Pajaczkowska: Solid State Comm. **78** (1991) 407

5.26 S. Katsuyama, Y. Ueda, K. Kosuge: Mat. Res. Bull. **24** (1989) 603

5.27 W.E. Pickett: Rev. Mod. Phys. **61** (1989) 433

5.28 H. Eskes, L.H. Tjeng, G.A. Sawatzky: Phys. Rev. B **41** (1990) 288

5.29 H. Eskes, G.A. Sawatzky, L.F. Feiner: Physica C **160** (1989) 424

5.30 Y. Ohta, T. Tohyama, S. Maekawa: Mod. Phys. Lett. B **5** (1991) 1315; Phys. Rev. B **43** 1991 2968

5.31 A.K. McMahan, R.M. Matrin, S. Satpathy: Phys. Rev. B **38** (1988) 6650

5.32 M.S. Hybertsen, M. Schlüter, N.E. Christensen: Phys. Rev. B **39** (1989) 9028

5.33 M. Hybertsen, E.B. Stechel, M. Schlüter, D.R. Jenninson: Phys. Rev. B **41** (1990) 11068

5.34 L.F. Mattheis: Phys. Rev. Lett. **58** (1987) 1028

5.35 J.-H. Xu, T.J. Watson-Yang, L. Yu, A.J. Freeman: Phys. Lett. A. **120** (1987) 489

5.36 R.E. Cohen, W.E. Pickett, H. Krakauer: Phys. Rev. Lett. **62** (1989) 831

5.37 W.E. Pickett, R.E. Cohen, H. Krakauer: Phys. Rev. Lett. **67** (1991) 228

5.38 L.F. Mattheis, D.R. Hamann: Phys. Rev.B **60** (1988) 2681

5.39 B.H. Brandow: Proc. 19th Intl. Conf. Rare Earth (1991), J. Less-Common Metals (1991)

5.40 V.I. Anisimov, J. Zaanen, O.K. Anderson: Phys. Rev. B **44** (1991) 943; V.I. Anisimov, M.A. Korotin, J. Zaanen, O.K. Andersen: Phys. Rev. Lett. **68** (1992) 345

5.41 N.N. Bogolubov: *Selected Works* (Naukova Dumka, Kiev 1970) (in Russian) v.2, p. 390

5.42 N.N. Bogolubov, V.L. Aksenov, N.M. Plakida: Physica C **153–155** (1988) 99; Theor. and Math. Phys. **93** (1992) 371

5.43 V. Emery: Phys. Rev. Lett. **58** (1987) 2794

5.44 C.M. Varma, S. Schmitt-Rink, E. Abrahams: Solid State Commun. **62** (1987) 681

5.45 F.C. Zhang, T.M. Rice: Phys. Rev. B **37** (1988) 3759

5.46 V.V. Nemoshkalenko, V.G. Aleshin: *Electron Spectroscopy of Crystals* (Naukova Dumka, Kiev 1983)

5.47 A. Fujimori, E. Takayama-Muromachi, Y. Uchida, B. Okai: Phys. Rev. B **35** (1987) 8814; A. Fujimori, E. Tokayama-Muromachi, Y. Uchida: Solid State Commun. **63** (1987) 857

5.48 N. Nücker, J. Fink, B. Renker, D. Ewert, C. Politis, P.J.W. Weijs, J.C. Fugle: Z. Phys. B **67** (1987) 9

5.49 N. Nücker, J. Fink, J.C. Fuggle, P.J. Durham, W.M. Temmerman: Phys. Rev. B **37** (1988) 5158

5.50 H. Romberg, M. Alexander, N. Nücker, P. Adelmann, J. Fink: Phys. Rev. B **42** (1990) 8768

5.51 M. Alexander, H. Romberg, N. Nücker, P. Adelmann, J. Fink, J.T. Markert, M.P. Maple, S. Uchida, H. Takagi, Y. Tokura, A.C.W.P. James, D.W. Murphy: Phys. Rev. B **43** (1991) 333

5.52 N. Nücker, H. Romberg, X.X. Xi, J. Fink, B. Gegenheimer, Z.X. Zhao: Phys. Rev.B **39** (1989) 6619

5.53 C.T. Chen, F. Sette, Y. Ma, M.S. Hybertsen, E.B. Stechel, W.M.C. Foulkes, M. Schlüter, S.-W. Cheong, A.S. Cooper, L.W. Rupp, B. Batlogg, Y.L. Soo, Z.H. Ming, A. Krol, Y.H. Kao: Phys. Rev. Lett. **66** (1991) 104

5.54 G. Mante, R. Claessen, T. Buslaps, S. Harm, R. Manzke, M. Skibowski, J. Fink: Z. Phys. B **80** (1990) 181

5.55 R. Claessen, R. Manzke, H. Carsten, B. Burandt, T. Buslaps, M. Skibowski, J. Fink: Phys. Rev. B **39** (1989) 7316

5.56 C.G. Olson, R. Liu, D.W. Lynch, R.S. List, A.J. Arko, B.W. Veal, Y.C. Chang, P.Z. Jiang, A.P. Paulikas: Phys. Rev. B **42** (1990) 381

5.57 G.A. Sawatzky: Nature **342** (1989) 480

5.58 R. Manzke, T. Buslaps, R. Claessen, J. Fink: Europhys. Lett. **9** (1989) 477

5.59 C.G. Olson, R. Liu, A.-B. Yang, D.W. Lynch, A.J. Arko, R.S. List, B.W. Veal, Y.C. Chang, P.Z. Jiang, A.P. Paulikas: Science **245** (1989) 731

5.60 Z.-X. Shen, D.S. Dessau, B.O. Wells, D.M. King, W.E. Spicer, A.J. Arko, D. Marshall, L.W. Lombardo, A. Kapitulnik, P. Dickinson, et al.: Phys. Rev. Lett. **70** (1993) 1553; R.J. Kelly, Jian Ma, G. Margaritondo, M. Onellion: Phys. Rev. Lett. **71** (1993) 4051

5.61 M. Peter, L. Hoffmann: In Proc. Intl. Seminar High−T_c Superconductivity (Dubna 1989), ed. by N.N. Bogolubov, V.L. Aksenov, N.M. Plakida (World Scientific, Singapore 1990) p. 130

5.62 H. Haghighi, J.H. Kaiser, S. Rayner, R.N. West, J.Z. Liu, R. Shelton, R.H. Howell, F. Solal, M.J. Fluss: Phys. Rev. Lett. **67** (1991) 382

5.63 J.C. Campuzano, L.C. Smedskjaer, R. Benedeck, G. Jennings, A. Banzil: Phys. Rev. B **43** (1991) 2788

5.64 C.M. Fowler, B.L. Freeman, W.L. Hults, J.C. King, F.M. Müller, J.L. Smith: Phys. Rev. Lett. **68** (1991) 534

5.65 T. Timusk, D.B. Tanner: In *Physical Properties of High-Temperature Superconductors*, ed. by D.M. Ginsberg (World Scientific, Singapore 1989) v.1, p. 339; D.B. Tanner, T. Timusk: Ibid. (1990) v.2, p. 363

5.66 Z. Schlesinger, L.D. Rotter, R.T. Collins, F. Holtzberg, C. Feild, U. Welp, G.W. Crabtree, J.Z. Liu, Y. Fang, K.G. Vandervoort: Physica C **185–189** (1991) 57

5.67 O.V. Dolgov, E.G. Maksimov, S.V. Shulga: Physica C **178** (1991) 266

5.68 S. Uchida: In Proc. Intl. Seminar High−T_c Superconductivity (Dubna 1989), ed. by N.N. Bogolubov, V.L. Aksenov, N.M. Plakida (World Scientific, Singapore 1990) p. 142; S. Uchida et al.: Phys. Rev. B **43** (1991) 7942

5.69 J. Orenstein, G.A. Thomas, A.J. Millis, S.L. Cooper, D.H. Rapkine, T. Timusk, L.F. Schneemeyer, J.V. Waszczak: Phys. Rev. B **42** (1990) 6342

5.70 I. Terasaki, T. Nakahashi, S. Takebayashi, A. Maeda, K. Uchinokura: Physica C **165** (1990) 152

5.71 Z. Schlesinger, R.T. Collins, F. Holtzberg, C. Feidl, S.H. Blanton, U. Welp, G.W. Crabtree, Y. Fang, J.Z. Liu: Phys. Rev. Lett. **65** (1990) 801

5.72 L.D. Rotter, Z. Schlesinger, R.T. Collins, F. Holtzberg, C. Feild, U.W. Welp, G.W. Crabtree, J.Z. Liu, Y. Fang, K.G. Vandervoort, S. Fleshler: Phys. Rev. Lett. **67** (1991) 2741

5.73 D.B. Romero, C.D. Porter, D.B. Tanner, L. Forro, D. Mandrus, L. Mihaly, G.L. Carr, G.P. Williams: Phys. Rev. Lett. **68** (1992) 1590

5.74 M. Cardona: Physica C **185–189** (1991) 65

5.75 D. Reznik, M.V. Klein, W.C. Lee, D.M. Ginsberg, S.-W. Cheong: Phys. Rev. B **46** (1992) 11725

5.76 P.B. Allen, Z. Fisk, A. Migliori: In *Physical Properties of High-Temperature Superconductors*, ed. by D.M. Ginsberg (World Scientific, Singapore 1989) v.1, p. 213

5.77 N.P. Ong: Ibid. (1990) v.2, 459

5.78 P.B. Allen, W.E. Pickett, H. Krakauer: Phys. Rev. B **37** (1988) 7482

5.79 T. Ito, Y. Nakamura, H. Takagi, S. Uchida: Physica C **185–189** (1991) 1267

5.80 M. Gurvich, A.T. Fiory: Phys. Rev. Lett. **59** (1987) 1337; In *Novel Superconductivity*, ed. by S.A. Wolf, V.Z. Kresin (Plenum, New-York 1987) p. 663

5.81 T.A. Friedmann, M.W. Rabin, J. Giapintzakis, J.P. Rice, D.M. Ginsberg: Phys. Rev. B **42** (1990) 6217

5.82 J.H. Kim, I. Bozovic, J.S. Harris, Jr., W.Y. Lee, C.B. Eom, T.H. Geballe, E.S. Hellman: Physica C **185–189** (1991) 1019

5.83 I.I. Mazin, O.V. Dolgov: Phys. Rev. B **45** (1992) 2509

5.84 D. Ihle, N.M. Plakida: Physica C **185–189** (1991) 1637; J. Magn. Magn. Mat. **104–107** (1992) 511

5.85 S. Martin, A.T. Fiory, R.M. Fleming, L.F. Schneermeyer, J.V. Waszczak: Phys. Rev. Lett. **60** (1988) 2194

5.86 T. Manako, Y. Shimakawa, Y. Kubo, H. Igarashi: Physica C **185–189** (1991) 1327

5.87 Y. Kubo, Y. Shimakawa, T. Manako, T. Kondo, H. Igarashi: Physica C **185–189** (1991) 1253

5.88 S. Uchida, H. Takagi, Y. Tokura: Physica C **162–164** (1989) 1677

5.89 N.P. Ong, T.W. Jing, T.R. Chien, Z.Z. Wang, T.V. Ramakrishnan, J.M. Tarascon, K. Remschnig: Physica C **185–189** (1991) 34

5.90 P.W. Anderson: Physica C **185–189** (1991) 11

5.91 A.F. Barabanov, L.A. Maksimov, A.V. Mikheyenkov: Superconductivity: Phys. Chem. Technol. **4** (1991) 3

5.92 H. Ushio, T. Schimizu, H. Kamimura: J. Phys. Soc. Jpn. **60** (1991) 1445

5.93 A.B. Kaiser, C. Uher: In *Studies of High Temperature Superconductors*, ed. by A. Navlikar (Nova Science Publisheres, New York 1990) v.7

5.94 A.B. Kaiser, G. Mountjoy: Phys. Rev. B **43** (1991) 6266

5.95 J.L. Cohn, S.A. Wolf, V. Selvamanickam, K. Salama: Phys. Rev. Lett. **66** (1991) 1098

5.96 V.V. Florentiev, A.V. Inyushkin, A.N. Taldenkov, O.K. Melnikov, A.B. Bykov: Superconductivity: Phys., Chem., Techn. **3** (1990) 2302; A.V. Inyushkin et al.: Ibid. **6** (1993) 985

5.97 J.F. Annet, N. Goldenfeld, S.R. Renn: In *Physical Properties of High-Temperature Superconductors*, ed. by D.M. Ginsberg (World Scientific, Singapore 1990) v.2, p. 571

5.98   J.R. Kirtley: Int. J. Mod. Phys. B **4** (1990) 201

5.99   M.R. Beasley: Physica C **185–189** (1991) 227

5.100   R.C. Dynes, F. Sharifi, A. Pargellis, E.S. Hellman, B. Miller, E. Hartford, Jr., J. Rosamilia: Physica C **185–189** (1991) 234

5.101   C.E. Gough, M.S. Colclough, E.M. Forgan, R.G. Jordan, M. Keene, C.M. Muirhead, A.I.M. Rae, N. Thomas, J.S. Abell, S. Sutton: Nature **326** (1987) 855

5.102   M. Sigrist, T.M. Rice: Z. Phys. B **68** (1987) 9; M. Sigrist : Rev. Mod. Phys. **63** (1991) 239

5.103   Ph.B. Allen, D. Rainer: Nature **349** (1991) 396

5.104   M. Shirai, N. Suzuki, K. Motizuki: J. Phys.: Condens. Matter **2** (1990) 3553

5.105   M. Nantoh, S. Heike, H. Ikuta, T. Hasegawa, K. Kitazawa: Physica C **185–189** (1991) 861

5.106   K. Ichimura, K. Nomura, F. Minami, S. Takekawa: Physica C **185–189** (1991) 861

5.107   W. Wei, M. Nantoh, H. Ikuta, T. Hasegawa, K. Kitazawa: Physica C **185–189** (1991) 863

5.108   T. Matsumoto, T. Kawai, K. Kitahama, S. Kawai, I. Shigaki, Y. Kawate: Physica C **185–189** (1991) 1907

5.109   N. Tsuda, D. Shimada, N. Miyakawa: Physica C **185–189** (1991) 1903

5.110   L.N. Bulaevskii, O.V. Dolgov, I.P. Kawakov, S.N. Maksimovskii, M.O. Ptitsyn, V.A. Stepanov, S.I. Vedeneev: Supercond. Sci. Technol. **1** (1988) 205

5.111   J.F. Zasadzinski, N. Tralshawala, Q. Huang, K.E. Gray, D.G. Hinks: In *Electron-phonon interaction in oxide superconductors* , ed. by R. Baquero (World Scientific, Singapore 1991) p. 46

5.112   B.A. Aminov, A.A. Bush, A.R. Kaul', et al.: JETP Lett. **54** (1991) 52; S.I. Vedeneev, K.A. Kuznetsov, V.A. Stepanov, A.A. Tsvetkov: JETP Lett. **57** (1993) 352

5.113   R.O. Anderson, R. Claessen, J.W. Allen, C.G. Olson, et al.: Phys. Rev. Lett. **70** (1993) 3163; D.M. King, Z.-X. Shen, et al.: Ibid. **70** (1993) 3159

5.114   M.J. Sumner, J.-T. Kim, Th.R. Lemberger: Phys. Rev. B **47** (1993) 12248; D. Mandrus, M.C. Martin, C. Kendziora, D. Koller, L. Forro, L. Mihaly: Phys. Rev. Lett. **70** (1993) 2629

5.115   B. Bücher, P. Steiner, J. Karpinski, E. Kaldis, P. Wachter: Phys. Rev. Lett. **70** (1993) 2012; T. Ito, K. Takeno, S. Uchida: Ibid. **70** (1993) 3955.

5.116   G. Xiao, M.Z. Cieplak, J.Q. Xiao, C.L. Chien: Phys. Rev. B **42** (1990) 8752

5.117   K. Sekizawa, Y. Nakano: Jpn. J. Appl. Phys., Ser. **7** (1992) 106

5.118   J. Fink, N. Nücker, E. Pellegrin, H. Romberg, M. Alexander, M. Knupfer: J. Electron Spectroscopy, HTSC Special Issue (1993)

5.119   R. Nemetschek, O.V. Misochko, B. Stadlober, R. Hackl: Phys. Rev. B **47** (1993) 3450; B. Stadlober, R. Nemetschek, O.V. Misochko et al.: In Proc. Intl. Conf. Low Temperature Physics LT-20 (Eugene, OR 1993), Physica B (1994)

5.120   T.P. Devereaux, D. Einzel, B. Stadlober, R. Hackl, D.H. Leach, J.J. Nemeier: Phys. Rev. Lett. **72** (1994) 396

Chapter 6

6.1   Ph.B. Allen: *Dynamical Properties of Solids*, ed. by G.K. Horton and A.A. Maradudin (North Holland, Amsterdam 1980) v.3, p. 95; Ph.B. Allen, B. Mitrović: Solid State Physics **37** (1982) 1

6.2   L. Pintschovius: Advances in Solid State Phys. **30** (1990) 183

6.3   L. Pintschovius, N. Pyka, W. Reichardt, Yu.A. Rumiantsev, N.L. Mitrofanov, A.S. Ivanov, G. Collin, P. Bourges: Physica C **185–189** (1991) 156

6.4   B. Renker, F. Gompf, E. Gering, D. Ewert, H. Rietschel, A. Dianoux: Z. Phys. B **73** (1989) 309

6.5 P. Böni, J.D. Axe, G. Shirane, R.J. Birgeneau, D.R. Gabbe, H.P. Jenssen, M.A. Kastner, C.J. Peters, P.J. Picone, T.R. Thurston: Phys. Rev. B **38** (1988) 185

6.6 A. Migliori, W.M. Visscher, S. Wong, S.E. Brown, I. Tanaka, H. Kojima, P.B. Allen: Phys. Rev. Lett. **64** (1990) 2458

6.7 W. Weber: Phys. Rev. Lett. **58** (1987) 1371

6.8 R. Feile: Physica C **159** (1989) 1

6.9 C. Thomsen, M. Cardona: In *Physical Properties of High-Temperature Superconductors*, ed. by D.M. Ginsberg (World Scientific, Singapore 1989) v.1, p. 409

6.10 C. Thomsen: *Light Scattering in Solids VI*, ed. by M. Cardona and G. Guntherodt (Springer, Berlin, Heidelberg 1991) p. 285

6.11 C. Thomsen, M. Cardona, B. Friedl, C.O. Rodriguez, I.I. Mazin, O.K. Anderson: Solid State Commun. **75** (1990) 219

6.12 B. Friedl, C. Thomsen, M. Cardona: Phys. Rev. Lett. **65** (1990) 915

6.13 R. Zeyher, G. Zwicknagl: Z. Phys. B **78** (1990) 175

6.14 B. Friedl, C. Thomsen, E. Schonherr, M. Cardona: Solid State Commun. **76** (1990) 1107

6.15 A.P. Litvinchuk, C. Thomsen, M. Cardona: Physica C **185–189** (1991) 987

6.16 C. Thomsen, B. Friedl, M. Cieplak, M. Cardona: Solid State Commun. **78** (1991) 727

6.17 E.T. Heyen, M. Cardona, J. Karpinski, E. Kaldis, S. Rusiecki: Phys. Rev. B **43** (1991) 12958

6.18 J. Mustre-de Leon, S.D. Conradson, I. Batistic, A.R. Bishop: Phys. Rev.Lett. **65** (1990) 1675; Phys. Rev.B **44** (1991) 2422

6.19 D. Mihailovic, C.M. Foster, K.F. Voss, T. Mertelj, I. Poberaj, N. Herron: Phys. Rev. B **44** (1991) 237

6.20 D. Mihailovic, C.M. Foster, K. Voss, A.J. Heeger: Phys. Rev. B **42** (1990) 7989

6.21 G. Ruani, R. Zamboni, C. Taliani, V.N. Denisov, V.M. Burlakov, A.G. Mal'shukov: Physica C **185–189** (1991) 963

6.22 H.S. Obhi, E.K.H. Salje: Physica C **171** (1990) 547

6.23 S. Sugai: Physica C **185–189** (1991) 76

6.24 A.D. Kulkarni, F.W. de Wette, J. Prade, U. Schroder, W. Kress: Phys. Rev. B **41** (1990) 6409

6.25 S. Barisic: Int. J. Modern Phys. B **5** (1991) 2439

6.26 T. Jarlborg: Solid State Commun. **67** (1988) 297; ibid. **71** (1989) 669

6.27 R.E. Cohen, W.E. Pickett, H. Krakauer: Phys. Rev. Lett. **62** (1989) 831

6.28 R.E. Cohen, W.E. Pickett, H. Krakauer: Phys. Rev. Lett. **64** (1990) 2575

6.29 O.K. Andersen, A.I. Liechtenstein, O. Rodriquez, I.I. Mazin, O. Jepsen, V.P. Antropov, O. Gunnarsson, S. Gopalan: Physica C **185–189** (1991) 147

6.30 S. Barisić, E. Tutiś: In Proc. Intl. Seminar on High Temperature Superconductivity (JINR E17-90-472, Dubna 1990) p. 59

6.31 G. Kotliar, P.A. Lee, N. Read: Physica C **153–155** (1988) 538

6.32 J. Mustre-de Leon, A. Batistic, A.R. Bishop: Phys. Rev. Lett. **68** (1992) 3236

6.33 K.A. Müller: Z. Phys. B **80** (1990) 193

6.34 S. Flach, N.M. Plakida, V.L. Aksenov: Int. J. Modern Phys. B **4** (1990) 1955

6.35 N.M. Plakida, S.E. Krasavin: Phys. Lett. A. **158** (1991) 313; S.E. Krasavin, N.M. Plakida: Superconductivity: Phys., Chem., Techn. **5** (1992) 1555

6.36 N.M. Plakida: Physica Scripta **T29** (1989) 77

6.37 N. Pyka, W. Reichardt, L. Pintschovius, G. Engel, J. Rossat-Mignod, J.Y. Henry: Phys. Rev. Lett. **70** (1993) 1457; W. Reichardt, L. Pintschovius, L. Pyka, et al.: In Proc. Intl. Conf. Low Temperature Physics LT-20 (Eugene, OR 1993), Physica B (1994)

6.38 R.A. Evarestov, Yu.E. Kitaev, M.F. Limonov, A.G. Panfilov: phys. stat. sol. (b) **179** (1993) 249

6.39 A.P. Litvinchuk, C. Thomsen, M. Cardona: Solid State Commun. **83** (1992) 343

Chapter 7

7.1 P.W. Anderson: Science **235** (1987) 1196; In *Frontiers and Borderlines in Many-Particle Physics*, Intl. School of Physics "Enrico Fermi", Course CIV, ed. by R.A. Broglia and J.R. Schrieffer (North-Holland, Amsterdam 1988)

7.2 P.W. Anderson: Phys. Rev. Lett. **64** (1990) 1839

7.3 N. Kawakami, S.-K. Yang: In *Physics of High-Temperature Superconductors*, ed. by S. Maekawa, M. Sato, (Springer, Berlin, Heidelberg 1992) p. 103

7.4 C. Varma: Int. J. Mod. Phys. B **3** (1989) 2083; A.E. Ruckenstein, C.M. Varma: Physica C **185–189** (1991) 134

7.5 G. Dopf, A. Muramatsu, W. Hanke: Phys. Rev. Lett. **68** (1992) 353

7.6 R. Hayn, V. Yushankhai, S. Lovtsov: Phys. Rev. B **47** (1993) 5253

7.7 P. Fulde: *Electronic correlations in molecules and solids* (Springer, Berlin, Heidelberg 1991)

7.8 H.B. Schutler, A.J. Fedro: Phys. Rev. B **45** (1992) 7588; J.H. Jefferson, H. Eskes, L.F. Feiner: Phys. Rev. B **45** (1992) 7959; J.H. Jefferson: Physica B **165 – 166** (1990) 1013; V.I. Belinicher, A.L. Chernyshev: Phys. Rev. B **47** (1993) 390;

7.9 Yu.A. Izyumov: Sov. Phys. Uspekhi **161** (1991) 1

7.10 E. Dagotto: Int. J. Mod. Phys. B **5** (1991) 77

7.11 L.N. Bulaevskii, E.L. Nagaev, D.I. Khomskii: Soviet Phys. JETP **27** (1968) 83

7.12 E.L. Nagaev: Soviet Phys. JETP **31** (1970) 682

7.13 W.F. Brinkman, T.M. Rice: Phys. Rev. B **2** (1970) 1324

7.14 S. Schmitt-Rink, C.M. Varma, A.E. Ruckenstein: Phys. Rev. Lett. **60** (1988) 2793

7.15 C.L. Kane, P.A. Lee, N. Read: Phys. Rev. B **39** (1989) 6880

7.16 F. Marsiglio, A. Ruckenstein, S. Schmitt-Rink, C. Varma: Phys. Rev. B **43** (1991) 10882

7.17 G. Martinez, P. Horsch: Phys. Rev. B **44** (1991) 317

7.18 Z. Liu, E. Manousakis: Phys. Rev. B **45** (1992) 2425

7.19 K.J. von Szczepanski, P. Horsch, W. Stephan, M. Ziegler: Phys. Rev. B **41** (1990) 2017

7.20 P. Horsch, W. Stephan: In Proc. of the NATO Advanced Research Workshop on Dynamics of Magnetic Fluctuations in High Temperature Superconductors, ed. by G. Reiter, P. Horsch, G. Psaltakis (Plenum Press, New York 1990)

7.21 W. Stephan, P. Horsch: Phys. Rev. Lett. **66** (1991) 2258

7.22 P. Horsch, W. Stephan: In *Electronic properties and mechanisms of high T_c superconductors*, ed. by T. Oguchi, K. Kadowaki, T. Sasaki (Elsevier Science Publishers B.V., Amsterdam 1992) p. 241

7.23 I. Bonca, P. Prelovsek, I. Sega: Europhys. Lett. **10** (1989) 87

7.24 A.F. Barabanov, R.O. Kuzian, L.A. Maksimov: J. Phys. C **3** (1991) 9129

7.25 H. Eskes, G.A. Sawatzky: Phys. Rev. B **43** (1991) 119

7.26 T. Tohyama, S. Maekawa: Physica C **191** (1992) 193

7.27 G. Dopf, J. Wagner, P. Dieterich, A. Muramatsu, W. Hanke: Phys. Rev. Lett. **68** (1992) 2082

7.28 K.W. Becker, W. Brenig, P. Fulde: Z. Phys. B **81** (1990) 165

7.29 P. Unger, P. Fulde: Phys. Rev. B **47** (1993) 8947

7.30 M. Sasaki, H. Matsumoto, M. Tachiki: Phys. Rev. B **46** (1992) 3022

7.31 S. Sorella: Phys. Rev. B **46** (1992) 11670

7.32 D.J. Scalapino: Physica C **185–189** (1991) 104

7.33 S.A. Trugman: Phys. Rev. B **41** (1990) 892

7.34 A. Kampf, J.R. Schrieffer: Phys. Rev. B **42** (1990) 7969

7.35 A. Virosztek, J. Ruvalds: Phys. Rev. B **42** (1990) 4064

7.36 N. Bulut, D. Hone, D.J. Scalapino, N.E. Bickers: Phys. Rev. B **41** (1990) 1797; N. Bulut, D.J. Scalapino: Phys. Rev. B **45** (1992) 2371; Phys. Rev. Lett. **68** (1992) 706; Phys. Rev. B **47** (1993) 3419

7.37 S. Wermbter, L. Tewordt: Phys. Rev. B **43** (1991) 10530; Ibid. **48** (1993) 10514

7.38 T. Moria, Y. Takahashi, K. Ueda: Physica C **185–189** (1991) 114; J. Phys. Soc. Jpn. **59** (1990) 2905; T. Moria, Y. Takahashi: Ibid. **60** (1991) 776

7.39 T. Tanamoto, K. Kuboki, H. Fukuyama: J. Phys. Soc. Jpn. **60** (1991) 3072, 4395(E)

7.40 Yu. Izyumov, B.M. Letfulov: J. Phys.: Condensed Matter. **2** (1990) 8905

7.41 T. Izuyama, D. Kim, R. Kubo: J. Phys. Soc. Jpn. 1963. **18** (1963) 1025

7.42 G. Baskaran, P.W. Anderson: Phys. Rev. B **37** (1988) 580; G. Baskaran: Phys. Scripta **T27** (1989) 53

7.43 L.B. Ioffe, A.I. Larkin: Phys. Rev. B **39** (1989) 8988; L.B. Ioffe, G. Kotliar: Phys. Rev. B **42** (1990) 10348

7.44 X.G. Wen, F. Wilczek, A. Zee: Phys. Rev. B **39** (1989) 11413

7.45 A. Zee: In *High Temperature Superconductivity* ed. by K. Bedell, D. Coffey, D.E. Meltzer, D. Pines, J.R. Schrieffer (Addison Wesley, 1990) Y.G. Chen, F. Wilczek, E. Witten, B.I. Halperin: Int. J. Mod. Phys. B **3** (1989) 1001

7.46 T.W. Lawrence, A. Szoke, R.B. Laughlin: Phys. Rev. Lett. **69** (1992) 1439

7.47 P. Lederer, D. Poilblanc, T.M. Rice: Phys. Rev. Lett. **63** (1989) 1519; D. Poilblanc, Y. Hasegawa, T.M. Rice: Phys. Rev. B **41** (1990) 1949

7.48 M. Luchini, M. Ogata, W. Putikka, T.M. Rice: Physica C **185–189** (1991) 141

7.49 N. Nagaosa, P.A. Lee: Phys. Rev. Lett. **64** (1990) 2450; Phys. Rev. B **43** (1991) 1233; Physica C **185–189** (1991) 130

7.50 R. Hlubina, W.O. Putikka, T.M. Rice, D.V. Khveshchenko: Phys. Rev. B **46** (1992) 11224

7.51 S. Chakravarty, B.I. Halperin, D. Nelson: Phys. Rev. Lett. **60** (1988) 1057; E. Manousakis: Rev. Mod. Phys. **63** (1991) 1

7.52 D. Ihle, M. Kasner: Phys. Rev. B **42** (1990) 4760

7.53 V.J. Emery, S.A. Kivelson, H.Q. Lin: Phys. Rev. Lett. **64** (1990) 475; S.A. Kivelson, V.J. Emery, H.Q. Lin: Phys. Rev. B **42** (1990) 6523

7.54 W.O. Puttika, M.U. Luchini, T.M. Rice: Phys. Rev. Lett. **68** (1992) 538

7.55 Yu.A. Izyumov, Yu.A. Skriabin: *Statistical Mechanics of Magnetically Ordered Sistems* (Nauka, Moscow 1987)

7.56 N.M. Plakida, V.Yu. Yushankhai, I.V. Stasyuk: Physica C **160** (1989) 80; V.Yu. Yushankhai, N.M. Plakida, P. Kalinay: Ibid. **174** (1991) 401

7.57 N.N. Bogolubov, S.V. Tyablikov: Soviet Phys. Doklady. **4** (1959) 604; D.N. Zubarev: Soviet Phys. Uspekhi. **3** (1960) 320

7.58 R.O. Zaitsev, V.A. Ivanov: Soviet Phys. Solid State **29** (1987) 2554, 3111; Zaitsev R.O.: Phys. Lett.A. **134** (1988) 199; V.A. Ivanov, R.O. Zaitsev: Intern. J. Mod. Phys. B **3** (1989) 1403

7.59 G. Baskaran, Z. Zou, P.W. Anderson: Solid State Commun. **63** (1987) 973

7.60 A.E. Ruckenstein, P.J. Hirschfeld, J. Appel: Phys. Rev. B **36** (1987) 857; Y. Suzumura, Y. Hasegawa, H. Fukuyama: J. Phys. Soc. Jpn. **57** (1988) 401

7.61 M. Cyrot: Solid State Commun. **62** (1987) 821; Ibid. **63** 1015.

7.62 Yu.A. Izyumov, B.M. Letfulov: Int. J. Mod. Phys. B **6** (1992) 321

7.63 M. Imada, Y. Hatsugai: J. Phys. Soc. Jpn. **58** (1989) 3752; M. Imada: Physica C **185–189** (1991) 1447; A. Moreo, D. Scalapino: Phys. Rev. B **43** (1991) 8211

7.64 M. Frick, W. von der Linden, I. Morgenstern, H. de Raedt: Z. Phys. B **81** (1990) 327; M. Frick, I. Morgenstern, W. von der Linden: Z. Phys. B **82** (1991) 339

7.65 N. Bulut, D.J. Scalapino, R.T. Scalettar: Phys. Rev. B **45** (1992) 5577; E. Dagotto, J. Riera, D. Scalapino: Ibid. P.5744.

7.66 I. Morgenstern, Th. Husslein, J.M. Singer, H.-G. Matuttis: Preprint HLRZ, Julich (1992)

7.67 J. Bonca, P. Prelovsek, I. Sega, H.Q. Lin, D.K. Campbell: Phys. Rev. Lett. **69** (1992) 526; Phys. Rev. B **47** (1993) 12224

7.68 E. Dagotto, J.R. Schrieffer: Phys. Rev. B **43** (1991) 8705

7.69 J. Spalek: Phys. Rev. B **38** (1988) 208; Physica C **153–155** (1999) 1267

7.70 S.V. Lovtsov, N.M. Plakida, V.Yu. Yushankhai: Z. Phys. B **82** (1991) 1

7.71 J.R. Schrieffer: Physica C **185–189** (1991) 17

7.72 D. Pines: Physica C **185–189** (1991) 120; Monthoux P., Balatsky A.V., Pines D.: Phys. Rev. Lett. **67** (1991) 3448; Phys. Rev. B **46** (1992) 14803; P. Monthoux, D. Pines: Phys. Rev. Lett. **69** (1992) 961; Phys. Rev. B **47** (1993) 6069

7.73 D.J. Scalapino, E. Loh, Jr., J.E. Hirsch: Phys. Rev. B **34** (1986) 8190; K. Miyake, S. Schmitt-Rink, C.M. Varma: Phys. Rev. B **34** (1986) 6554

7.74 N.E. Bickers, D.J. Scalapino, R.T. Scalettar: Int. J. Mod. Phys. B **1** (1987) 687; N.E. Bickers, D.J. Scalapino, S.R. White: Phys. Rev. Lett. **62** (1989) 961; N.E. Bickers, S.R. White: Phys. Rev. B **43** (1991) 8044; H. Shimahara, S. Takada: J. Phys. Soc. Jpn. **57** (1988) 1044.

7.75 I.E. Dzyaloshinskii: Soviet Phys. JETP **66** (1987) 848

7.76 S. Lenck, S. Wermbter, L. Tewordt: J. Low Temp. Phys. **80** (1990) 269

7.77 J.R. Schrieffer, X.-G. Wen, S.C. Zhang: Phys. Rev. B **39** (1989) 11663; A. Kampf, J.R. Schrieffer: Ibid. **41** 6399

7.78 V.S. Babichenko, Yu. Kagan: JETP Letters **56** (1992) 305

7.79 W. Weber: Adv. in Solid State Phys. **28** (1988) 141

7.80 W. Jin, M.H. Degami, R.K. Kalia, P. Vashishta: Phys. Rev. B **45** (1992) 5535

7.81 Ph.B. Allen: Nature. **335** (1988) 419; Carbotte J.P.: Rev. Mod. Phys. **62** (1990) 1027

7.82 M.K. Crawford, W.E. Farneth, R. Miao, R.L. Harlow, C.C. Torardi, E.M. McCarron: Physica C **185–189** (1991) 1345; M.K. Crawford, W.E. Farneth, E.M. McCarron, R.L. Harlow, E.H. Moudden: Science. **250** (1990) 1390; M. Ronay, M.A. Frisch, T.R. McGuire: Phys. Rev. B **45** (1992) 355

7.83 J.P. Franck, S. Gygax, G. Soerensen, E. Altshuler, A. Hnatiw, J. Jang, M.A.-K. Mohamed, M.K. Yu, G.I. Sproule, J. Chrzanowski, J.C. Irwin: Physica C **185 – 189** (1991) 1379; J.P. Franck et al. Phys. Rev. B **44** (1991) 5318; J.P. Franck, S. Havker, J.H. Brewer: Phys. Rev. Lett. **71** (1993) 283

7.84 N. Babushkina, A. Inyushkin, V. Ozhogin et al.: Physica C **185–189** (1991) 901; Appl. Superconductivity **1** (1993) 359

7.85 B. Batlogg, S.-W. Cheong, G.A. Thomas, S.L. Cooper, L.W. Rupp, Jr., D.H. Rapkine, A.S. Cooper: Physica C **185–189** (1991) 1385

7.86 J.P. Carbotte, M. Greeson, A. Perez-Gonzales: Phys. Rev. Lett. **66** (1991) 1789; J.P. Carbotte, E.J. Nicol: Physica C **185–189** (1991) 162

7.87 R. Zeyher: Z. Phys. B **80** (1990) 187

7.88 I.I. Mazin, O.K. Anderson, A.I. Liechtenstein, O. Jepsen, V.P. Antropov, S.N. Rashkeev, V.I. Anisimov, J. Zaanen, R.O. Rodriguez, M. Methfessel: In Proc. Intl. Conf. Lattice effects in High-T_c Sureconductors (Santa Fe 1992) ed. by Y. Bar-Yam, T. Egami, J. Mustre-de Leon, A.R. Bishop (Wold Scientific, Singapore 1992) p. 235

7.89 H. Krakauer, W.E. Pickett, R.E. Cohen: Phys. Rev. B **47** (1993) 1002

7.90 Ju.H. Kim, K. Levin, R. Wentzcovitch: Phys. Rev. B **44** (1991) 5148

7.91 G.M. Vujicic, V.L. Aksenov, N.M. Plakida, S. Stamenkovic : Phys. Lett. A. **73** (1979) 439; J. Phys. C **14** (1981) 2377

7.92 N.M. Plakida, V.L. Aksenov, S.L. Drechsler: Europhys. Lett. **4** (1987) 1309

7.93 J.C. Phillips: Phys. Rev. B **36** (1987) 861; Phys. Rev. Lett. **59** (1987) 1856; Solid State Commun. **65** (1988) 227; Ibid. **68** 769

7.94 A. Bussman-Holder, H. Butner, A. Simon: Phys. Rev. B **39** (1989) 207; A. Bussman-Holder, A.R. Bishop, I. Batistic : Ibid. **43** (1991) 137; A. Bussman-Holder, A. Migliori, Z. Fisk, J.L. Sarrao, R.G. Leisure, S.-W. Cheong: Phys. Rev. Lett. **67** (1991) 512.

7.95 J.R. Hardy, J.W. Flocken: Phys. Rev. Lett. **60** (1988) 2191
7.96 T. Galbaatar, S.L. Drechsler, N.M. Plakida, G.M. Vujicic: Physica C **176** (1991) 496
7.97 M. Arai, K. Yamada, A.C. Hannon, Y. Hidaka, D. Sugimoto, A.D. Taylor: Phys. Rev. Lett. **69** (1992) 359
7.98 J. Ranninger: Z. Phys. B **84** (1991) 167
7.99 A.S. Alexandrov, J. Ranninger, S. Robaszkiewicz: Phys. Rev. B **33** (1986) 4526
7.100 A.S. Alexandrov, J. Ranninger: Phys. Rev. B **23** (1981) 1796; Ibid. **24** (1981) 1164
7.101 A.S. Alexandrov: Phys. Rev. B **38** (1988) 925; Sov. Phys. JETP. **68** (1989) 167; Physica C **158** (1989) 337
7.102 R. Micnas, J. Ranninger, S. Robaszkiewicz: Rev. Mod. Phys. **62** (1990) 113
7.103 D. Emin, M.S. Hillery: Phys. Rev. B **39** (1989) 6575; D. Emin: Phys. Rev. Lett. **62** (1989) 1544; Physica C **185–189** (1991) 1593
7.104 H. Eschrig, S.-L. Drechsler: Physica C **173** (1991) 80; H. Eschrig, S.-L. Drechsler, J. Malek: Physica C **185–189** (1991) 1461
7.105 L.J. De Jongh: Physica C **152** (1988) 171; H. Kamimura: In *Physics of High-Temperature Superconductors*, ed. by S. Maekawa, M. Sato, (Springer, Berlin, Heidelberg 1992) 63
7.106 I.O. Kulik: Int. J. Mod. Phys. B **1** (1988) 851
7.107 Y. Bar-Yam: Phys. Rev. B **43** (1991) 359, 2601
7.108 A. Ramsak, P. Horsch, P. Fulde: Phys. Rev. B **46** (1992) 14305
7.109 J. Zhong, H.-B. Schuttler: Phys. Rev. Lett. **69** (1992) 1600
7.110 R. Combescot: Phys. Rev. Lett. **67** (1991) 148
7.111 J.E. Hirsch, D.J. Scalapino: Phys. Rev. Lett. **56** (1986) 2732
7.112 J. Friedel: J. Phys.: Condens. Matter. **1** (1989) 7757
7.113 C.C. Tsuei, D.M. Newns, C.C. Chi, P.C. Pattnaik: Phys. Rev. Lett. **65** (1990) 2724; D.M. Newns, P.C. Pattnaik, C.C. Tsuei: Phys. Rev. B **43** (1991) 3075; D.M. Newns et al.: Phys. Rev. Lett.: **69** (1992) 1264; R. Combescot: Ibid. **68** (1992) 1089
7.114 R.S. Markiewicz: J. Phys. Cond. Matter **2** (1990) 665; Physica C **183** (1991) 303; Int. J. Mod. Phys. B **5** (1991) 2037
7.115 A.A. Gorbatsevich, V.Ph. Elesin, Yu.V. Kopaev: Phys. Lett. **125A** (1987) 149
7.116 N.M. Plakida, V.S. Udovenko: Mod. Phys. Lett. B **6** (1992) 541
7.117 A.N. Das, J. Konior, D.K. Ray, A.M. Oles: Phys. Rev. B **44** (1991) 7680
7.118 Chan Minh-Tien, N.M. Plakida: Physica C **206** (1993) 90
7.119 N.M. Plakida, R. Hayn: Zs. Phys. **93** (1994) 313
7.120 W.A. Little: Phys. Rev. **134** (1964) A1416
7.121 V.I. Ginsburg: Phys. Lett. **13** (1964) 101; Sov. Phys. JETP **20** (1965) 1549
7.122 W.A. Little: In *Physics of High-Temperature Superconductors*, ed. by S. Maekawa, M. Sato, (Springer, Berlin, Heidelberg 1992) p. 113
7.123 E.A. Pashitskii: Superconductivity: Phys., Chem., Techn. **3** (1990) 2669; JETP Letters **57** (1993) 648
7.124 V.Z. Kresin: Phys. Rev. B **35** (1987) 8716; V.Z. Kresin, H. Moravitz: Phys. Rev. B **37** (1988) 7854; J. Superconductivity **1** (1988) 89
7.125 J. Ruvalds: Phys. Rev. B **35** (1987) 8869; In *Novel Superconductivity* ed. by A. Wolf, V.Z. Kresin (Plenum Publ. Corp. 1987) p. 455
7.126 M. Tachiki, S. Takahashi: Phys. Rev. B **39** (1988) 218
7.127 C.M. Varma, S. Schmitt-Rink, E. Abrahams: Solid State Commun. **62** (1987) 681
7.128 Y. Kuroda, C.M. Varma: Phys. Rev. B **42** (1990) 8619; E.J. Nicol, J.P. Carbotte, T. Timusk: Ibid. **43** (1991) 473; Solid State Commun. **76** (1990) 937
7.129 A.R. Bishop, R.L. Martin, K.A. Müller, Z. Tesanovic: Z. Phys. B **76** (1989) 17; Solid State Commun. **68** (1988) 337; A.R. Bishop: In *Earlier and Recent Aspects of Superconductivity*, ed. by J.G. Bednorz, K.A. Müller (Springer, Berlin, Heidelberg 1990) p. 482
7.130 Yu.B. Gaididei, V.M. Loktev: phys. stat. sol.(b). **147** (1988) 307

7.131 W.A. Weber: Z. Phys. B **70** (1988) 323; W.A. Weber, A.L. Shelankov, X. Zotos: Physica C **162–164** (1989) 307

7.132 M. Georgiev, M. Borissov: Phys. Rev. B **39** (1989) 1124 M. Georgiev, M. Borissov, A. Vavrek: Physica C **162–164** (1989) 1519

7.133 K.H. Johnson, D.P. Clougherty, M.E. McHenry: Mod. Phys. Lett. B **3** (1989) 1367

7.134 V.A. Moskalenko: Fizika Metallov i Metallov. (USSR) **8** (1959) 503

7.135 A.A. Gorbatsevich, Yu.V. Kopaev: Physica C **185–189** (1991) 1711; JETP Letters. **51** (1990) 327

7.136 K.B. Efetov: Phys. Rev. B **47** (1991) 5538

7.137 P. Konsin, N. Kristoffel, T. Ord: Phys. Lett. **A129** (1988) 339; Ibid. **137** (1989) 420; Ibid. **143** (1990) 83

7.138 J.E. Hirsch: Phys. Lett. **A134** (1989) 451; Physica C **158** (1989) 326; Phys. Lett. **A138** (1989) 83; J.E. Hirsch, F. Marsiglio: Phys. Rev. B **39** (1989) 11515

7.139 F. Marsiglio: Physica C **160** (1989) 305

7.140 Qimiao Si, Y. Zha, K. Levin, J.P. Lu: Phys. Rev. B **47** (1993) 9055; Y. Zha, K. Levin, Qimiao Si: Ibid. **47** (1993) 9124

7.141 M.L. Kulić, R. Zeyher: Phys. Rev. B **49** (1994) 4395

7.142 A.A. Abrikosov, J.C. Campuzano, K. Gofron: Physica C **214** (1993) 73

7.143 J.D. Perkins, J.M. Graybeal, M.A. Kastner, R.J. Birgenau, J.P. Falck, M. Greven: Phys. Rev. Lett. **71** (1993) 1621

Conclusion

8.1 R.C. Haddon et al.: Nature **350** (1991) 321; A.E. Hebard et al.: Ibid. 600.

8.2 D.W. Murphy, M.C. Rosseinsky, R.C. Haddon, A.P. Ramirez, A.E. Hebard, R. Tycko, R.M. Fleming, G. Dabbagh: Physica C **185–189** (1991) 403

8.3 K. Tanigaki, I. Hirosawa, T.W. Ebbesen, J. Mizuki, Y. Shimakawa, Y. Kubo, J.S. Tsai, S. Kuroshima: Nature **356** (1992) 419

Subject Index

Springer-Verlag
and the Environment

We at Springer-Verlag firmly believe that an international science publisher has a special obligation to the environment, and our corporate policies consistently reflect this conviction.

We also expect our business partners – paper mills, printers, packaging manufacturers, etc. – to commit themselves to using environmentally friendly materials and production processes.

The paper in this book is made from low- or no-chlorine pulp and is acid free, in conformance with international standards for paper permanency.